DOPPLER RADAR PHYSIOLOGICAL SENSING

WILEY SERIES IN BIOMEDICAL ENGINEERING AND MULTI-DISCIPLINARY INTEGRATED SYSTEMS

KAI CHANG, SERIES EDITOR

Advances in Optical Imaging for Clinical Medicine
Nicusor Iftimia, William R. Brugge, and Daniel X. Hammer (Editors)

Antigen Retrieval Immunohistochemistry Based Research and Diagnostics
Shan-Rong Shi and Clive R. Taylor

Introduction to Nanomedicine and Nanobioengineering
Paras N. Prasad

Biomedical Image Understanding
Joo-Hwee Lim, Sim-Heng Ong, and Wei Xiong (Editors)

Doppler Radar Physiological Sensing
Olga Boric-Lubecke, Victor M. Lubecke, Amy D. Droitcour, Byung-Kwon Park, and Aditya Singh (Editors)

DOPPLER RADAR PHYSIOLOGICAL SENSING

Edited by

OLGA BORIC-LUBECKE

VICTOR M. LUBECKE

AMY D. DROITCOUR

BYUNG-KWON PARK

ADITYA SINGH

IEEE

WILEY

Published by John Wiley & Sons, Inc., Hoboken, New Jersey

Published simultaneously in Canada

Library of Congress Cataloging-in-Publication Data:

Doppler radar physiological sensing / edited by Olga Boric-Lubecke, Victor M. Lubecke,
Amy D. Droitcour, Byung-Kwon Park, Aditya Singh.
p. ; cm.
Includes bibliographical references and index.
ISBN 978-1-118-02402-7 (cloth)
I. Boric-Lubecke, Olga, 1966-, editor. II. Lubecke, Victor M., editor. III. Droitcour, Amy D., editor.
IV. Park, Byung-Kwon, editor. V. Singh, Aditya, 1984-, editor.
[DNLM: 1. Heart Rate. 2. Monitoring, Physiologic–methods. 3. Respiratory Rate.
4. Signal Processing, Computer-Assisted. 5. Ultrasonography, Doppler–methods. WG 140]
QP113
612.1′71–dc23

2015028401

Typeset in 10/12pt TimesLTStd by SPi Global, Chennai, India

Printed in the United States of America

10 9 8 7 6 5 4 3 2 1

1 2016

CONTENTS

LIST OF CONTRIBUTORS

Olga Boric-Lubecke, Department of Electrical Engineering, University of Hawaii at Manoa, Honolulu, Hawaii, United States

Amy D. Droitcour, Wave 80 Biosciences, Inc., San Francisco, California, United States

Aly Fathy, Department of Electrical Engineering and Computer Science, University of Tennessee, Knoxville, Tennessee, United States

John Kiriazi, QCT RF Systems, Qualcomm Inc., San Diego, California, United States

Jenshan Lin, Department of Electrical and Computer Engineering, University of Florida, Gainesville, Florida, United States

Victor M. Lubecke, Department of Electrical Engineering, University of Hawaii at Manoa, Honolulu, Hawaii, United States

Wansuree Massagram, Department of Computer Science and Information Technology, Naresuan University, Phitsanulok, Thailand

Isar Mostafanezhad, Department of Physics and Astronomy, University of Hawaii at Manoa, Honolulu, Hawaii, United States

Byung-Kwon Park, DAS Sensor SW Engineering Team, Hyundai Mobis Mechatronics R&D Center, Gyeonggi-Do, South Korea

Aditya Singh, University of Hawaii Neuro-science and MRI research Program, John A. Burns School of Medicine, Honolulu, Hawaii, United States

Alex Vergara, Theranova LLC, San Francisco, California, United States

Yazhou Wang, Boston Design Center, RF Micro Devices, Inc., Billerica, Massachusetts, United States

Shuhei Yamada, Department of Electrical Engineering, University of Hawaii at Manoa, Honolulu, Hawaii, United States

Ehsan Yavari, Department of Electrical Engineering, University of Hawaii at Manoa, Honolulu, Hawaii, United States

1

INTRODUCTION

AMY D. DROITCOUR[1], OLGA BORIC-LUBECKE[2], SHUHEI YAMADA[2], AND VICTOR M. LUBECKE[2]

[1]*Wave 80 Biosciences, Inc., San Francisco, California, United States*
[2]*Department of Electrical Engineering, University of Hawaii at Manoa, Honolulu, Hawaii, United States*

Noncontact detection and monitoring of human cardiopulmonary activity is one of the most promising solutions for sleep monitoring, postsurgery monitoring, home health care, and search and rescue applications. Without contact or subject preparation (special clothing, attachments, etc.), this could facilitate health monitoring to the chronically ill, enable sleep monitoring outside of sleep laboratories, detect survivors under rubble, and deliver warnings of emergencies or changes in conditions of patients. Doppler radar remote sensing of physiological signatures has shown promise to this end. The development of Doppler radar for remote sensing of vital signs, with proof of concept demonstrated for various applications [Li et al., 2013], could offer a platform for unobtrusive, noncontact, yet continuous physiological monitoring systems.

Cardiopulmonary monitoring is typically carried out with contact sensors such as electrocardiogram (ECG) electrodes. The use of contact sensors is neither possible nor desirable in many situations, due to, for example, skin irritation, or simply lack of access for direct contact. The long-term, continuous use of contact sensors is also limited by degradation in contact quality over time. Some examples of long-term health-care monitoring that would benefit from noncontact sensing include monitoring postsurgery patients, chronic and elderly patients, and patients with sleep disorders. Premature infants and burn victims will also clearly benefit

Doppler Radar Physiological Sensing, First Edition.
Edited by Olga Boric-Lubecke, Victor M. Lubecke, Amy D. Droitcour, Byung-Kwon Park, and Aditya Singh.

from noncontact sensing due to compromised skin integrity. A failure to respond to patient deterioration promptly and appropriately can lead to increased morbidity and mortality, increased requirement for intensive care, and elevated costs [Tarassenko et al., 2006]. Early identification of patient deterioration is important, as it can prevent subsequent cardiopulmonary arrest and reduce mortality. Early recognition of physiological abnormalities coupled with the rapid intervention of suitably trained staff may result in an improvement in the functional outcome or mortality rate. Early recognition relies on the physiological observations being measured accurately and at intervals appropriate to the condition of the patient. However, many patients are not monitored regularly, and some vital signs such as respiratory rate are measured significantly less frequently than other vital signs. There is a need for straightforward, automated, continuous physiological monitoring technology.

Noncontact physiological monitoring may make a significant impact beyond health-care applications, especially in situations where direct access to the subject is not available. Such situations include, for example, occupancy sensors for energy efficiency [Yavari et al., 2014], search and rescue operations for survivor detection under rubble [Chuang et al., 1990], and detection of adversaries through walls [Lubecke et al., 2007].

1.1 CURRENT METHODS OF PHYSIOLOGICAL MONITORING

Assessment of cardiopulmonary functions is most often performed with contact sensors when direct access to the subject is available. ECG is the gold standard for heart monitoring that is often used in hospital and ambulatory settings, whereas there is no equivalent gold standard for respiratory monitoring. There are many different approaches used for respiratory monitoring; however, none of them are easily applied. Even though respiratory rate is a key early indicator of physiological instability that may lead to a critical event, respiratory rate is measured significantly less frequently than other vital signs, such as blood pressure, pulse rate, and arterial oxygen saturation. Among the vital signs, respiratory rate is the only sign that is typically measured manually, via visual assessment, with a nurse counting chest excursions.

The current practices to measure respiration are divided into three categories: measurement of oxygen saturation, measurement of airflow, and measurement of respiratory effort/movement [Webster, 2010]. Pulse oximetry measures the percentage of hemoglobin (Hb) that is saturated with oxygen. A source of light originates from the probe at two wavelengths (650 and 805 nm). The light is partly absorbed by hemoglobin, at amounts which differ depending on whether it is saturated or desaturated with oxygen. Direct measurement of airflow typically uses a spirometer with a mouth piece or a face mask. It contains a precision differential pressure transducer for the measurements of respiration flow rates. The spirometer records the volume and rate of air that is breathed in and out over a specified time. These spirometers are rarely used continuously because they have large dead volumes and high resistance, which make them unpleasant to use. Indirect measurement of airflow, such as with a thermocouple or capnograph, has less adverse effects, but still requires the placement of sensors in front of the nose and/or mouth. Respiratory effort/movement measurement can be monitored by measuring body volume changes; transthoracic inductance and impedance plethysmographs, strain gauge measurement of thoracic circumference, pneumatic respiration transducers, and whole-body plethysmographs are examples of indirect techniques. Each respiratory measurement method has unique advantages and disadvantages. Pulse oximetry measurements indicate that a respiratory disturbance has occurred, but do not provide respiratory rate. Airflow measurements are the most accurate, but interfere with normal respiration. The whole-body plethysmograph can be highly accurate and does not interfere with respiration, but requires immobilization of the patient. The performance of commonly used transducers (belts or electrodes) for ambulatory respiration monitoring significantly degrades over time with wear and tear. Impedance plethysmography, performed through ECG electrodes, is the most common method of continuously measuring respiratory rate in the hospital.

The electrocardiograph (ECG) is traditionally considered the standard way to measure the cardiac activity. It records the electrical activity of the heart over time. Electrical waves cause the heart muscle to contract. These waves pass through the body and can be measured at electrodes attached to the skin. Electrodes on different sides of the heart measure the activity of different parts of the heart muscle. An ECG displays the voltage between pairs of these electrodes, and the muscle activity that they measure, from different directions. This display indicates the overall rhythm of the heart, and weaknesses in different parts of the heart muscle. The other approach is pulse measurement of changes in blood volume in the skin. Pulse measurements, such as a photoplethysmograph (PPG) or piezoresistance, use optical or pressure sensors to identify pulses of blood driven by heartbeats. These are less invasive and simpler than ECG, yet both of these methods require patients to be tethered to the sensing devices.

1.2 NEED FOR NONCONTACT PHYSIOLOGICAL MONITORING

The ability to remotely detect vital signs such as heart beat and respiration is particularly useful in situations where direct contact with the subject is either impossible or unwanted. Avoidance of problems such as skin irritation, restriction of breathing, and electrode contacts is desirable in a number of health-care applications, including monitoring of patients with compromised skin, and sleep monitoring. Beyond health care, the potential applications that could benefit from remote sensing of physiological signals include fatigue monitoring, border crossing monitoring, occupancy sensors, sense through the wall, and search and rescue operations.

1.2.1 Patients with Compromised Skin

Development of reliable noninvasive physiological monitoring is an important goal in modern health-care research. Knowledge of routinely monitored heart and respiratory patterns would be clinically useful in many situations. In neonatal intensive care units, infants often suffer skin damage from adhesive tape, electrocardiogram electrodes, electroencephalogram electrodes, and transcutaneous probes, with some lesions leaving scars [Colditz et al., 1999]. Monitoring the cardiac state of burn victims can be challenging because it is sometimes difficult to find enough skin on which to apply an ECG electrode. Sometimes the electrode is stapled to the skin, or to an undebrided burn area [Loo et al., 2004]. Often an esophageal ECG must be used because adequate skin area cannot be located. A wireless heart and respiration rate monitor could fill the needs for both neonates and burn victims, by enabling the monitoring of these vital signs without contacting the skin with electrodes.

1.2.2 Sleep Monitoring

Cardiopulmonary activity is the main parameter used in the study of sleep disorders. Sleep is widely understood to play a key role in physical and mental health. The quality and quantity of sleep that an individual gets can have a significant impact on learning and memory, metabolism and weight, safety, mood, cardiovascular health, disease, and immune system function. Obstructive sleep apnea (OSA) is the most common sleep disorder with an estimated 12 million Americans suffering from it. Risk factors include gender, weight and age (being male, overweight, and over the age of 40), but sleep apnea can strike anyone at any age, even children. One of every five adults has at least mild OSA, and one of fifteen has at least moderate OSA [Young et al., 2002]. OSA has many negative effects, including excessive daytime sleepiness, increased risk of motor vehicle accidents, hypertension, psychological distress, and cognitive impairment. Apnea is the cessation of airflow for 10 s or longer, and OSA is apnea that occurs in spite of respiratory effort. To differentiate between central and obstructive apneic events, measurements of respiratory movement must be made in addition to measurements of airflow [Phillips et al., 1998]. Current laboratory polysomnography (PSG) is cumbersome, inconvenient, and expensive, causing considerable interest in portable monitoring of the condition. A Doppler radar monitoring system could identify respiratory movement, without the difficulties that accompany a full polysomnographic recording [Singh et al., 2013].

The gold standard for the clinical diagnosis of obstructive sleep apnea syndrome (OSAS) is PSG, consisting of simultaneous recordings of electrophysiological and respiratory signals, and overnight monitoring of the patient in a specially equipped sleep laboratory [Kryger et al., 2000]. However, the scarcity of sleep clinics and the expense associated with standard PSG allows treatment of small numbers of OSAS cases. The lack of awareness among the public and health-care professionals results in the vast majority of sufferers remaining undiagnosed and untreated, despite the fact that this serious disorder can have significant consequences. Untreated sleep apnea can cause high blood pressure and other cardiovascular disease, memory problems, depression, weight gain, and headaches. Moreover, untreated sleep apnea may be responsible for job impairment and motor vehicle collisions. A simple, less costly, noninvasive, reliable and ambulatory screening method for OSAS is desirable.

Sudden infant death syndrome (SIDS) is believed to be attributable to sleep apnea. Although rates of SIDS have declined sharply in the past 15 years, SIDS is still the third leading cause of infant mortality and the leading cause of postneonatal infant mortality [Arias et al., 2003; Hunt and Hauck, 2004]. In 2001, 8.1% of all infant deaths were caused by SIDS, affecting 55.5 of every 100,000 live births. Apparent life-threatening events (ALTEs), defined as an episode that is characterized by some combination of apnea, color change, marked change in muscle tone, choking, or gagging, are experienced by 2.46 of every 1000 infants [Kiechl-Kohlendorfer et al., 2005]. Although home electronic surveillance does not reduce the risk of SIDS at this time, this may be due to limits of current home-monitoring devices, or high false-negative rates. If obstructed breathing, central apnea, bradycardia, or oxygen saturation could be reliably detected, intervention could save infants' lives [Hunt and Hauck, 2004]. A Doppler radar device could detect central apneic events, where there is no respiratory motion, and bradycardia, where the heart rate slows. Doppler radar could be an integral part of a combination of sensors that could provide reliable home SIDS monitoring.

1.2.3 Elderly Monitoring

The population share of the elderly around the globe has been steadily increasing, due to improvements in health care and decrease in birth rates. The global percentage of people aged 60 years or above increased from 9.2% in 1990 to 11.7% in 2013 and will continue to grow as a proportion of the world population, projected to reach 21.1% by 2050 [United Nations, 2013]. This has resulted in an increasing need for health-care equipment specialized for routine in home monitoring of the elderly. With the reduced mobility, elderly adults may be at high risk of gait or balance disorders, which are the major causes of fall in this population and risk factors for increasing morbidity and mortality. Therefore, more specialized health-care equipment is needed for long-term monitoring of gait for the elderly.

The Doppler radar has demonstrated potential for human gait monitoring [Wang and Fathy, 2011]. A Doppler radar-based approach for gait monitoring and fall detection was proposed, with good accuracy in distinguishing common fall events from normal movements [Mercuri et al., 2013]. Such a system could be potentially linked to medical monitoring personnel to provide a prompt alert in the event of emergencies.

1.3 DOPPLER RADAR POTENTIAL FOR PHYSIOLOGICAL MONITORING

The development of Doppler radar for sensing of cardiopulmonary sensing, with proof of concept demonstrated for various applications [Li et al., 2013], could offer a platform to establish noncontact, continuous, physiological monitoring systems.

Doppler radar systems can perform noncontact sensing of respiratory and cardiac signatures at a distance, through clothing, walls, or debris. A particular advantage of Doppler radar is its ability to detect both heart and respiratory signals simultaneously, but independently, and without contact with the subject. This may be particularly useful in chronic disease management, sleep studies, heart rate variability (HRV) and energy balance studies, and for obtaining biometric signatures for security applications.

1.3.1 Principle of Operation and Power Budget

Radar, an acronym for RAdio Detection And Ranging, describes a system that transmits an electromagnetic signal and senses the echo from reflecting objects, thereby gaining information about those objects. The time delay between the transmitted and received signals indicates the distance to the target; the frequency shift of the received signal due to Doppler effect enables calculation of the target's velocity; and the strength of the signal gives information about the target's radar cross section, which provides information about its size, geometry, and composition. A major advantage of radio and microwave frequency radar systems is that these waves can penetrate through some objects that light cannot penetrate, allowing detection of objects that cannot be seen. However, radar systems developed for different applications may operate at many different frequencies, varying from a few megahertz to optical frequencies.

The Doppler effect can be observed as the change of frequency or pitch when a wave source moves either toward or away from the observer. This principle was discovered by the Austrian physicist Christian Doppler in 1842, and it applies to all wave motion. Doppler radar uses this principle to measure target velocity from the frequency shift between the transmitted electromagnetic wave and the wave reflected from the moving object.

Radar systems were originally developed for military applications including surveillance and weapon control. Radar now has many civil applications, including navigation of aircraft, ships, and spacecraft, remote sensing of the environment (including weather), and law enforcement. Depending on the radar system hardware and the type of signal sent, it may be possible to detect the range and/or angle to the target, the size and shape of the target, and the linear and/or rotational velocity of the target [Skolnik, 1990]. Depending on which of these parameters is most important to sense, as well as the range to and the nature of the target, different radar topologies may be used. A pure continuous-wave (CW) system can readily detect moving targets via the Doppler shift of the received signal, although it cannot detect the range. Frequency-modulated continuous-wave (FMCW) radar systems can detect both the range to and the velocity of the target. Altimeters and Doppler navigation devices use FMCW radar systems [Saunders, 1990]. Pulsed radar allows transmitting and receiving to occur at different times, and it is used when the return signal is much smaller than the transmitted signal and, therefore, difficult to sense the received signal in the presence of the transmitted signal [Skolnik, 1990]. Different types of Doppler radar and their applications are discussed in more detail in Chapter 2.

A low-power Doppler radar system can be used to sense physiological movement, enabling the monitoring of vital signs parameters without contact, and through clothing and bedding [Li et al., 2013]. A low-power radio frequency (RF) signal is transmitted, and as it reflects from the patient's body, the echo is modulated by physiological motion. The Doppler shift theory states that a reflected signal from an object with periodic movement but zero net velocity is phase modulated. This phase is proportional to the displacement of the subject. If the subject is the human body, the reflected signal will contain information on the positional variations on the surface due to cardiopulmonary activity. A combination of hardware and software compares the echo signal with the transmitted signal, and extracts a physiological motion signal.

CW, FMWC, and pulsed Doppler radar can be used to sense physiological movement. A CW radar topology is the simplest radar topology for two reasons: a single oscillator can be used for both the transmitter and the receiver, and the extremely narrow signal bandwidth avoids interference and rejects stationary clutter. A pure CW radar system can measure targets at any range (subject to the signal-to-noise ratio (SNR)) that are moving at any velocity (subject to the receiver bandwidth) without ambiguity, unlike pulsed or modulated systems that have limited velocity resolution. However, CW radar systems cannot detect range without modulation, and when modulated, the same range ambiguities found in pulsed radar systems are present.

When the goal of the measurement is target motion rather than distance to the target, a pure CW radar system is very effective. When the CW signal is directed at a target, it is reflected and frequency-modulated by the target motion. According to Doppler theory, a target with a time-varying position but no net velocity will reflect the signal, modulating its phase in proportion to the time-varying position of the target. A stationary person's chest has a periodic movement with no net velocity, and a CW radar with the chest as the target will, therefore, receive a signal similar to the transmitted signal, with its phase modulated by the time-varying chest position. Demodulating the phase will then provide a signal directly proportional to the chest position, which contains information about movement due to heartbeat and respiration, from which heart and respiration rates can be determined. Noncontact heart and respiration monitors have been developed based on this principle [Lin, 1992].

The most significant issues with CW radar are linked to its nature of constantly transmitting and receiving, which results in the inability to separate reflections temporally. A portion of the transmitted signal leaks from the transmitter to the receiver, either through coupling between the transmit and receive circuitry, or directly through the antenna(s). In addition, clutter reflects some of the signal and its noise sidebands back to the receiver, adding to the signal power at the transmit frequency due to leakage. These unwanted signals result in a DC offset and low-frequency noise if they are not eliminated before the signal is detected.

1.3.1.1 Frequency and Power Considerations The peak-to-peak chest motion due to respiration in adults ranges from 4 to 12 mm [DeGroote et al., 1997; Kondo et al, 1997], while the peak-to-peak motion due to the heartbeat is about 0.5 mm [Ramachandran and Singh, 1989]. The amount of phase modulation due to chest motion will be proportional to displacement and the operating frequency. At 2.4 GHz, 1 cm of displacement corresponds to about 1 rad of phase change, and the phase change increases proportionally to increasing frequency. The ability of the system to discern changes in phase will depend on the overall SNR that is determined by the size of the moving surface, displacement amplitude, range to target, and electrical properties of the radar system.

The electrical properties of biological tissue affect the amount of signal that is reflected and transmitted, both at the skin–air interface and at interfaces between different tissues within the body. The electric properties of biological tissue depend on frequency of operation, and are largely governed by the percentage of water content. Tissues with high water content, such as skin, muscle, and blood, are more lossy and more readily absorb electromagnetic waves. As the frequency of operation increases, the losses increase, whereas tissues with low water content, such as bone and fat, are largely transparent to electromagnetic waves. The tissue contrast based on different absorption and propagation characteristics has been used for microwave imaging.

Doppler radar detects all motion in the radar field of view. If the antenna is placed in contact with the skin, internal organ motion may be measured, assuming that there is enough penetration for the electromagnetic wave to reach the internal organs, and propagate back to the receiver after the reflection. If the Doppler radar is placed at some standoff distance from a subject, at frequencies of 2.4 GHz and above, about 50% of the incident power will be reflected at the air–skin interface, and more than 90% of the reflected power will come from the surface reflection. Thus, in this case, for all practical purposes, we can assume that noncontact Doppler radar physiological measurements are measurements of skin surface motion.

1.3.2 History of Doppler Radar in Physiological Monitoring

Physiological motion detection with CW Doppler radar has been known since the 1970s [Lin, 1975], and with FMCW [Sharpe, 1990] and ultra-wideband (UWB) [Staderini, 2002] Doppler radar since the 1980s. Understanding of microwave noninvasive physiological sensing has advanced tremendously in the past few decades, and recent advances in wireless technology have enabled further progress in medical radar, culminating with recent FDA approvals. Widespread use of microwave technology and digital processors in common household communications devices has driven down costs, making it possible to develop practical radar monitors that cost significantly less than conventional cardiopulmonary assessment instruments.

Microwave Doppler radar monitoring of respiratory and cardiac movements was first demonstrated in the late 1970s, when respiration [Lin, 1975; Lin et al., 1977] and heart beats [Lin et al., 1979] were measured separately, with a breath-hold required for the heart measurement [Lin et al., 1979]. X-band sweep oscillators were used, with horn antennas directing the microwave energy toward the upper torso of the subjects. A nonanesthetized rabbit's respiration was measured from a distance of 30 cm [Lin, 1975]. In [Lin et al., 1977], the same system was used with an apnea-detector circuit, and was tested on a rabbit and two cats, all of which were anesthetized and intubated. Both hyperventilation and apnea were induced in the animals, and both states were clearly detected by the microwave monitor. Microwave apexcardiography was demonstrated by using a continuous-wave 2-GHz microwave signal with an antenna placed 3 cm above the apex, and the precordial motions were easily detected [Lin et al., 1979].

From the mid-1980s through the late 1990s, radar transceivers were developed that incorporated analog and digital signal processing to separate the small heart signal from the much larger respiration signal, so the subject did not need to hold his/her breath for the heart rate to be measured, and heart and respiration could be measured simultaneously [Chan and Lin, 1987; Chen et al., 2000; Chuang et al., 1990; Greneker, 1997; Seals et al., 1986]. These transceivers were used for the detection of heart and respiration rates of persons in rubble, persons behind walls, and Olympic athletes. An analog amplification and filtering for separation of heart and respiration signatures was combined with 8-bit digitization and digital signal processing to detect heart and respiration rates [Chan and Lin, 1987]. An automatic clutter-cancellation circuit was developed to facilitate measurement of the heart and respiration signatures through seven layers of brick [Chuang et al., 1990] and through 10 ft of rubble [Chen et al., 2000]. Heart and respiration rates of athletes were successfully detected at ranges exceeding 10 m [Greneker, 1997]. At 100 m standoff distance, the limit was moving background clutter, not the system sensitivity. A quadrature receiver was used to avoid phase-demodulation null points [Seals et al., 1986].

More recently, Doppler radar vital signs monitoring was explored to detect hypovolemic states and shock in persons under rubble or in biochemical hazard conditions that could pose danger to health-care providers [Matsui et al., 2004a, 2004b]. Hypovolemic rabbits and rabbits in shock could be reliably distinguished from the control rabbits based on the Doppler radar information by using linear discriminant analysis on the heart and respiration rates. The hypovolemic rabbits have higher heart and respiration rates. Doppler radar was also used to estimate arterial blood pH without contact using heart and respiration rate monitoring coupled with an infrared thermographic temperature measurement and an exhaled gas (CO and CO_2) analyzer [Matsui et al., 2006]. This measurement was successful at estimating blood pH using linear regression analysis on hypovolemic rabbits.

The connection of this technology to the existing wireless communications infrastructure was also investigated [Boric-Lubecke and Lubecke, 2002; Lubecke et al., 2002; Boric-Lubecke et al., 2003]. A modified wireless LAN PCMCIA card was used to detect heart and respiration [Boric-Lubecke et al., 2003], and a module that combines the transmitted and reflected signals from any wireless communication device, such as a cordless telephone, was used to detect heart and respiration [Lubecke et al., 2002]. Using this technology to directly connect Doppler measurement of heart and respiration rate to health-care providers has been proposed [Boric-Lubecke and Lubecke, 2002].

In addition, UWB radar has been used for the measurement of heart and respiration rates. Using 0.4-W pulses and a 1-GHz central frequency, heart rates were detected through 1 m of air and a 0.4-m brick wall [Immoreev and Samkov, 2005] and respiration was measured at up to 5 m [Ossberger et al., 2004].

Research efforts in the last decade have been moving the technology development toward lower power, lighter weight, smaller form factor, better accuracy, longer detection range, and more robust operation for practical portable and handheld applications. Among many possible applications this technology can be used for, health care seems to be drawing the most interest. As an example, a baby monitor using this technology was demonstrated to monitor SIDS [Hafner et al., 2007; Li et al., 2009]. The Doppler radar embedded into the baby monitor detects tiny baby movements induced by breathing. If no movement is detected within 20 s, an alarm goes off to warn the parents. Operating in similar ways, biomedical Doppler radar is also being investigated as a cost-effective solution for long-term monitoring of sleep apnea [Singh et al., 2013]. Human studies in clinical environment have validated this technology as a potential substitute for conventional respiratory monitors [Droitcour et al., 2009; Massagram et al., 2009] and a useful tool for precise assessment of key parameters relating to cardiopulmonary activity including body orientation [Kiriazi et al., 2012]. Furthermore, recent studies have demonstrated that Doppler radar could help a medical linear accelerator to track the location of a mobile tumor during radiotherapy with the help of advanced signal-processing algorithms [Li et al., 2011; Gu et al., 2012]. Doppler radar has also been applied to monitor the health and behavior of land and sea animals including lizards and fish [Singh et al., 2012b, 2012a; Hafner et al., 2012].

With growing interests in health and life sciences in the engineering community, many researchers have been contributing to the technology advancement in this field. This has led to various radar front-end architectures including the conventional homodyne/heterodyne [Xiao et al., 2006], self-/mutual-injection locking [Wang et al., 2011], and coherent low-IF [Mostafanezhad and Boric-Lubecke, 2014]. Each of these architectures shows its specific advantage in certain environments. Signal conditioning and processing methods such as adaptive DC calibration [Vergara et al., 2008b; Gu and Li, 2012], arctangent demodulation [Park et al., 2007], and noise cancellation [Li and Lin, 2008; Oum et al., 2008; Fletcher and Jing, 2009; Wiesner, 2009] have been proposed to enable practical applications of the biomedical radar. Various techniques, including blind source separation (BSS) and the use of passive RF tags, have been applied to isolate multiple targets and clutter noise [Vergara et al., 2008a; Singh and Lubecke, 2013]. Hardware implementations from benchtop fast prototyping using fundamental RF/microwave instruments [Gu et al., 2010] to radar-on-chip application-specific integrated circuits (ASICs) [Droitcour et al., 2002, 2004; Li et al., 2008, 2010] have been demonstrated to make this technology available to both researchers and general users.

The selection of published works outlining Doppler radar physiological monitoring history from 1975 to date are described briefly in Table 1.1 with the year of publication, reference, description, and results.

TABLE 1.1 Doppler Radar Physiological Monitoring from 1975 to 2014

References	Description	Results
Lin [1975]	X-band sweep oscillator, rectangular horn antenna	Respiration measured on rabbit and human at 30-cm range
Lin et al. [1977]	Microwave apnea detector proposed for low-birth-weight infants 3-mW at 30-cm, 3-cm by 3-cm horn	Respiration measured on cats and rabbit at 30-cm range Detection of apnea and hyperventilation
Lin et al. [1979]	"Apexcardiography" heart measurements made while breath held Signal amplitude and phase vary with antenna location	Heart measurement made 3-cm from apex
Seals et al. [1986]	Analog signal processing separates heart and respiration signs Digital rate detection algorithm 10- and 3-GHz quadrature radar systems	Measured heart and respiration simultaneously with digital rate detection algorithm
Chen et al. [2000]	An X-band microwave life-detection system has been developed for vital signs	Measured heartbeat and breathing of human subjects lying on the ground at a distance of about 30 m or located behind a cinder block wall
Chan and Lin [1987]	Heart and respiration separated with analog and digital signal processing: amplification, filtering, 8-bit ADC sampled at 80 Hz 10.5-GHz, 10-mW, horn antenna a few centimeter from subject	Heart and respiration obtained simultaneously at 5–7-cm range
Chuang et al. [1990]	Heart and respiration for detecting victims in clutter 10-GHz system does not penetrate wet bricks, but 2-GHz system does Automatic clutter cancellation algorithm introduced	Successful measurement with subject both faceup and facedown, through 4–7 layers of brick, with 2- and 10-GHz radar systems

(*Continued*)

TABLE 1.1 ***(Continued)***

References	Description	Results
Greneker [1997]	0.6-m dish aimed at subject's thorax 24-GHz, 30-mW output signal, 40-dB antenna gain	Heartbeat and respiration measured at ranges exceeding 10 m
Chen et al. [2000]	Life-detection system for victims under rubble or behind barriers including 6 in. of steel, bricks, and cylinder blocks Heart and respiration were measured	450-MHz signal penetrates deepest into concrete rubble without metal 1150-MHz signal penetrates rubble with metallic wire mesh
Droitcour et al. [2002]	The first CMOS and BiCMOS Doppler radar chips	A fully integrated direct conversion Doppler radar that detects heart and respiration movement at a distance of 50 cm. The 1.6 GHz transceiver is implemented in both CMOS and BiCMOS technologies
Lubecke et al. [2002]	Add-on module uses signals from existing wireless devices to measure heart and respiration rates	Heart and respiration were measured with a 2.4-GHz cordless phone and a 2.4-GHz signal generator
Boric-Lubecke et al. [2003]	Modified wireless LAN PCMCIA cards are used to sense heart and respiration rates	Heart and respiration were obtained successfully at 40-cm
Droitcour et al. [2004]	Range correlations and I/Q performance benefits in single-chip Doppler radars	The range-correlation effect on residual phase noise is a critical factor when detecting small phase fluctuations with a high-phase-noise on-chip oscillator. Phase-noise reduction due to range correlation was experimentally evaluated
Matsui et al. [2004a]	Determining hypovolemic and shock states using linear discriminant analysis Heart and respiration rates of hypovolemic rabbits 1215 MHz, 70-mW output power	Linear discriminant analysis effectively predicted hypovolemic state of 10 rabbits
Ossberger et al. [2004]	Ultra-wideband pulse radar Wavelet signal processing	Respiration measured at 1–5-m and through a wall at 85-cm

Immoreev and Samkov [2005]	Ultra-wideband radar, 1-GHz central frequency	Measured respiration through 1-m air and 0.4-m brick wall on one subject
Matsui et al. [2006]	Used 1215-MHz radar, infrared thermography, and exhaled CO/CO_2 analyzer to estimate blood pH	Significant correlation of measured and calculated blood pH on rabbits
Xiao et al. [2006]	Heterodyne Ka-band radar	Double-sideband transmission is used for avoiding null detection point
Park et al. [2007]	Used 2.4-GHz quadrature receiver. Arctangent demodulation with DC offset compensation to detect heart and respiration rate	Robust and accurate output data obtained from arctangent demodulation. Unwanted DC offset was successfully eliminated
Hafner et al. [2007]	Vital signs sensing using a baby monitor and a passive sensor node	Signal from a baby monitor, with an addition of a passive sensor is used to detect vital signs
Li and Lin [2008]	4–7-GHz quadrature transceiver with the complex and arctangent demodulation for random movement cancellation	The complex signal demodulation is more favorable for random body movement cancellation. The arctangent demodulation has the advantage of eliminating the harmonic and intermodulation interference at high frequencies using high gain antennas
Li et al. [2008]	A 5 GHz double-sideband radar sensor chip	The differential architecture has the advantage of reducing local oscillation leakage. The 5 GHz double-sideband system can avoid null detection point by frequency tuning
Vergara et al. [2008b]	DC offset cancellation and DC information preservation	Digitally controlled voltage feedback and center finding preserves the important DC information for optimal extraction of phase information in the quadrature system
Vergara et al. [2008a]	Blind source separation of human motion	The successful separation of cardiopulmonary motion and hand motion for a single subject

(*Continued*)

TABLE 1.1 ***(Continued)***

References	Description	Results
Wiesner [2009]	CW vital sign radar with enhanced target detection probability in a highly cluttered environment with range and angle-of-arrival estimation	A respiration signature around 0.2 Hz and higher harmonics along with heartbeat signature around 1.3 Hz are clearly discernible
Massagram et al. [2009]	2.4-GHz quadrature system with linear demodulation method to detect heart rate variability (HRV) and respiratory sinus arrhythmia (RSA)	The data were obtained from 12 human subjects in seated and supine positions. High accuracy in extracting the HRV and RSA indices was achieved
Droitcour et al. [2009]	The first clinical validation of the respiratory rate accuracy	A human study in a clinical environment validated Doppler radar technology as a potential substitute for conventional respiratory monitors
Li et al. [2009]	Infant vital sign monitor	The Doppler radar embedded into the baby monitor detects tiny baby movements induced by breathing. If no movement is detected within 20 s, an alarm goes off to warn the parents
Fletcher and Jing, 2009	Dual-antenna differential radar front-end system for short-range heart rate and background motion noise removal	Designed two helical directional antennas at 2.46-GHz and 2.510-GHz to detect heart rate at 0.5-m distance with motion noise cancellation
Li et al. [2010]	A direct-conversion 5.8-GHz radar sensor chip with 1-GHz bandwidth	Analyses on sensitivity and link budget guide the design of high-sensitivity noncontact vital sign detector. This radar sensor chip is software configurable to set the operation point and detection range for optimal performance
Gu et al. [2010]	Fast prototyping using fundamental RF/microwave instruments	Heterodyne digital quadrature demodulation architecture that helps mitigate quadrature channel imbalance and eliminate the complicated DC offset calibration required for arctangent demodulation

Wang et al. [2011]	Single antenna self-injection-locked (SIL) radar presented to detect vital signs with random body motion cancellation	Two-radar array placed 2-m away at 2.4 GHz is implemented with random body motion cancellation for monitoring subject jogging on treadmill. Provided theoretical basis for determining signal-to-noise spectral density ratio
Gu et al. [2012]	DC-coupled adaptive tuning radar sensor to precisely measure respiratory movement in motion-adaptive radiotherapy	Submillimeter accuracy was achieved in measurement for respiratory gating and has potential tumor tracking
Kiriazi et al. [2012]	Dual-frequency (2.4 and 5.8-GHz) radar for cardiopulmonary effective radar cross section (ERCS) and displacement in the direction of incidence	Results showed that an ERCS is larger for the back of the torso and smaller for the side than to the front, while the respiration depth is smaller in the prone position than in supine
Singh et al. [2012a]	Lizard activity monitoring and motion classification	Lizard monitoring in a nonconfined laboratory environment and uses multiple Doppler radars operating at 10.525 GHz
Hafner et al. [2012]	Fish heart rate monitoring	The heart rate detected by the body contact radar matched the heart rate detected by ECG
Singh et al. [2013]	Doppler radar for clinical sleep monitoring	Doppler radar is being investigated as a cost-effective solution for long-term monitoring of sleep apnea
Singh and Lubecke [2013]	2.45-GHz system with a frequency doubling passive planar harmonic tag	The harmonic radar system isolates the respiration of a tagged human subject from an untagged large scattering object
Mostafanezhad and Boric-Lubecke [2014]	Coherent low-IF receiver architecture	Measurements on a mechanical target and a human subject demonstrate a signal-to-noise ratio improvement of 7 dB, which can increase the range of operation by 50%

REFERENCES

Arias E, MacDorman MF, Strobino DM, Guyer B. Annual summary of vital statistics – 2002. Pediatrics 2003;112(6):1215–1230.

Boric-Lubecke O, Atwater G, Lubecke VM. Wireless LAN PC card sensing of vital signs. Proceedings of IEEE Topical Conference on Wireless Communications Technology; 2003. p 206–207.

Boric-Lubecke O, Lubecke VM. Wireless house calls: using communications technology for health care monitoring. IEEE Microwave Mag 2002;3(3):43–48.

Chan KH, Lin JC. Microprocessor-based cardiopulmonary rate monitor. Med Biol Eng Comput 1987;25(9):41–44.

Chen K-M, Huang Y, Zhang J, Norman A. Microwave life-detection systems for searching human subjects under earthquake rubble or behind barrier. IEEE Trans Biomed Eng 2000;47(1):105–114.

Chuang H-R, Chen Y-F, and Chen K-M. Microprocessor-controlled automatic clutter-cancellation circuits for microwave systems to sense physiological movements remotely through the rubble. Proceedings of the Instrumentation and Measurement Technology Conference; 1990. p 177–181.

Colditz PB, Dunster KR, Joy GJ, Robertson IM. Anetoderma of prematurity in association with electrocardiographic electrodes. J Am Acad Dermatol 1999;41(3):479–481.

DeGroote A, Wantier M, Cheron G, Estenne M, Pavia M. Chest wall motion during tidal breathing. J Appl Physiol 1997;83(5):1531–1537.

Droitcour AD, Boric-Lubecke, O, Lubecke VM, Lin, J. 0.25 μm CMOS and BiCMOS single-chip direct-conversion Doppler radars for remote sensing of vital signs. IEEE International Solid-State Circuits Conference 2002, ISSCC 2002. Digest of Technical Papers; 2002. Vol. 1, p 348–349.

Droitcour AD, Boric-Lubecke O, Lubecke VM, Lin J, Kovacs GTA. Range correlation and I/Q performance benefits in single-chip silicon Doppler radars for noncontact cardiopulmonary monitoring. IEEE Trans Microwave Theory Tech 2004;52:838–848.

Droitcour AD, Seto TB, Byung-Kwon P, Yamada S, Vergara A, El Hourani C, Shing T, et al., Non-contact respiratory rate measurement validation for hospitalized patients. Engineering in Medicine and Biology Society, 2009, EMBC 2009. Annual International Conference of the IEEE; 2009. p 4812–4815.

Fletcher R, Jing H. Low-cost differential front-end for Doppler radar vital sign monitoring. IEEE MTT-S International Microwave Symposium Digest, 2009, MTT'09; 2009. p 1325–1328.

Greneker EF. Radar sensing of heartbeat and respiration at a distance with applications of the technology. Radar 97 Conference Proceedings; 1997. p 150–154.

Gu C, Li C. DC coupled CW radar sensor using fine-tuning adaptive feedback loop. Electron Lett 2012;48:344–345.

Gu C, Li C, Lin J, Long J, Huangfu J, Ran L. Instrument-based noncontact Doppler radar vital sign detection system using heterodyne digital quadrature demodulation architecture. IEEE Trans Instrum Meas 2010;59:1580–1588.

Gu C, Li R, Zhang H, Fung A, Torres C, Jiang S, Li C. Accurate respiration measurement using DC-coupled continuous-wave radar sensor for motion-adaptive cancer radiotherapy. IEEE Trans Biomed Eng 2012;59:3117–3123.

Hafner N, Drazen JC, Lubecke VM. Fish heart rate monitoring by body-contact Doppler radar. IEEE Sensors J 2012;13:408–414.

Hafner N, Mostafanezhad I, Lubecke VM, Boric-Lubecke O, Host-Madsen A. Non-contact cardiopulmonary sensing with a baby monitor. Engineering in Medicine and Biology Society, 2007, EMBS 2007. 29th Annual International Conference of the IEEE; 2007. p 2300–2302.

Hunt CE, Hauck FR. Sudden infant death syndrome. In: Behrman RE, Kliegman RM, Jenson HB, editors. Nelson Textbook of Pediatrics. 17th ed. Philadelphia: Saunders; 2004. p 1380–1385.

Immoreev IY, Samkov S. Short-distance ultrawideband radars. IEEE Aerospace Electron Syst Mag 2005;20(6):9–14.

Kiechl-Kohlendorfer U, Hof D, Peglow UP, Traweger-Ravanelli B, Kiechl S. Epidemiology of apparent life threatening events. Am Acad Pediatr Grand Rounds 2005;90(3):297–300.

Kiriazi JE, Boric-Lubecke O, Lubecke VM. Dual-frequency technique for assessment of cardiopulmonary effective RCS and displacement. IEEE Sensors J 2012;12:574–582.

Kondo T, Uhlig T, Pemberton P, Sly PD. Laser monitoring of chest wall displacement. Eur Resp J 1997;10:1865–1869.

Kryger MH, Roth T, Dement WC. Principles and Practice of Sleep Medicine. 3rd ed. Philadelphia: W.B. Sunders Co.; 2000. p 869.

Li C, Cummings J, Lam J, Graves J, Wu W. Radar remote monitoring of vital signs. IEEE Microwave Mag 2009;10(1):47–56.

Li C, Gu C, Li R, Jiang SB. Radar motion sensing for accurate tumor tracking in radiation therapy. 12th Annual IEEE Wireless and Microwave Technology Conference, Clearwater, FL; 2011.

Li C, Lin J. Random body movement cancellation in Doppler radar vital sign detection. IEEE Trans Microwave Theory Tech 2008;56:3143–3152.

Li C, Lubecke VM, Boric-Lubecke O, Lin J. A review on recent advances in Doppler radar sensors for noncontact healthcare monitoring. IEEE Trans Microwave Theory Tech 2013;61(5):2046–2060.

Li C, Xiao Y, Lin J. A 5 GHz double-sideband radar sensor chip in 0.18 μm CMOS for non-contact vital sign detection. IEEE Microwave Wireless Compon Lett 2008;18:494–496.

Li C, Yu X, Lee C-M, Li D, Ran L, Lin J. High-sensitivity software-configurable 5.8-GHz radar sensor receiver chip in 0.13-μm CMOS for noncontact vital sign detection. IEEE Trans Microwave Theory Tech 2010;58:1410–1419.

Lin JC. Non-invasive microwave measurement of respiration. Proc IEEE 1975;63(10):1530.

Lin JC. Microwave sensing of physiological movement and volume change: a review. Bioelectromagnetics 1992;13(6):557–565.

Lin JC, Dawe E, Majcherek J. A noninvasive microwave apnea detector. Proceedings of the San Diego Biomedical Symposium; 1977. p 441–443.

Lin JC, Kiernicki J, Kiernicki M, Wollschlaeger PB. Microwave apexcardiography. IEEE Trans Microwave Theory Tech 1979;27(6):618–620.

Loo S, Kuo T, Waters G, Muller M, Brown TLH. A mobile electrode for ECG monitoring. Burns 2004;30(2):203.

Lubecke VM, Boric-Lubecke O, Beck E. A compact low-cost add-on module for Doppler radar sensing of vital signs using a wireless communications terminal. The IEEE Microwave Theory and Techniques Symposium Digest; 2002. p 1767–1770.

Lubecke VM, Boric-Lubecke O, Host-Madsen A, Fathy AE. Through-the-wall radar life detection and monitoring. IEEE/MTT-S International Microwave Symposium, 2007; 3–8 June 2007. p 769–772.

Massagram W, Lubecke VM, Host-Madsen A, Boric-Lubecke O. Assessment of heart rate variability and respiratory sinus arrhythmia via Doppler radar. IEEE Trans Microwave Theory Tech 2009;57(10):2542–2549.

Matsui T, Hagisawa K, Ishizuka T, Takase B, Ishihara M, Kikuchi M. A novel method to prevent secondary exposure of medical and rescue personnel to toxic materials under biochemical hazard conditions using microwave radar and infrared thermography. IEEE Trans Biomed Eng 2004b;51(12):2184–2188.

Matsui T, Hattori H, Takase B, Ishihara M. Non-invasive estimation of arterial blood pH using exhaled CO/CO_2 analyzer, microwave radar, and infrared thermography for patients after massive hemorrhage. J Med Eng Technol 2006;20(2):97–101.

Matsui T, Ishizuka T, Takase B, Ishihara M, Kikuchi M. Non-contact determination of vital sign alterations in hypovolemic states induced by massive hemorrhage: an experimental attempt to monitor the condition of injured persons behind barriers or under disaster rubble. Med Biol Eng Comput 2004a;42(6):807–811.

Mercuri M, Soh PI, Pandey G, Karsmakers P, Vandenbosch GAE, Leroux P, Schreurs D. Analysis of an indoor biomedical radar-based system for health monitoring. IEEE Trans Microwave Theory Tech 2013;61(5):2061–2068.

Mostafanezhad I, Boric-Lubecke O. Benefits of coherent low-IF for vital signs monitoring using Doppler radar. IEEE Trans Microwave Theory Tech 2014;62(10):2481–2487.

Ossberger G, Buchegger T, Schimback E, Stetzler A, Weigel R. Non-invasive respiratory movement detection and monitoring of hidden humans using ultra wideband pulse radar. Proceedings of the International Workshop on Ultrawideband Technologies; 2004. p 395–399.

Oum JH, Kim D-W, Hong S. Two frequency radar sensor for non-contact vital signal monitor. 2008 IEEE MTT-S International Microwave Symposium Digest; 2008. p 919–922.

Park BK, Boric-Lubecke O, Lubecke VM. Arctangent demodulation with DC offset compensation in quadrature Doppler radar receiver systems. IEEE Trans Microwave Theory Tech 2007;55:1073–1079.

Phillips BA, Anstead MI, Gottlieb DJ. Monitoring sleep and breathing: methodology; Part I: Monitoring breathing. Clin Chest Med 1998;19(1):203–212.

Ramachandran G, Singh M. Three-dimensional reconstruction of cardiac displacement patterns on the chest wall during the P, QRS, and T-segments of the ECG by laser speckle interferometry. Med Biol Eng Comput 1989;27(5):525–530.

Saunders WK. CW and FM radar. In: Skolnik MI, editor. Radar Handbook. 2nd ed. San Francisco: McGraw-Hill, Inc.; 1990. p 14.1–14.45.

Seals J, Crowgey SR, Sharpe SM. Electromagnetic vital signs monitor. Georgia Tech Research Institute Biomedical Division, Atlanta, GA, Final Report Project A-3529-060; 1986.

Sharpe S, Seals J, MacDonald A, Crowgey S. Non-contact vital signs monitor. US patent# 4,958,638. 1990.

Singh A, Baboli M, Gao X, Yavari E, Padasdao B, Soll B, Boric-Lubecke O, Lubecke VM. Considerations for integration of a physiological radar monitoring system with gold standard clinical sleep monitoring systems. 35th Annual International Conference of the IEEE Engineering in Medicine and Biology Society (EMBC), 2013, 3–7 July 2013. p 2120–2123.

Singh A, Hafner N, Butler M, Lubecke VM. A data efficient method for characterization of Chameleon Tongue motion using Doppler radar. Annual International Conference of the IEEE Engineering in Medicine and Biology Society, 2012, EMBC 2012; 2012b.

Singh A, Lee S, Butler M, Lubecke VM. Activity monitoring and motion classification of the Lizard Chamaeleo Jacksonii using multiple Doppler radars. Annual International Conference of the IEEE Engineering in Medicine and Biology Society, 2012, EMBC 2012; 2012a.

Singh A, Lubecke VM. Adaptive noise cancellation for two frequency radar using frequency doubling passive RF tags. IEEE Trans Microwave Theory Tech 2013;61(8):2975–2981.

Skolnik MI. An introduction to radar. In: Skolnik MI, editor. Radar Handbook. 2nd ed. San Francisco: McGraw-Hill, Inc.; 1990. p 1.1–1.21.

Staderini EM. UWB radars in medicine. IEEE Aerospace Electron Syst Mag 2002;17(1):13–18.

Tarassenko L, Hann A, Yound D. Integrated monitoring and analysis for early warning of patient deterioration. Br J Anaesth 2006;97(1):64–68.

United Nations. World Population Ageing. New York, NY: Department of Economic and Social Affairs, Population Division; 2013.

Vergara A, Boric-Lubecke O, Lubecke VM. DC information preservation for cardiopulmonary monitor utilizing CW Doppler radar. 30th Annual International Conference of the IEEE Engineering in Medicine and Biology Society, 2008, EMBS 2008; 20–25 August 2008b. p 1246–1249.

Vergara A, Petrochilos N, Boric-Lubecke O, Host-Madsen A, Lubecke V. Blind source separation of human body motion using direct conversion Doppler radar. IEEE MTT-S International Microwave Symposium Digest; 2008a. p 1321–1324.

Wang Y, Fathy AE. Micro-Doppler signatures for intelligent human gait recognition using a UWB impulse radar. 2011 IEEE International Symposium on Antennas and Propagation (APSURSI); 3–8 July 2011. p 2103–2106.

Wang F-K, Horng T-S, Peng K-C, Jau J-K, Li J-Y, Chen C-C. Single-antenna Doppler radars using self and mutual injection locking for vital sign detection with random body movement cancellation. IEEE Trans Microwave Theory Tech 2011;59(12):3577–3587.

Webster J. Medical Instrumentation. Wiley; 2010.

Wiesner A. A multifrequency interferometric CW radar for vital signs detection. 2009 IEEE Radar Conference; 2009. p 1–4.

Xiao Y, Lin J, Boric-Lubecke O, Lubecke VM. Frequency-tuning technique for remote detection of heartbeat and respiration using low-power double-sideband transmission in the Ka-band. IEEE Trans Microwave Theory Tech 2006;54(5):2023–2032.

Yavari E, Song C, Lubecke VM, Boric-Lubecke O. Is there anybody in there?: Intelligent radar occupancy sensors. IEEE Microwave Mag 2014;15(2):57–64.

Young T, Peppard PE, Gottlieb DJ. Epidemiology of obstructive sleep apnea: a population health perspective. Am J Respir Crit Care Med 2002;165(9):1217–1239.

2

RADAR PRINCIPLES

EHSAN YAVARI, OLGA BORIC-LUBECKE, AND SHUHEI YAMADA

Department of Electrical Engineering, University of Hawaii at Manoa, Honolulu, Hawaii, United States

Radar is an acronym for *Ra*dio *D*etection *A*nd *R*anging; a radar system transmits an electromagnetic signal and observes the echo reflected from one or more objects, to obtain information about the presence, position, and motion of the object(s). In the early days of radar (1920s–1950s), the main functions of radar systems were limited to the detection of a target and estimation of its range. Since then, advances in radar system hardware and software have enabled radar systems to detect, differentiate, classify, image, and track the range, altitude, direction, or velocity of multiple moving or fixed targets simultaneously.

2.1 BRIEF HISTORY OF RADAR

Radar systems were originally developed for military applications including surveillance and weapon control. Current applications of radar are very broad and include navigation of aircraft, ships, and spacecraft, automotive anticollision systems, meteorological precipitation monitoring, radio astronomy, and geological observations, in addition to military applications such as air-defense systems, antimissile systems, and guided-missile systems. The use of electromagnetic waves was pioneered by James Clerk Maxwell, who developed classical electromagnetic theory in the 1860s, and Heinrich Hertz who was the first to demonstrate the transmission and reflection of radio waves in the late 1880s. The use of reflected electromagnetic waves to

Doppler Radar Physiological Sensing, First Edition.
Edited by Olga Boric-Lubecke, Victor M. Lubecke, Amy D. Droitcour, Byung-Kwon Park, and Aditya Singh.

detect objects was not explored until the 1900s, when the Doppler effect, which was described by Christian Andreas Doppler in 1842 [Coman, 2005], was used to detect distant moving ships objects by Christian Huelsmeyer and Nikola Tesla. In the 1920s, Dr. A. Hoyt Taylor of the Naval Research Laboratory developed a radar for ship tracking that was first installed on a ship in 1937 [Skolnik, 1990]. In 1924, Sir Edward Victor Appleton used what is now known as frequency-modulated continuous-wave (FMCW) radar to prove the existence of and measure the distance to the ionosphere [Saunders, 1990]. Sir Robert Alexander Watson Watt developed a radar system to detect storms while working at the British Meteorological Office. In 1935, Sir Watt developed a radar system for detecting enemy aircraft before they were visible, and he received a patent for the first pulsed radar system. By 1939, Great Britain had a chain of radar stations along its coasts to detect enemy arrivals by air and by sea, which was instrumental in World War II. During World War II, imaging radars and sweep displays were developed.

After World War II was over, radar was applied to several civilian applications and was further advanced for additional military applications, as described in Section 2.5.

2.2 RADAR PRINCIPLE OF OPERATION

When a radar system transmits a pulse of radio frequency (RF) energy, a small portion of that energy is reflected by objects in the path of the transmitted pulse. Because the speed of electromagnetic waves in free space is the same as the speed of light, it is possible to calculate the range of the objects from the time between when the pulse was transmitted and when the reflection was received. If the radar system is configured to measure the frequency of the reflected electromagnetic signal, the object's velocity can be calculated from the shift in frequency from the transmitted signal to the received signal. The power of the reflected signal provides information about the size, geometry, and composition of the object. A major advantage of radio and microwave radar systems is that the electromagnetic waves in this frequency range can penetrate clouds, fog, and dust, enabling detection of the objects that are not visible [Skolnik, 1990].

As shown in Fig. 2.1, a radar system typically consists of a transmitter, an antenna, a receiver, and a signal-processing unit. The transmitter generates the electromagnetic wave and amplifies it to the required power. A directional antenna both concentrates the wave in the direction of the target and enables determination of the direction of the target; electronically tunable antenna arrays are often used for this purpose. The receiver converts the signal from the transmission frequency to either an intermediate frequency or baseband, separates the signal from both noise and interference, and amplifies the signal enough for digitization and/or display. Signal processing is used to reject clutter and other noise, while discriminating the desired signal from interference, and to extract information from the signal.

In principle, radar can operate at any radio frequency; however, practical considerations of antenna size, transmitter power, range to target, and radar resolution will determine constrains on operation frequency. Radio frequency range is usually

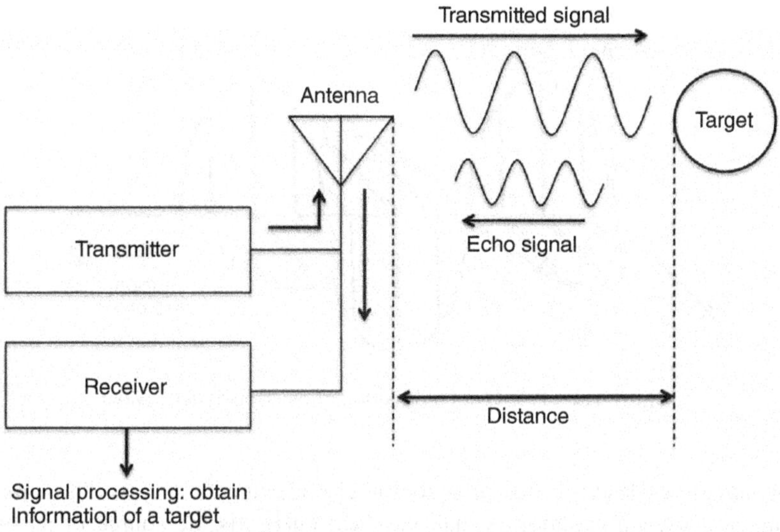

Figure 2.1 Basic principle of radar. A target will reflect an echo signal; the echo signal's power, phase delay, and frequency depends on the target's distance, radar cross section, and velocity.

defined as frequency range from 3 kHz to 300 GHz. Standard radar-operating frequencies are in the range of 3 MHz–300 GHz, as defined by IEEE Standard for Letter Designation for Radar-Frequency Bands.

2.2.1 Electromagnetic Wave Propagation and Reflection

Electromagnetic energy travels through air at the speed of light of 3×10^8 m/s. The electromagnetic waves are reflected if they encounter a material with a different dielectric constant. A radar system has a receiver intended to detect these reflected waves, indicating an object with a different dielectric constant in the propagation direction. The maximum range of a radar system is slightly affected by atmospheric and weather conditions, since the signal's power loss increases as water vapor absorbs electromagnetic power. In general, as the frequency of the electromagnetic wave increases, the power loss increases, except for several frequency regions resonant with water vapor.

The properties of electromagnetic field are described by Maxwell's equations:

$$\begin{aligned}
\nabla \cdot E &= \frac{\rho}{\varepsilon} \\
\nabla \cdot H &= 0 \\
\nabla \times E &= -\mu \frac{\partial H}{\partial t} \\
\nabla \times H &= \varepsilon \frac{\partial E}{\partial t} + j
\end{aligned} \tag{2.1}$$

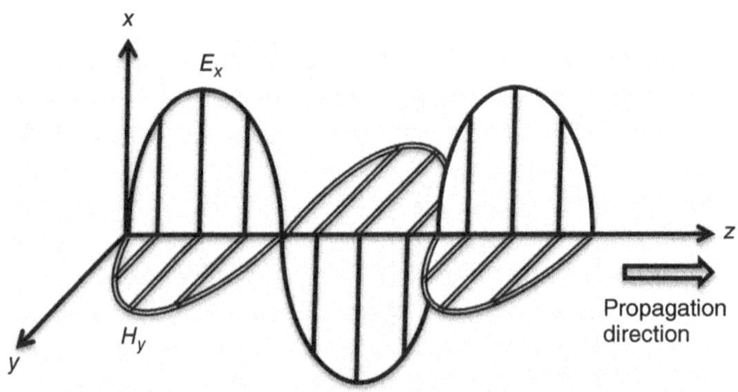

Figure 2.2 Propagation direction of electromagnetic wave.

These equations describe the behavior of electromagnetic fields. The simplified explanation of each equation is that electrical charge is the source of an electric field (Gauss's law), there are no magnetic monopoles (Gauss's law for magnetism), time-varying magnetic field induces an electric field (Faraday's law), and time-varying electric field and/or current induces a magnetic field (Ampere's law). In conjunction, Faraday's law and Ampere's law indicate that time-varying electric and magnetic fields mutually induce each other, generating a time-varying electromagnetic field that propagates as a wave. The propagation direction of the electromagnetic wave is orthogonal to the plane of both the electrical and magnetic fields, as shown in Fig. 2.2. In this case, electric field is the x component E_x, magnetic field is the y component H_y, and the wave is propagated along z axis.

If there is a discontinuity in the dielectric constant of the medium the electromagnetic wave is propagating through, some of the wave will be transmitted and some will be reflected, depending on the properties of the two materials. This situation is illustrated in Fig. 2.3.

2.2.2 Radar Cross Section

The radar cross section (RCS) is the property of a scattering object, or target, which represents the fraction of the echo signal returned to the radar by the target compared with the echo that would be reflected by a perfectly conducting sphere with a 1 m^2 cross-sectional area.

A definition of the RCS also can be expressed in terms of electromagnetic scattering as

$$\sigma = \frac{\text{Power reflected toward source/Unit solid angle}}{\text{Incident power density}/4\pi} = 4\pi R^2 \frac{|E_r|^2}{|E_i|^2} \tag{2.2}$$

where R is the range to the target, E_r is the electric field strength of the echo signal at the radar, and E_i is the electric field strength incident on the target. It is assumed in

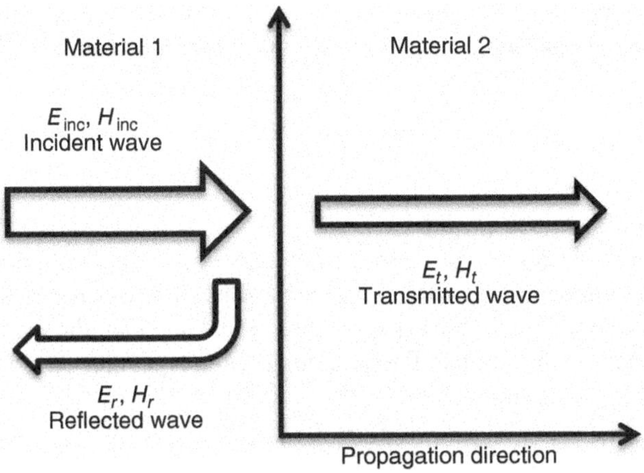

Figure 2.3 Incident wave transmission and reflection at a planar boundary between two materials.

this equation that the target is far enough from the radar that the incident wave can be considered to be planar rather than spherical. Sometimes the RCS is described as the cross-sectional area at the target location required to intercept the amount of power, which, if scattered uniformly in all directions, would produce an echo power at the radar receiver equal to that produced at the radar by the real target. Real targets, of course, do not scatter the incident energy uniformly in all directions. In general, there are two extreme conditions of RCS in terms of wavelength. First, if the wavelength is large compared with the object's dimensions, scattering is in Rayleigh region. The RCS in Rayleigh region is determined more by the volume of the scatterer than by its shape. At the other extreme, where the wavelength is small compared with the object's dimensions, scattering is in the optical region. Scattering from aircraft or ships at microwave frequencies generally is in the optical region where the RCS is affected more by the shape of the object than by its projected area.

In between the Rayleigh and the optical region is the resonance region where the radar wavelength is comparable to the object's dimensions.

2.2.3 Radar Equation

The radar equation is used to estimate the received power in a radar system, for a target with given properties and range from the transceiver, and it can be used to help determine the system's theoretical limits. The estimated received power is based on the transmitted power, the range to the target, and the properties of the transmit antenna, the target, and the receive antenna.

If an isotropic antenna radiates an electromagnetic wave, the wave is propagated uniformly in all directions, and the energy density spreads spherically. Therefore, at a point at the distance R from the radar, the total power will be distributed across the surface A:

$$A = 4\pi R^2 \tag{2.3}$$

Therefore, the power density is inversely proportional to the radius of the sphere. The power density of the isotropic antenna S_i at range R_1 can be calculated as

$$S_i = \frac{P_t}{4\pi R_1^2} \ [\mathrm{W/m^2}] \tag{2.4}$$

where P_t is the transmitted power.

While a spherical segment or isotropic antenna emits equal radiation in all directions (at constant transmitted power), if the transmitted power is focused to provide more radiation in one direction by using a directional antenna, the increase of power density in the focused direction of the radiation is called antenna gain. The directional power density, S_g, is given as

$$S_g = S_i \cdot G \tag{2.5}$$

where G is antenna gain.

The radar target intercepts a portion of the radiated power and reflects it, partially in the direction of the radar receiving antenna. The RCS, σ, is determined by the amount of power incident on the target that is re-radiated toward the antenna. This quantity depends on several factors, but it is generally true that a bigger area reflects more power than a smaller area. The reflected power P_d at the target can be expressed by the power density S_i, the antenna gain G, and the RCS σ:

$$P_d = S_i \cdot G \cdot \sigma = \frac{P_i \cdot G \cdot \sigma}{4\pi R_1^2} \ [\mathrm{W}] \tag{2.6}$$

In order to simplify the analysis, the target can be regarded as a source that radiates the reflected power. Since the reflected power can be considered with the same conditions as the transmitted power, the power density yielded at the receiver S_r is given by

$$S_r = \frac{P_d}{4\pi R_2^2} \ [\mathrm{W/m^2}] \tag{2.7}$$

where S_r is the power density at the receiver, P_d is the reflected power by the target in watts, and R_2 is the range between target and receiving antenna in meters. At the radar antenna, the received power depends on the power density at the receiver S_r and the effective antenna area A_e:

$$P_r = S_r \cdot A_e \tag{2.8}$$

where P_r is the power at the receiver in watts, and A_e is the effective antenna area in square meters. The effective antenna area determines the portion of the radiated energy the receiving antenna can capture. The power received is equal to the power density at the antenna, multiplied by the effective capture area of the receiving antenna. The differences between the actual antenna area and effective antenna area are caused by losses, such that the received power at the antenna is not equal to the input power. The effective antenna area can be calculated from the geometric antenna area as

$$A_e = A \cdot K_a \tag{2.9}$$

where A_e is the effective antenna area in square meters, A is the geometric antenna area in square meters, and K_a is the antenna efficiency. From Equations 2.7 and 2.8, the received power, P_r, is then calculated as

$$P_r = \frac{P_d A_e}{4\pi R_2^2} \text{ [W]} \tag{2.10}$$

From Equations 2.10 and 2.6, received power can be expressed as

$$P_r = \frac{P_t G \sigma A_e}{(4\pi)^2 R_1^2 R_2^2} \text{ [W]} \tag{2.11}$$

Now, assuming that range R_2 (from target to antenna) and range R_1 (from antenna to target) are equal, Equation 2.11 can be expressed as

$$P_r = \frac{P_t G \sigma A_e}{(4\pi)^2 R^4} \text{ [W]} \tag{2.12}$$

where G is the antenna gain in terms of the wavelength λ and the effective antenna area:

$$G = \frac{4\pi A_e}{\lambda^2} \tag{2.13}$$

Solving for A, and substituting Equation 2.13 into 2.12, after simplification it yields

$$P_r = \frac{P_t G^2 \sigma \lambda^2}{(4\pi)^3 R^4} \text{ [W]} \tag{2.14}$$

Finally, solving Equation 2.14 for range R, the radar range equation is given as

$$R = \sqrt[4]{\frac{P_t G^2 \lambda^2 \sigma}{P_r (4\pi)^3}} \text{ [m]} \tag{2.15}$$

where, in general, parameters P_t, G, and λ can be regarded as constants in a given radar system.

One application of the radar equation is determining the maximum distance at which a target can be detected. The smallest received power that can be detected by the radar is called $P_{r\min}$ and is a function of the receiver and digitization electronics. The maximum range $R_{\max}$ is calculated as

$$R_{\max} = \sqrt[4]{\frac{P_t G^2 \lambda^2 \sigma}{P_{r\max} (4\pi)^3}} \text{ [m]} \tag{2.16}$$

2.3 DOPPLER RADAR

Doppler radar is typically used to detect moving targets, and estimate their velocity. Common applications of Doppler radar include imaging weather fronts, and detecting speeding vehicles. Security systems motion detectors and door openers are other common uses of Doppler radar-based motion detectors.

2.3.1 Doppler Effect

The Doppler effect, or Doppler shift, is the change in the frequency of received waves when an observer is moving relative to the source of the wave. This phenomenon was discovered by the German physicist Christian Doppler and it applies to all wave motion, including sound, light, and electromagnetic waves. If the frequency of a sound from a source is held constant, and both the source and the observer of the sound remain stationary, the sound stays at the same frequency because the observer is receiving the same number of waves per second as the source is producing. If either the source or the observer is moving toward the other, the observer will perceive the sound at a higher frequency than that at which it was generated because the observer captures more waves per second. Alternatively, if the source and the observer are moving away from each other, the observer will perceive a lower frequency because the observer captures fewer waves per second.

If the range to the target is R, then the total number of wavelengths λ in the two-way path from radar to target and return is $2R/\lambda$. Each wavelength corresponds to a phase change of 2π radians. The total phase change in the two-way propagation path is then

$$\phi = 2\pi \times \frac{2R}{\lambda} = 4\pi R/\lambda \tag{2.17}$$

If the target is in motion relative to the radar, R is changing and so will the phase. Differentiating (2.17) with respect to time provides the rate of change of phase, which is the angular frequency:

$$\omega_d = \frac{d\phi}{dt} = \frac{4\pi}{\lambda}\frac{dR}{dt} = \frac{4\pi v_r}{\lambda} = 2\pi f_d \tag{2.18}$$

where $v_r = dR/dt$ is the radial velocity (m/s), or rate of change of range with time. If, as shown in Fig. 2.4, the angle between the target's velocity vector and the radar line of sight to the target is θ, then $v_r = v\cos\theta$, where v is the speed, or magnitude of the vector velocity. The rate of change of ϕ with time is the angular frequency $\omega_d = 2\pi f_d$, where f_d is the Doppler frequency shift. Thus from (2.18),

$$f_d = \frac{2v_r}{\lambda} = \frac{2f_t v_r}{c} \tag{2.19}$$

where f_t is radar frequency and c is the speed of light or 3×10^8 m/s.

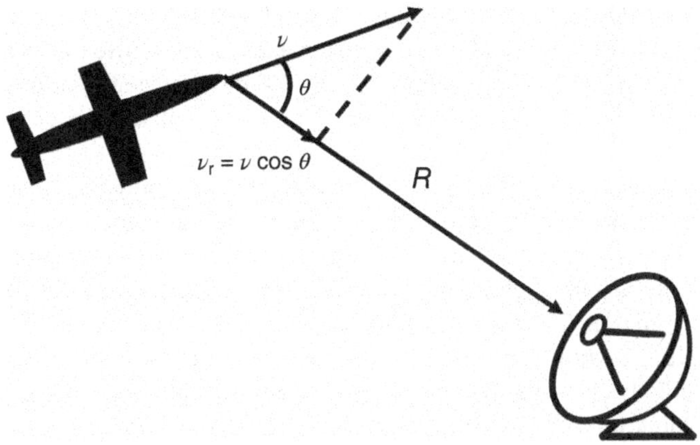

Figure 2.4 Geometry of a radar and a target in deriving the Doppler frequency shift.

2.3.2 Doppler Radar Waveforms: CW, FMCW, Pulsed

2.3.2.1 Continuous Wave A continuous wave (CW) radar system constantly transmits and receives a very narrow bandwidth signal. The CW radar transceiver has a simple topology (Fig. 2.5), consisting of a signal source used for both transmitting and receiving, and either a heterodyne or homodyne receiver. Since a CW system constantly transmits and receives, there is no need for a switch to control a transmit/receive terminal, as is required in pulsed radar system [Banks, 1975]. A CW radar either uses a single antenna with a duplexer or circulator to isolate the transmit and receive signals, or uses separate antennas for transmit and receive.

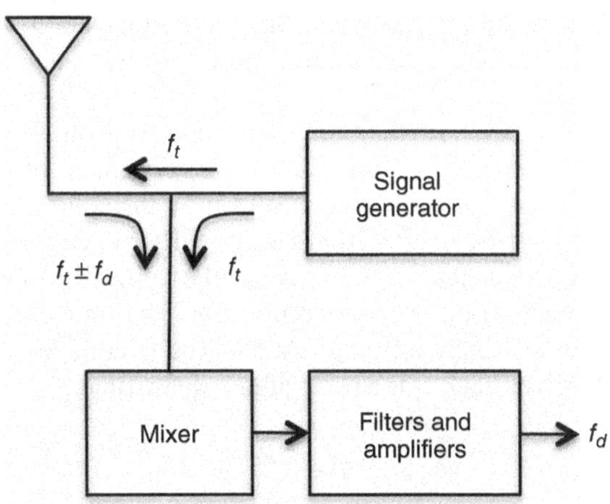

Figure 2.5 CW Doppler radar block diagram.

Due to its extremely narrow bandwidth, simple filters can be used at each stage of the receiver. The narrow bandwidth signal makes it straightforward to detect the Doppler shift in a CW radar system; a pure CW radar system can unambiguously measure the velocity of targets at any range and moving at any velocity.

The main disadvantage of CW radar results is leakage from the transmitter to the receiver, which is a side effect caused by both the transmitter and receiver being on constantly throughout operation. A portion of the transmitted signal leaks from the transmitter to the receiver, either through coupling between transmit and receive circuitry, or directly through the antenna(s). This leakage injects the radar receiver with a large signal at the transmit frequency that has not reflected off the target. In addition, clutter, or nontarget, nonmoving objects, reflects some of the signal and noise sidebands back to the receiver, injecting the receiver with more signal at the transmit frequency. These unwanted signals result in a DC offset and low-frequency noise if they are not eliminated before the signal is detected.

A CW radar system with a single-ended receiver similar to that illustrated in Fig. 2.5 cannot distinguish approaching and receding targets because both positive and negative Doppler shifts fold into one frequency band after signal downconversion to baseband. Either a coherent heterodyne or a quadrature homodyne receiver must be used to avoid spectrum folding and distinguish the direction or target motion [Saunders, 1990]. A quadrature homodyne receiver is discussed in more detail in Chapter 4.

2.3.2.2 Frequency-Modulated Continuous Wave A CW Doppler radar system can determine the velocity of moving target because the system can detect the frequency shift in the received CW signal. In order to detect the range to the target, the signal must have some type of timing marker such that the transit time can be measured. Both velocity and range can be measured with radar systems that modulate the CW signal with any of several methods, including FMCW, stepped frequency continuous wave (SFCW), coded modulation (CM), noise modulation (NM), synthesized pulse modulation (SPM), holographic modulation (HM), or amplitude modulation (AM) [Daniels, 2010].

The frequency modulation is commonly triangular for FMCW systems, such that the frequency varies gradually. Figure 2.6 shows an example of the modulation in an FMCW radar. The transmitted signal is shown by the solid triangular waveform and the dashed curve represents the frequency of the received echo signal from a stationary target. The frequency excursion, Δf, or the bandwidth of the FMCW signal, determines the accuracy of the range measurement. The frequency modulation at a rate, f_m, determines the maximum range that can be unambiguously detected. The transmitted signal arrives back at the radar after a delay time:

$$T = \frac{2R}{c} \tag{2.20}$$

where R is the range to the target. The received signal and the transmitted signal are multiplied in a mixer to produce the frequency difference. In Fig. 2.6, it is assumed

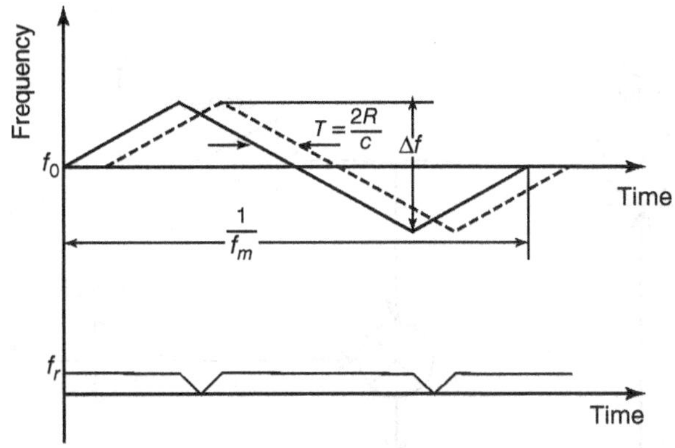

Figure 2.6 (Top) Frequency–time relation in an FMCW radar with linear triangular frequency modulation. Solid lines represent the transmitted signal, dashed lines represent the received signal delayed by a time $T = 2R/c$; Δf = frequency excursion, f_m = modulation frequency. (Bottom) Absolute value of the frequency difference between the transmitted and received signals.

that the only frequency shift is that due to the target range, f_r. From the geometry of Fig. 2.6, the frequency shift due to range, f_r can be shown to be

$$\frac{4Rf_m\Delta f}{c} \tag{2.21}$$

If there is a Doppler frequency shift f_d from the target motion, the total frequency difference is $f_r + f_d$ during half the modulation period and $f_r - f_d$ during the other half of the modulation period. The target range can be obtained by averaging these two frequency differences over the period $1/f_m$. The range resolution of an FMCW radar is inherently given by

$$\Delta R = \frac{c}{2\Delta f} \tag{2.22}$$

where Δf is the bandwidth of the FM sweep.

An example of a radar system for that uses FMCW for range detection is an altimeter to detect the altitude of an aircraft above the Earth [Skolnik, 2000].

2.3.2.3 Pulse Doppler Radar Pulse radar is the most commonly used radar system. A pulse radar system transmits narrow pulses with a large peak power at a constant pulse repetition frequency (PRF) and analyses the time-delayed received echoes reflected from target objects. Pulse radar that uses the Doppler shift for detecting moving targets is either a moving target indication (MTI) radar or a pulse Doppler radar [Skolnik, 2000]. Traditional pulse radar, shown in Fig. 2.7, has the advantage of measuring range, in addition to velocity information. The common method to measure range with a radar is to measure the time delay between transmission and reception of a pulse. Since the RF energy travels at the speed of light, $c \approx 3 \times 10^8$ m/s, the total

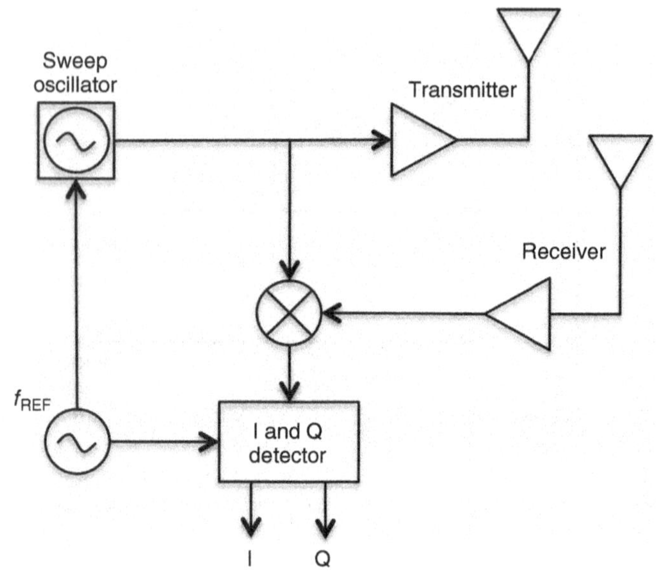

Figure 2.7 Block diagram of FMCW radar with homodyne receiver [Komemou, 2009].

round-trip delay between transmission and reception of the pulse is

$$\tau = \frac{2R}{c} \tag{2.23}$$

where R is the range of the target and τ is the round-trip time. The main advantage of pulsed radar over CW or FMCW radar is its time discrimination between transmit and receive, such that leakage from the transmitter and strong echoes from short-range clutter are separated temporally from the weaker echoes of long-range targets.

If the radar pulse width is long enough and if the target's velocity is high enough, it may be possible to detect the Doppler frequency shift on the basis of the frequency change within a single pulse in case of pulse radar. Figure 2.8 illustrates the baseband demodulated signal when there is a recognizable Doppler frequency shift. To detect a Doppler shift on the basis of a single pulse of width, T generally requires that there be at least one cycle of the Doppler frequency f_d within the pulse; or that $f_d T > 1$. Otherwise, the Doppler shift is shown sampled at the PRF. In this case, the amplitude of the received pulse is modulated by the Doppler effect, and more than one pulse is needed to recognize a change in the echo frequency due to the Doppler effect.

2.4 MONOSTATIC AND BISTATIC RADAR

There are two basic radar configurations based on the spatial relationship between the transmitting and receiving antennas: monostatic and bistatic. As shown in Fig. 2.9, in a true monostatic configuration, one antenna provides the path through the air for

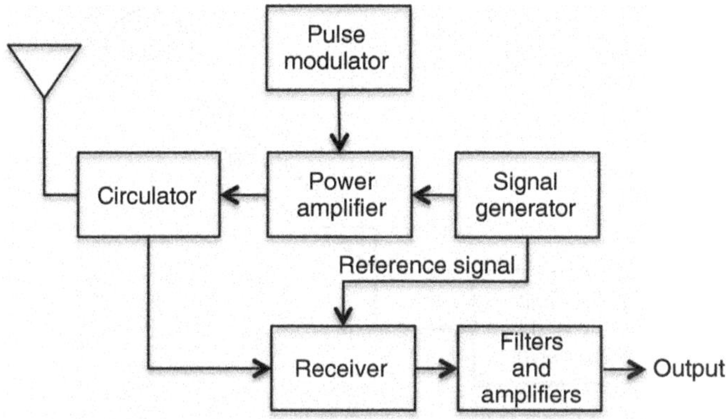

Figure 2.8 Block diagram of pulse Doppler radar.

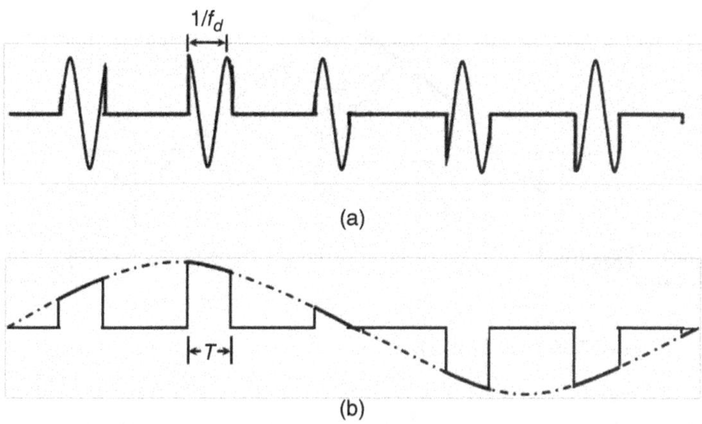

Figure 2.9 (a) Baseband demodulated signal when the Doppler frequency $f_d > 1/T$; (b) baseband signal for the Doppler frequency $f_d < 1/T$.

both the transmitter and receiver [Barton et al., 1997a]. In a bistatic configuration, the transmit and receive antennas are separated by a distance on the order of the distance from one of the antennas to the target [Barton et al., 1997b]. In cases where there are two separate antennas but they are not separated by a large angle, the system may be referred to as "pseudo-monostatic." A multistatic radar has at least two transmitters and one receiver or at least one transmitter and two receivers, or multiple transmitters and multiple receivers [Doughty et al., 2006]. Figure 2.10 shows the configuration of a multistatic radar.

A radar transmitter usually transmits a high-power electromagnetic signal to ensure the reflected signal from the target will be large enough to detect; for long-distance applications, electromagnetic wave power levels can be up to hundreds of kilowatts or even megawatts. Furthermore, the receiver is a very sensitive device, designed to detect small signals of the order of milliwatts to nanowatts that come

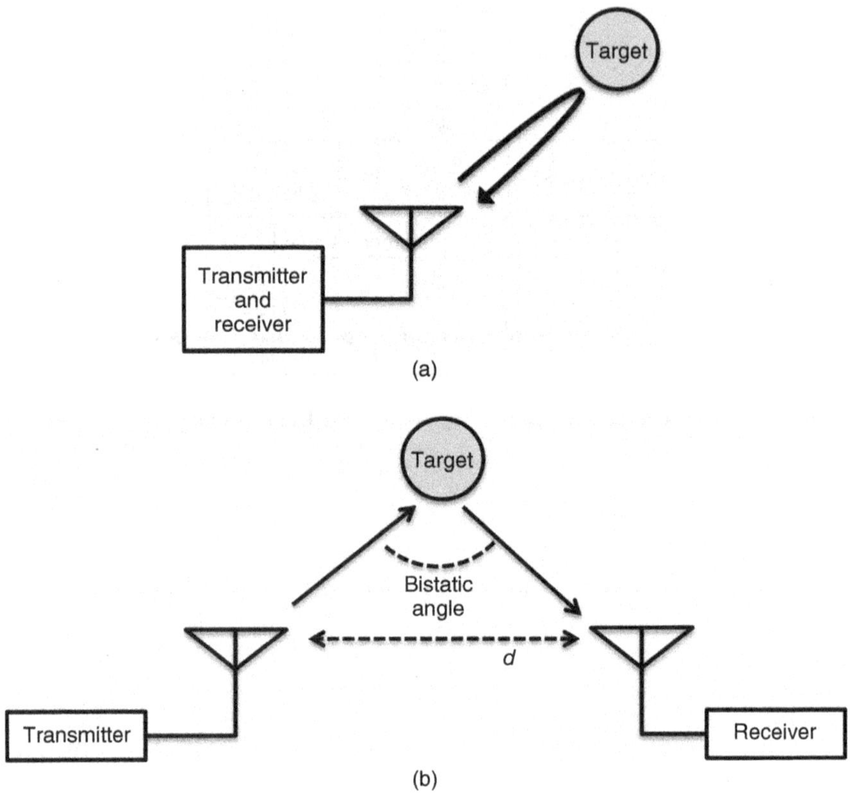

Figure 2.10 (a) Monostatic and (b) bistatic radar configuration.

from small or distant targets. In monostatic radar systems, the high transmitting power is sometimes directly coupled to the receiver, causing self-jamming, saturation, or damage to the receiver. A bistatic configuration is more efficient than monostatic in terms of self-jamming since either physical distance or circuitry isolation can provide sufficient isolation between transmitting and receiving paths.

Despite these challenges, many modern radars are monostatic due to the simple architecture enabled by a monostatic system. To obtain enough isolation between the transmit and receive circuitry, some special devices are used including ferrite circulators and resonant cavity or waveguide duplexers. In pulsed radar systems, the transmitter and receiver are not operated at the same time, which provides a large degree of additional isolation.

There are several applications for the bistatic configurations. For example, a semi-active missile has only the receiver portion on the cruising system, while the transmitter is on another platform. The transmitter is radiating primary signal to a target, while the missile is receiving the reflected signal to estimate a target's location. The bistatic radar also can be used for increasing the capability of detecting stealth targets. One of the techniques for creating stealth targets is making a target in special shape to maximize wave dispersion and minimize the backscatter signal toward the monostatic radar. However, that detoured reflected signal is usually providing a large RCS in some bistatic direction, which may be used for detecting that stealthy target [Richards, 2010]. Bistatic and multistatic radar systems are also often used for covert operation of the receiver, portability of the receiver, or to avoid electronic countermeasures. Multistatic systems can use spatial diversity to improve detection, resolution, and rejection of multipath effects.

2.5 RADAR APPLICATIONS

Radar has been employed to detect targets on the ground, on or under the sea, in the air, in space, and below ground. The major areas of radar application are briefly described in the following, including military defense and weapons systems, remote monitoring of the Earth's surface, the ocean, and other planets, reconnaissance imaging, ground-penetrating radar for archeological expeditions, weather surveillance, air traffic control, and others.

Radar is an important part of military air-defense systems, missile guidance systems, and reconnaissance imaging [Griffiths and Willis, 2010]. Air-defense radar systems can detect and recognize aircraft and airborne weapons, and track their position, course, and speed. Weapon guidance radar systems guide a missile to its intended target. High-resolution imaging radars, such as synthetic aperture radar, have been used for reconnaissance purposes and for detecting fixed and moving targets on the battlefield. The military has been the major user of radar and the major means by which new radar technology has been developed.

While all radar systems sense something about a target at a distance, the term "remote sensing" is typically used to describe the use of aerial sensors (on aircraft or satellites) to provide information about the Earth's atmosphere, surface, and oceans, the use of sensors to determine information about other planets, and the use of surface-based systems to detect items below the surface of the Earth. Interferometric synthetic aperture radar (InSAR) is used to product precise digital elevation models of large-scale terrain, and to detect centimeter-scale changes in deformation over time spans of days and years [Rodriguez and Martin, 1992]. It is used for geophysical monitoring of earthquakes, volcanoes, and landslides. InSAR is also used for monitoring of ground subsidence caused by depletion of aquifers, oil drilling, and mining to determine the stability of structures. Radar altimeters on satellites are used to measure the height and wavelength of ocean waves to determine the speed and direction of wind speeds and surface ocean currents [Xu et al., 2010]. Radar remote sensing systems have also been used for planetary observation, such as the use of a radar altimeter to map Venus beneath its visually opaque clouds. Ground-penetrating radar is used to locate buried structures and utility lines, to map archeological features, and to identify land mines and tunnels. Similar systems have been used for oil and gas exploration. Entomologists and ornithologists have applied radar to study the movements and migrations of insects and birds, which cannot be easily achieved by other methods.

Weather surveillance radar is a regular part of TV weather reporting. These pulse Doppler systems are used to locate precipitation (including its height above the surface), calculate its motion, estimate its type (rain, snow, hail), and to forecast its future position and intensity. Radars have been employed around the world to safely control air traffic in the vicinity of airports (Air Surveillance Radar, or ASR) [Skolnik, 1999], in routes between airports (Air Route Surveillance Radar, or ARSR), as well as ground vehicular traffic and taxiing aircraft on the ground (Airport Surface Detection Equipment, or ASDE). The ASR also maps regions of poor weather so that aircraft can be directed around them. There are also radar systems specifically used to detect wind shears that are potentially hazardous for aircraft, called Terminal Doppler Weather Radar or TDWR [Barry, 2000]. The Air Traffic Control Radar Beacon System (ATCRBS and Mode-S) is widely used for the control of air traffic; although it is not a true radar system, it uses radar-like technology.

Radar is found on ships and boats for collision avoidance and to observe navigation buoys, especially when the visibility is poor. Similar shore-based radars are used for surveillance of harbors and river traffic. Low-flying military aircraft depend on terrain avoidance and terrain following radars to avoid colliding with obstructions or high terrain. Military aircraft employ ground-mapping radars to image a scene. Radio altimeters are radars used to indicate the height of an aircraft above the terrain and as a part of self-contained guidance systems over land. Space vehicles have used radar for rendezvous and docking, and for landing on the moon [Kayton and Fried, 1997].

Large ground-based radars are used for the detection and tracking of satellites and other space objects. In the field of radar astronomy, the use of Earth-based radar systems has helped in understanding the nature of meteors and comets, establishing an accurate measurement of the Astronomical Unit, and observing the moon and nearby planets, all before adequate space vehicles were available to explore these entities at close distances.

There are now many everyday uses of radar systems. Most people are familiar with radar speed guns. These are small radar units used to measure the speed of moving objects, including vehicles (typically for enforcing speed limits) and pitched baseballs. Radar systems are now included in many automobiles to make the vehicles safer by warning of impending collision or warning of obstructions or people behind a vehicle or in the side blind zone. Radar motion-detection systems are also employed for intruder-detection in many home security systems.

REFERENCES

Banks DS. Continuous wave (CW) radar. Electron Prog 1975;17(2):34–41.

Barry AS. 2000. Available at: http://ieeexplore.ieee.org/xpl/articleDetails.jsp?tp=&arnumber=884925&url=http%3A%2F%2Fieeexplore.ieee.org%2Fxpls%2Fabs_all.jsp%3Farnumber%3D884925.

Barton DK, Leonov AI, Morozov IA, Hamilton PC. Radar Technology Encyclopedia. Norwood, MA: Artech House; 1997a. p 343.

Barton DK, Leonov AI, Morozov IA, Hamilton PC. Radar Technology Encyclopedia. Artech House; 1997b. p 328–329.

Coman IM. Christian Andreas Doppler – the man and his legacy. Eur J Echocardiogr 2005;6(1):7–10.

Daniels DJ. EM Detection of Concealed Targets. Hoboken, NJ: Wiley; 2010.

Doughty S,Woodbridge K,Baker C. Characterisation of a multistatic radar system. 3rd European Radar Conference; 2006. p 5–8.

Griffiths H, Willis N. Klein Heidelberg – The first modern bistatic radar system. IEEE Trans Aerosp Electron Syst 2010;46(4):1571–1588.

Kayton M, Fried WR. Avionics Navigation Systems. John Wiley & Sons; 1997.

Komemou G. Radar Technology. In-Tech; 2009.

Richards MA. Principles of Modern Radar: Basic Principles. SciTech Publishing; 2010.

Rodriguez E, Martin JM. Theory and design of interferometric synthetic aperture radars. IEE Proc F Radar Signal Process 1992;139(2):147–159.

Saunders K. CW and FM radar. In: Skolnik MI, editor. Radar Handbook. 2nd ed. San Francisco: McGraw-Hill, Inc.; 1990. p 14.1–14.45.

Skolnik M. Improvements for air-surveillance radar. IEEE Radar Conference; 1999. p 18–21.

Skolnik MI. An introduction to radar. In: Skolnik MI, editor. Radar Handbook. 2nd ed. San Francisco: McGraw-Hill, Inc.; 1990. p 1.1–1.21.

Skolnik MI. Introduction to Radar Systems. 3rd ed. McGraw-Hill; 2000.

Xu X, Liu H,Yang S. Mechanism and system design of satellite interferometric Synthetic Aperture Radar altimeter. 2010 Second IITA International Conference on Geoscience and Remote Sensing (IITA-GRS); 2010. Vol. 2, p 209–211.

3

PHYSIOLOGICAL MOTION AND MEASUREMENT

AMY D. DROITCOUR[1] AND OLGA BORIC-LUBECKE[2]

[1]*Wave 80 Biosciences, Inc., San Francisco, California, United States*
[2]*Department of Electrical Engineering, University of Hawaii at Manoa, Honolulu, Hawaii, United States*

Respiratory, heart, and circulatory movements that can be detected without contact by Doppler radar are concentrated not only in the thorax, where the lungs and heart lie, but they also occur in the abdomen, which moves during respiration, and at other points on the body where superficial pulses are present.

3.1 RESPIRATORY SYSTEM MOTION

3.1.1 Introduction to the Respiratory System

Although the main function of the respiratory system is gas exchange, it is also used to maintain the body's acid–base balance. The level of ventilation, typically measured as the respiratory rate and the tidal volume, is controlled by the central nervous system, central and peripheral chemoreceptors, and lung receptors. Ventilation is dependent on the arterial partial pressures of oxygen and carbon dioxide, and these are affected by the intake of oxygen, the absorption of oxygen, perfusion of the lungs with blood, transport of oxygen through the vascular system, oxygen demand in tissues, excretion of carbon dioxide, production of acids or bases, intake of acids or bases, excretion of acids or bases, and stimulation of respiratory centers.

Doppler Radar Physiological Sensing, First Edition.
Edited by Olga Boric-Lubecke, Victor M. Lubecke, Amy D. Droitcour, Byung-Kwon Park, and Aditya Singh.

Because such a broad range of factors can affect ventilation, the respiratory system's function can be altered by changes in a variety of organ systems, including the nervous system, the cardiovascular system, the respiratory system, and the excretory system. This makes the respiratory rate, rhythm, regularity, depth, and volume a broad indicator of imbalance in these organ systems, as well as an indicator of respiratory disorders.

Respiratory rate is a key vital sign for home monitoring of progression of diseases and for hospital monitoring to prevent medical crises. Trends in respiratory rate can indicate progression of cardiopulmonary illnesses, including acute respiratory distress syndrome, pulmonary edema, pulmonary embolism, pneumonia, chronic obstructive pulmonary disease (COPD), and severe heart failure. Changes in respiratory rate can also indicate sepsis, systemic inflammation, low blood volume, and malfunctions of the excretory system or central nervous system disorders, including intracranial pressure, neurogenic shock, pain, and opioid-induced respiratory depression.

Additional physiological information and information about the respiratory system can be obtained when the rhythm, regularity, and depth of respiration are measured in addition to the rate.

3.1.2 Respiratory Motion

The respiratory system's primary function is the exchange of carbon dioxide for oxygen in the lungs. For gas exchange to occur in the lungs, air with carbon dioxide needs to be removed from the lungs and air with oxygen needs to be inspired. In respiration, muscles contract to generate changes in thorax volume, which creates pressure differences between the thorax and the external environment, causing air to move in and out of the lungs, from areas of high pressure to areas of low pressure. The motions of the thorax and the abdomen cause significant displacements at the skin surface that are measurable with Doppler radar, allowing noncontact measurement of respiration rates. This section describes the motion associated with breathing and how this motion affects the skin surface motion.

Figure 3.1 shows the location of the muscles associated with breathing, the lungs and the ribs. As the diaphragm contracts, its dome descends into the abdominal cavity, causing the thorax to elongate and increase in volume and pushing the abdominal viscera out against the compliant abdominal wall. In normal inspiration, the diaphragm extends 1–2 cm into the abdominal cavity. In deep inspiration, the diaphragm can descend as much as 10 cm; at this point, the abdominal wall is stretched to its limit of compliance, and the abdominal pressure increases, limiting the downward motion of the diaphragm.

When abdominal displacement is prevented, for this or any other reason, further contraction of the diaphragm causes the lower ribs to elevate, further decreasing the thoracic pressure [Rodarte and Shardonofsky, 2000]. The external intercostal muscles contract simultaneously with the diaphragm for inspiration. If the diaphragm contracted alone, the decrease in pressure would pull the rib cage downward and

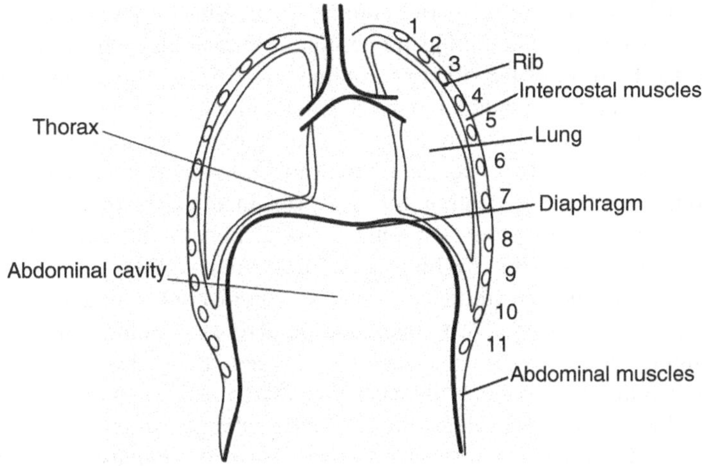

Figure 3.1 The thoracic wall, body cavities, and muscles of respiration. After Osmond [1995].

inward, decreasing the amount of air inspired. Contraction of the external intercostal muscles pulls the ribs upward and outward, further increasing the volume of the thorax and preventing the collapse of the ribcage. If the external intercostals contracted by themselves, the decrease in pleural pressure would cause the flaccid diaphragm to be displaced into the thorax rather than leading to inspiration. Joint action by the external intercostal muscles and the diaphragm is required for inspiration [Macklem, 1995; Macklem et al., 1978].

There are three types of rib movement at different points in the rib cage: the "pump-handle" motion of the upper ribs, the "bucket-handle" motion of the lower ribs, and the "caliper" motion of the lowest ribs, as shown in Fig. 3.2. The dominant motion of the upper ribs is rotation upward around their long axis, known as

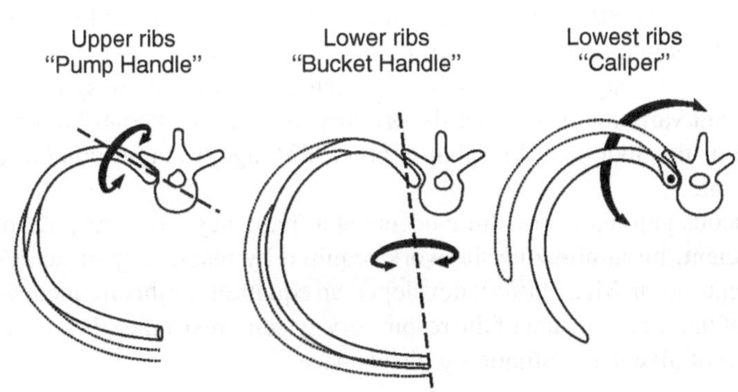

Figure 3.2 Motion of upper, lower, and lowest ribs. After Osmond [1995].

"pump-handle" motion. The lower ribs connect to the spine differently than the upper ribs so that they can glide as well as rotate. The combination of this motion and the rotation keeps the front of the rib at approximately a constant location, and the ribs effectively rotate upward while fixed at the front and the back in a "bucket-handle" motion. The lowest ribs are not connected to the sternum, and are known as floating ribs. These ribs tend to flare open and backward, rotating around their connection with the spine in a "caliper" motion. Since the ribs increase in size and curvature as they go downward, at any given horizontal cross section, the diameter of the thorax at that height increases as the ribs hinge upward [Osmond, 1995].

During normal quiet breathing, no muscles contract for expiration; the elastic recoil of the alveoli is sufficient to decrease the alveolar volume. During exercise, speech, singing, coughing, or sneezing, muscles are required for expiration. The abdominal wall muscles contract, increasing the abdominal pressure and pushing the contents of the abdomen up against the relaxed diaphragm, pushing the diaphragm into the thorax. Contraction of the abdominal muscles also depresses the lower ribs and pulls down the lower ribs, further decreasing the volume of the thorax. The internal intercostal muscles also contract in expiration, depressing the upper rib cage [Osmond, 1995]. Contraction of the external intercostal muscles raises and enlarges the rib cage, further increasing the volume of the thorax. Muscles in the abdominal wall are muscles of deep expiration: contraction of these muscles increases the abdominal pressure, elevating the diaphragm and depressing the ribs. Contraction of the internal intercostals pulls the ribs downward, decreasing the volume of the thorax for active expiration [Osmond, 1995].

Several respiratory parameters related to the breathing pattern can be collected, including respiratory rate, tidal volume, inspiratory time, expiratory time, rib cage excursions, and abdominal excursions. In Tobin et al. [1983], the respiratory parameters of 65 healthy test subjects were acquired. The average breathing frequency was 16.6 ± 2.8 breaths/min, with a tidal volume of 383 ± 91 mL and a minute volume of 6.01 ± 1.39 L/min. The inspiratory time of 1.62 ± 0.31 s was shorter than the expiratory time, with an inspiratory ratio of 0.421 ± 0.033. Typical breathing patterns in young adults were very regular, without sighs; older adults tended to have less regular breathing patterns. Some of the older adults had significant variation in tidal volumes and brief central apneas, or cessation of breathing. The subjects all had the rib cage and abdominal compartments moving in synchrony. There was significant variation in whether the primary displacement contribution was from the ribcage or the abdomen. In Tobin et al. [1983], age was not found to affect the respiratory rate.

Spontaneous automatic breathing occurs at a frequency where respiration is most energy efficient, by minimizing the work required by respiratory muscles to obtain adequate ventilation. Mead [1960] developed an equation for this frequency, which is a function of the time constant of the respiratory system (resistance times compliance) and the ratio of alveolar ventilation to dead space.

3.1.3 Chest Wall Motion Associated with Breathing

The chest surface motion associated with breathing is the combination of the abdominal and rib cage movements as described in the previous section. Kondo et al. [2000] present magnetic resonance imaging (MRI) data indicating a linear correlation between cross-sectional area of the thorax, displacement of the diaphragm, displacement of the rib cage, and lung volume. Wilson et al. [1987] present data showing that the pump-handle angle varies from 20° to 30° on the third rib, and from 30° to 37° on the seventh rib. This motion caused the rib radius to vary from 10.6 to 10.8 mm at rib three, and to vary from 13.7 to 14.2 mm at rib seven. DeGroote et al. [1997] measured the chest motion in the front/back, left/right, and up/down directions at 36 points. The largest motions were the sternum, which moved forward 4.3 mm with inspiration, and the navel, which moved forward 4.03 mm with inspiration. Kondo et al. [1997] measured the relationship between tidal volume and abdominal wall linear displacement with a laser displacement measuring device; they found that the abdomen distended 4 mm with a 400 mL inspiration, and 11 mm with an 1100 mL inspiration. They also showed a 12-mm abdominal displacement during spontaneous breathing in another subject. Overall, there is a 4–12 mm radial expansion of the thorax during breathing, depending on individual physiology and how much air is inspired.

3.1.4 Breathing Patterns in Disease and Disorder

3.1.4.1 Chronic Obstructive Pulmonary Disease COPD is an irreversible lung obstruction, such as chronic bronchitis or emphysema. COPD is characterized by the need to generate high intrathoracic pressures to exhale, due to loss of elastic recoil, which causes the airway to collapse easily and high pressures are required to exhale.

Patients with COPD have increased respiratory rate and tidal volume, leading to an increased minute volume. The inspiratory time is typically a very short fraction of the total respiratory time. Some COPD patients get a high level of carbon dioxide in their blood, a state known as hypercapnia. These patients may have a breathing pattern known as Cheyne–Stokes respiration. Cheyne–Stokes respiration is a repeating pattern of breathing that involves progressively deeper and sometimes faster breathing, followed by a gradual decrease in volume and sometimes rate that results in a temporary stop in breathing called an apnea. Some COPD patients exhibit asynchronous motion between the rib cage and abdomen.

According to American Thoracic Society COPD guidelines [American Thoracic Society and European Respiratory Society, 2004], tachypnea (a high respiratory rate) is not present in mild COPD, is likely to be present in moderate COPD, and is very likely to be present in severe COPD. The Global Strategy for the Diagnosis, Management, and Prevention of COPD guidelines (GOLD guidelines) [Global Initiative for Chronic Obstructive Lung Disease (GOLD), 2008] indicate that respiratory rate is often increased to more than 20 breaths/min in COPD. These same guidelines state that a respiratory rate greater than 25 breaths/min is an indication for noninvasive ventilation in COPD patients, and a respiratory rate greater than 35 breaths/min is an indication for invasive mechanical ventilation in COPD patients.

3.1.4.2 ***Restrictive Lung Disease*** Restrictive lung disease is a category of chronic disorders that cause a decrease in the ability to expand the lungs. Restrictive lung diseases are either interstitial or extrapulmonary. Interstitial lung disease causes inflammation of the lung parenchyma (the covering of the lungs) and the connective tissue that holds the air sacs together. When the inflammation is chronic, the tissues change, restricting breathing. Extrapulmonary restrictive lung disease can be caused by neuromuscular diseases and disordered, deformities of the thoracic cage, or pleurisy.

Patients with restrictive lung disease have a very high respiratory rate, with normal inspiratory time ratio, and normal tidal volume, leading to a high minute volume. These patients typically have rhythmic breathing without asynchrony of rib cage to abdominal movement.

3.1.4.3 ***Central and Obstructive Apnea*** Apnea is a pause in breathing for several seconds. There are there forms of apnea: central, obstructive, and mixed. Central apnea involves a lack of respiratory effort, while obstructive apnea involves a blockage of the airflow in spite of respiratory effort. The muscle tone of the body ordinarily relaxes during sleep, and at the level of the throat the human airway is composed of collapsible walls of soft tissue, which can obstruct breathing during sleep, causing what is known as obstructive sleep apnea. When the airway is obstructed, respiratory effort increases, causing higher thoracic pressures until either the airway opens enough to breathe, or there is a microarousal, and the muscles tighten, opening the airway.

During obstructive sleep apnea, with respiratory effort and a blocked airway, the thorax and abdomen are moving, but air is not flowing. This often results in "paradoxical breathing" where the abdomen expands while the thorax contracts, and vice versa, as air moves between the thorax and abdomen. The frequency and depth of the motion increase until the microarousal, after which point the respiratory rate returns to normal until the next apneic event.

3.1.4.4 ***Asthma*** Asthma is a lung disease characterized by reversible airway obstruction, airway inflammation, and airway hyperresponsiveness to a variety of stimuli. Asymptomatic asthmatics typically have breathing patterns similar to normal patients. During an attack, asthmatics have been found to have a markedly increased tidal volume and greater asynchronous motion between the ribcage and abdomen.

3.2 HEART SYSTEM MOTION

The heart drives blood through the lungs and to tissues throughout the body. When the heart contracts to generate the pressure that drives blood flow, it moves within the chest cavity, hitting the chest wall, and creating a measurable displacement at the skin surface. This section describes the location and anatomy of the heart, the electrical and mechanical events that cause contraction, the motion of the heart during contraction, and how that motion affects chest wall motion.

As the heart beats and drives blood into the arteries it rotates and its size changes, causing motion of the chest wall that can be detected at the skin surface, both by palpation and with noncontact sensors. The greatest motion occurs at the fourth and fifth intercostal space when the left ventricle strikes the chest wall as it contacts. More gradual motions due to filling of the heart occur in the left parasternal region. The maximum motion detected at the apex with noncontact sensors has an average of 0.6 mm, and this value is expected to vary widely over population due to differences in physiology, health, fitness, and age. However, this average motion is sufficient to provide detection with a Doppler radar system.

3.2.1 Location and Gross Anatomy of the Heart

The heart is located in the middle of the thorax, between and partially overlapped by the lungs. The sternum covers the front of the heart, as do the cartilages of the third, fourth, and fifth ribs, as shown in Fig. 3.3. Two-thirds of the heart is to the left of the midline. The heart rests on the diaphragm, tilted forward and to the left, so the apex is forward of the rest of the heart. Motion of the apex can be felt at the fourth or fifth intercostal space, near the left midclavicular line [Schlant et al., 1990].

The left side of the heart pumps blood to the organs and tissues, while the right side of the heart pumps blood to the lungs. A diagrammatic section of the heart is shown in Fig. 3.4. The vena cava, carrying blood from the peripheral tissues, enters at the upper right of the heart, into the right atrium. Blood from the right atrium enters the right ventricle, directly beneath the sternum, when the tricuspid valve opens. When the

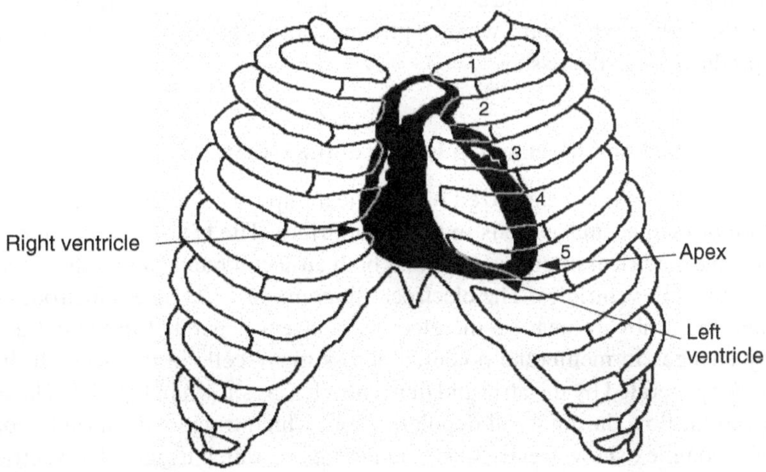

Figure 3.3 The location of the heart in the rib cage. The intercostal spaces are indicated by the numbers 1–5. The heart is beneath the sternum and the cartilage of the third, fourth, and fifth ribs. After Flint [1859].

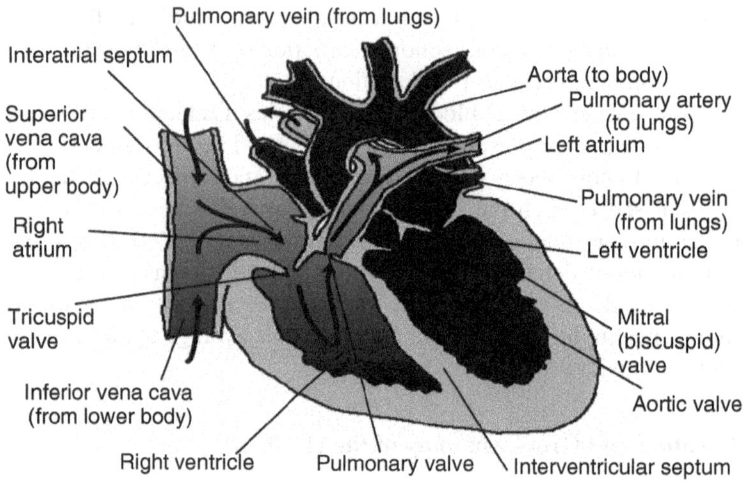

Figure 3.4 Diagrammatic section of the heart. The arrows indicate the direction of blood flow. After Vander et al. [1998].

right ventricle contracts, the pulmonary valve opens, and blood exits from the top of the right ventricle in the front of the heart and into the pulmonary artery, which takes blood to the lungs, where gas exchange removes carbon dioxide from and introduces oxygen to the blood. Blood from the lungs returns to the heart through two pulmonary veins, which enter the left atrium at the top and back of the heart, along the midline of the thorax. When the mitral valve is open, blood from the left atrium enters the left ventricle. When the left ventricle contracts, the aortic valve opens and blood exits from the top of the heart into the aorta, which begins the system of arteries that deliver blood to the tissues of the body, where it provides nutrients and oxygen to and removes waste products from the tissues.

3.2.2 Electrical and Mechanical Events of the Heart

The heart's beating is synchronized by electrical impulses that originate as the depolarization of the pacemaker cells in the right atrium. The heart's conduction system transmits the electrical impulses such that both atria contract at about the same time, followed by both ventricles. The electrocardiogram, or ECG, uses electrodes on the chest and the limbs to measure the electrical current generated in the extracellular fluid by changes in membrane potential across many cells in the heart. It displays waveforms generated by the atria and the ventricles, as shown in Fig. 3.5. The P wave shows current flow during atrial depolarization, which triggers the atria to contract. The QRS complex shows ventricular depolarization, which triggers the ventricles to contract. The T wave shows ventricular repolarization; atrial repolarization occurs at the same time as the QRS complex, so it is not visible in the ECG. The use of multiple combinations of recording locations on the limbs and the chest delivers information

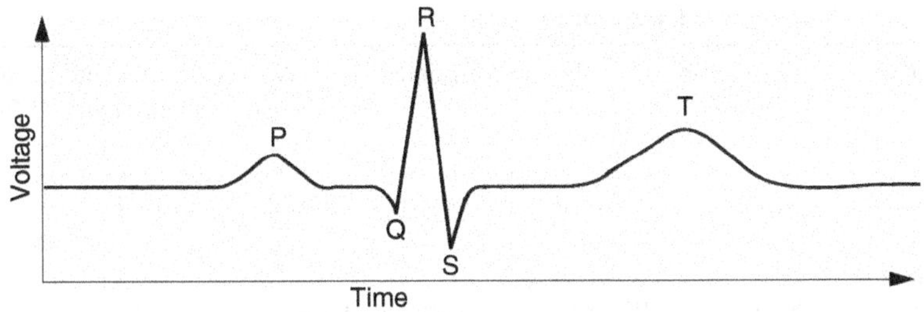

Figure 3.5 Example of an electrocardiogram. Atrial depolarization causes the P wave, ventricular depolarization causes the QRS complex, and ventricular repolarization causes the T wave.

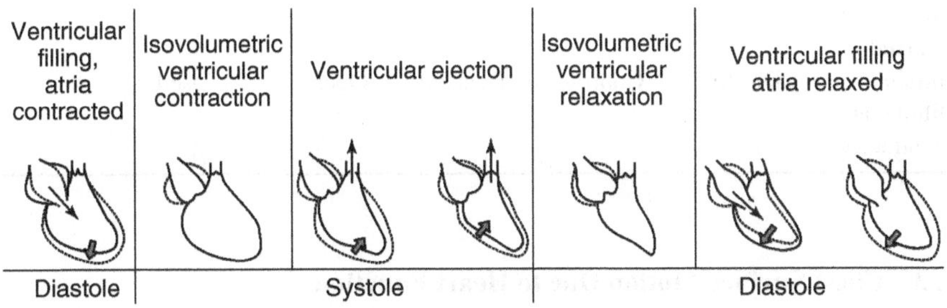

Figure 3.6 Motion of the heart throughout the cardiac cycle. After Opie [2001].

about different areas of the heart; the shapes and sizes of the P and T waves and the QRS complex vary with electrode placement.

The depolarization of the heart begins a cycle of atrial and ventricular contractions that cause chest wall motion, which is measurable by motion sensors such as Doppler radar. The motion of the left side of the heart is shown in Fig. 3.6, the phases of this cycle are outlined in Table 3.1, and the pressures and volumes in the left side of the heart during these cycles are illustrated in the Wiggers diagram in Fig. 3.7. During systole, the contracting ventricles eject blood, and during diastole, the relaxed ventricles fill with blood.

In systole, when the ventricles initially contract, the ventricular pressure is still less than that of the aorta so that the aortic valve is closed, and the ventricle maintains constant volume in isovolumetric ventricular contraction. Once the ventricular pressure is greater than the aortic pressure, the aortic valve opens, and ventricular ejection begins. When the ventricles stop contracting, they maintain a constant volume while the atrial pressure is less than the ventricular pressure in isovolumetric ventricular relaxation. Once the atrioventricular (AV) valves open, the ventricles begin to fill with blood from their respective atrium's; this initial filling is passive, with the atria relaxed. Then the atrial contraction starts and fills the ventricles until ventricular pressure is greater than atrial pressure, and then AV valves close [Awtry and Loscalzo, 2001a].

TABLE 3.1 Mechanical Events of the Heart

Mechanical Event	Systole or Diastole	Atria	Ventricles	AV Valves	Aortic and Pulmonary Valves	Stage, Fig. 3.5
Isovolumetric ventricular contraction	Systole	Relaxed	Contracted	Closed	Closed	1
Ventricular ejection	Systole	Relaxed	Contracted	Closed	Open	2
Isovolumetric ventricular relaxation	Diastole	Relaxed	Relaxed	Closed	Closed	3
Ventricular filling, atria relaxed	Diastole	Relaxed	Relaxed	Open	Closed	4
Ventricular filling, atria contracted	Diastole	Contracted	Relaxed	Open	Closed	4

3.2.3 Chest Surface Motion Due to Heart Function

The noncontact Doppler radar system operates at frequencies where it detects primarily skin surface motions. Changes in the shape and volume of the heart during systole move the ribs and soft tissue near the heart, causing the chest to pulse with each heartbeat. This section explores how heart motion translates to both palpable and visible motions.

The contraction and relaxation of the left ventricle causes a larger chest motion than other heart actions in healthy subjects. During isovolumetric contraction, the heart normally undergoes a partial rotation in a counterclockwise (when facing the patient) direction, causing the lower front part of the left ventricle to strike the front of the chest wall [Braunwald and Perlkoff, 2001]. The left ventricle also shortens as it contracts, making the heart more spherical, increasing its diameter and further adding to the impulse on the chest wall [Dressler, 1937]. The peak outward motion of the left ventricular impulse occurs either simultaneously with or just after the opening of the aortic valve (just before the upstroke of the carotid pulse); then the left ventricular apex moves inward [Braunwald and Perlkoff, 2001; Deliyannis et al., 1964]. The left ventricular motion causes the chest to pulse outward briefly, and the adjacent chest retracts during ventricular ejection [Gillam et al., 1964]. This impulse occurs at the lowest point on the chest where the cardiac beat can be seen, and it is normally above the anatomical apex, in the fourth and fifth intercostal spaces in the left midclavicular line [Awtry and Loscalzo, 2001b]. In healthy patients, this is usually the point of maximal impulse (PMI). It is typically palpable as a single brief outward motion, but it may not be palpable in as many as half of normal subjects over 50 years of age; obese, muscular, emphysematous, and elderly persons may have weakened or undetectable

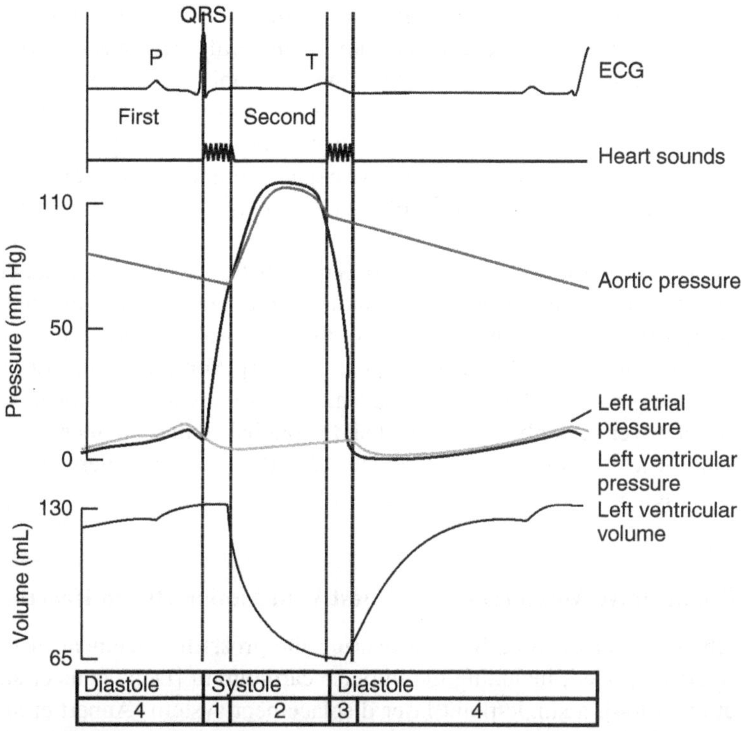

Figure 3.7 During the beginning of systole, the ventricles are contracting, but all the valves in the heart are closed; this is known as the isovolumetric ventricular contraction (1). The pressure in the ventricle increases, and when it is greater than the pressure in the aorta, the aortic valve opens, and ventricular ejection (2) begins. The pressure in the ventricle decreases and blood flows out of it, and when the pressure drops below that of the aortic valve, the aortic valve closes and diastole begins. Since all the valves in the heart are closed and the ventricle is relaxing, this is known as the isovolumetric ventricular relaxation period (3). When the left ventricular pressure drops below that of the atria, the mitral valve opens, and ventricular filling (4) begins. After Vander et al. [1998].

pulsations [Braunwald and Perlkoff, 2001]. Some studies found a second outward movement at the apex: the pre-ejection beat [Deliyannis et al., 1984].

There are also more gradual motions in the left parasternal region of the chest. There is an outward motion of the apex with left ventricular diastolic filling and an outward motion of the left parasternal region at the third intercostal space due to bulging of the left atrium at the end of systole [Braunwald and Perlkoff, 2001; Deliyannis et al., 1984]. An increase in pulmonary blood flow can cause a prominent systolic pulsation in the second intercostal space to the left of the sternum, caused by the closure of the pulmonic valve [Braunwald and Perlkoff, 2001]. Because left parasternal motion is smaller than apex motion and occurs over a wide area of the precordium rather than at a localized point, it is more difficult to palpate, although this motion is present in all healthy persons. Gillam et al. [1964] found that the left parasternal portion of the chest wall moves outwardly during early systole, followed by retraction in late systole in most normal subjects, but in some subjects, only the retraction occurs. Surface vibration from sounds measured in a stethoscopic exam causes negligible motion compared with the gross surface displacements caused by the heart striking the chest wall and the expansion and contraction of the heart [Awtry and Loscalzo, 2001b]. Motion of the right ventricle is not generally palpable in healthy patients.

Mechanical circuits have been proposed as models for the chest wall, but vibration measurement in soft tissues is not a well-studied topic. There have been no thorough studies on how the heart striking the inside of the chest wall couples to motion on the skin surface [Vermarien and van Vollenhoven, 1984]. Some quantitative measurements have been made of the chest displacement, but there are no known studies of how these vary over age or body type. Not all the published measurements have taken into account how their measurement device loads the chest and alters the measurement [Vermarien and van Vollenhoven, 1984].

3.2.4 Quantitative Measurement of Chest Wall Motion Due to Heartbeat

Many techniques for quantitatively measuring the gross displacement of the chest wall have been applied, including the impulse cardiogram [Deliyannis et al., 1984; Gillam et al., 1964], a single-point laser displacement system [Aubert et al., 1984; Ronaszeki et al., 1990], structured lights and the Moiré effect [Brandt et al., 1986], laser speckle interferometry [Ramachandran and Singh, 1989; Singh and Ramachandran, 1991], a capacitance transducer [Ramachandran et al., 1991], a magnetic displacement sensor [Mohri et al. [1987, 1985]], and a phonocardiographic microphone [Ikegaya et al., 1971].

The impulse cardiogram has been used to make quantitative measurements of chest motion due to heartbeat at the apex by Deliyannis et al. [1984] and at the left parasternal region by Gillam et al. [1964]. This impulse cardiogram consisted of a metal rod supported by springs that is affixed to the chest wall in order to measure chest wall motion. Displacement of the rod interrupts a beam of light on a photoelectric cell, which alters the resistance in an electrical current. In the measurement of normal subjects, Deliyannis et al. [1984] found that the largest impulse cardiogram measured had a 1-cm amplitude. Gillam et al. [1964] measured an average left parasternal deflection of 3.6 mm in 14 normal subjects. The outward movement did not last longer than two-thirds of systole in any of the normal subjects. In six of the subjects, an outward movement due to the atrial beat was detected with a maximum pulse amplitude of 5 mm. In all the normal subjects, the apical impulse displacement was larger than the left parasternal displacement.

Ramachandran et al. [1991] used a capacitance transducer to measure out-of-plane chest wall motion on five subjects. The subjects were asked to hold their breath during the measurement in order to isolate heart-related movement. A maximum displacement of 0.04 mm was measured at the apex during the T wave.

Magnetic displacement sensors were used to measure chest wall pulsation by Mohri et al. [1987, 1985]. A small magnet was placed on the skin at the measurement site, and a magnetic sensor determined changes in the magnetic field. The field sensor has an amorphous wire core as a component in a bridge circuit to sense changes in the magnetic field as the magnet moves while close to the core. The sensor with two cores has a 9-mm linear range and 1 μm resolution [Mohri et al., 1985], while the sensor with the star-shaped core has a 20-mm linear region and 0.2 μm resolution [Mohri et al., 1987]. The maximum measured chest displacement for the one subject in Mohri et al. [1987] was 0.21 mm, while in Mohri et al. [1985] a healthy subject had a maximum displacement of 0.035 mm, and an overweight subject had a maximum chest displacement of 0.012 mm.

Ikegaya et al. [1971] used phonocardiographic microphones to measure the motion of the chest wall in one subject. The microphones were calibrated to account for the coupling between the chest wall and the microphone using calculated chest wall impedance. The amount of measured motion in this study depended on the amount of force applied to the chest. When a mass of 100 g was applied, the chest motion was measured to be 0.05 mm, and when a mass of 200 g was applied, the chest motion was 0.08 mm.

Berson and Pipberger [1966] placed a lamp on the chest and used a detector with a photo-potentiometer to measure the chest motion, in three dimensions: normal to the chest, left-to-right, and head-to-foot. The three 30–40-year-old male subjects each held their breath while they were measured at three different points—the apex, the fourth intercostal space to the left of the sternum, and the fifth intercostal space to the right of the sternum. Normal measurements at the apex ranged from 0.10 mm to 0.84 mm, and the magnitudes of the displacements in directions other than normal to the chest wall were comparable with those normal to the chest wall.

Single-point laser displacement has been used to measure chest wall displacement [Aubert et al., 1984; Ronaszeki et al., 1990]. Aubert et al. [1984] used an infrared (850 nm) laser displacement measuring system and found a 0.6 to 0.2 mm displacement at the PMI at the apex on five normal male subjects, 20–40 years old. Ronaszeki et al. [1990] used a similar system to measure the apex motion in 16 men, but the absolute displacement was not recorded. In the data plot shown, a scale bar is given, and the measured peak-to-peak distance was 1.25 mm.

Brandt et al. [1986] evaluated chest wall motion with structured lights and the Moiré effect. This technique gives a contour map of distance from the source, so that the difference in plots must be assessed in order to determine the relative displacement. These images were recorded on one subject in Brandt et al. [1986]; the amplitude of the maximum displacement was estimated to be 1.7 mm, and the diameter of the quasi-spherical displaced area was about 8 mm.

Ramachandran and Singh [1989] used laser speckle interferometry to measure the displacement of the chest wall due to cardiac action on 10 healthy men of different builds. A thin layer of paint was applied to the chest to enhance reflectivity, and the seated subject was asked to hold his breath during recording. Since the scan time for the 3-D image was long compared with a cardiac cycle, the measurement was synchronized with the ECG so that the scan could be performed over several cardiac cycles. The maximum displacement was over the apex and the left ventricle during the QRS wave, or ventricular contraction, 0.57 ± 0.11 mm at the apex and 0.53 ± 0.10 mm at the left ventricle. The largest significant displacement during the P wave, when the atria contract and the ventricles fill, was above the left ventricle, 0.45 ± 0.07 mm. The largest significant displacement during the T wave, when the ventricles begin to refill, was at the apex, 0.45 ± 0.03 mm. No assessment of how the build of the subject affected the measurements was given.

Singh and Ramachandran [1991] used a similar technique to measure the in-plane cardiac displacement pattern of the area over the heart. The in-plane motion is expected to be much less than the motion perpendicular to the chest plane. Again, the cardiac movement was isolated by each subject holding his breath, and the exposure of the laser light was synchronized with the ECG so that different phases of the cardiac cycle could be measured when the scan time was greater than the cardiac cycle length. The maximum in-plane displacement was 0.09 mm, measured at the apex during the QRS complex. Right ventricular in-plane motion was also at its maximum during the QRS complex, with a displacement of 0.07 mm. In-plane motion would not be measured by Doppler radar if it is pointed perpendicular to the chest, but it could be measured from the side.

While these studies are useful for getting an idea of how much the chest wall moves with heartbeat, they leave many areas open to future research. First, none of these studies indicated the error due to the measurement, only the variation between subjects, so the accuracy of the data is unclear. Second, the number of subjects is small in all of these studies, with 20 being the greatest number of subjects, and some only giving quantitative data for a single subject. None of these studies compared males and females, and those that compared a healthy subject with an overweight subject or cardiac patient only used one of each, which does not provide reliable information about how body shape or heart condition affects the chest wall motion due to heartbeat. Third, in measurements that involved a sensor sitting on the chest, the sensor may have affected the chest motion, making those measurements unreliable. In Ikegaya et al. [1971], doubling the weight of the phonographic microphone sensor increased the measured motion by 60%, indicating that the pressure applied to the chest in contacting measurements can significantly affect the measurement. The only noncontact measurements of out-of-plane chest wall displacement due to the motions of the heart that had more than one person in the study were those using infrared laser displacement [Aubert et al., 1984] and laser speckle interferometry [Ramachandran and Singh, 1989]. They both found the maximum displacement to be over the apex, and approximately 0.6 mm. Fourth, none of the studies explored how the position of the subject affected the chest wall motion due to heartbeat. The subject position affects how easy it is to palpate motion at the chest wall, so the position likely also affects the amount of chest wall motion without pressure as well.

The values of motion at the skin surface due to heartbeat are expected to vary widely between individuals due to physiological difference, age differences, and body shape differences. It is expected that the amount of chest motion due to the heartbeat changes with age, since the amount of and speed of the motion of the heart within the chest changes with age. Yip et al. [2002] found the expected amplitude of motion of the mitral valve along the long axis of the heart was 1.49 cm at age 20 and 1.22 cm at age 84. The expected velocity was 7.48 cm/s at age 20 and 5.22 cm/s at age 84. Owen [1999] found that displacement of the septum decreased with age, but displacement of the left lateral wall and the posterior wall of the heart remained constant between ages 49 and 73.

Arcem et al. [2002] found that the absolute diastolic displacement of annular sites in children increased significantly with increasing body weight (which is expected since the size of the heart and thorax is increasing), but the percent displacement was inversely proportional to body weight.

As the heart beats and drives blood into the arteries it rotates and its size changes, causing motion of the chest wall that can be detected at the skin surface, both by palpation and with noncontact sensors. The greatest motion occurs at the fourth and fifth intercostal space when the left ventricle strikes the chest wall as it contacts. More gradual motions due to filling of the heart occur in the left parasternal region. The maximum motion detected at the apex with noncontact sensors has an average of 0.6 mm, and this value is expected to vary widely over population due to differences in physiology, health, fitness, and age. However, this average motion is sufficient to provide detection with a Doppler radar system.

3.3 CIRCULATORY SYSTEM MOTION

Blood vessels carry blood from the heart to the tissues and back. The blood is carried through the body by the pumping of the heart, the recoil of the arteries, the compression of veins by skeletal muscle, and the negative pressure in the thorax during inspiration. Blood pulses through the distensible arteries, which expand when the heart pumps blood into them during systole and contract during diastole, when the aortic valve is closed. As the arteries expand and contract, the skin above them moves; the skin surface motion is most prominent above superficial arteries. The following sections describe the location and structure of the arteries and veins, how they distend as the pressure of the blood in them varies during the cardiac cycle, and how this distension affects the skin surface motion. The skin surface moves measurably due to arterial pulsations at locations where the artery is near the skin surface. This motion does not necessarily occur at the same time as the maximum chest wall motion due to the heart beat. The delay between the R-wave, which causes ventricular contraction, and the posterior tibial pulse is approximately 0.26 s [Hong and Fox, 1994]. This delay from the chest wall motion to the motion at the furthest pulse points from the heart could cause some spreading in time of the heart signal when measured by Doppler radar, since it integrates overall motion. However, according to Mohri et al. [1987], the chest displacement is four times that of the largest carotid pulse, and therefore the much smaller pulse should not cause a major problem.

Although the relationship of skin surface displacement due to pulse and age has not been published, increases in arterial rigidity with age are well proven, and they cause decreased change in cross-sectional area with increasing age. Therefore, it is expected that any skin surface motion due to arterial pulses will decrease as the age of the measurement subject increases.

This section focuses on motion at the superficial pulse sites, which most likely create the largest displacement of arterial pulses. However, other areas of the body likely also pulse at the heart rate, although with a smaller amplitude. Ko et al. [2005] use interferometric holograms to measure cerebral pulsations on the scalps of patients with incomplete skulls. In the process, they also noticed motion in the eyes at the pulse rate. The amplitude of the pulsations was not provided.

3.3.1 Location and Structure of the Major Arteries and Veins

The diameter of the arterial vessels progressively decreases from the aorta to the capillaries, as shown in the model of the arterial system in Fig. 3.8. When arteries are near the skin surface, their pulses are palpable and sometimes visible. The main superficial arteries are the carotid artery in the neck, the brachial artery at the elbow, the radial artery in the wrist, the femoral artery in the upper thigh, the popliteal artery in the back of the knee, the posterior tibial artery in the inside ankle, and the dorsalis pedis artery on the top of the foot.

Elastic arterial tissue enables arteries to accept blood from the heart in impulses while delivering blood to capillaries by gradually stretching and recoiling. As shown in Fig. 3.9, during systole, when the heart pumps blood into the aorta, only one-third of the stroke volume (the volume of blood which leaves the left ventricle) leaves the

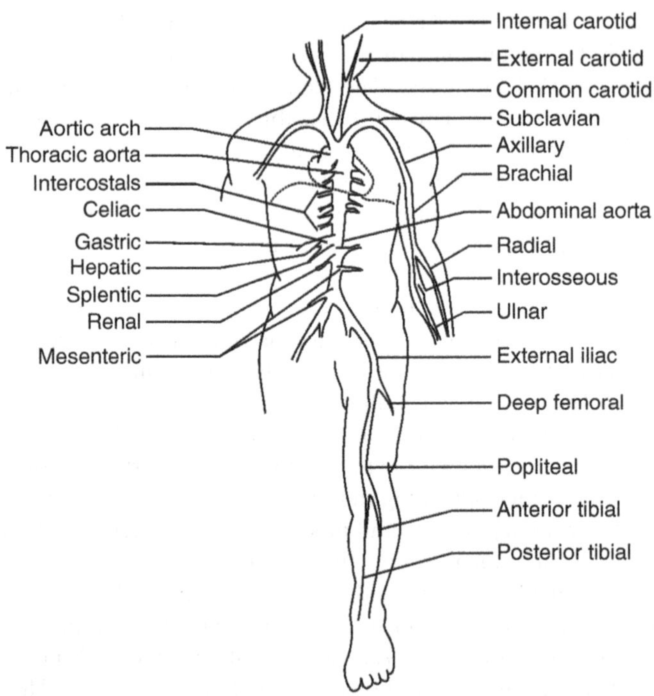

Figure 3.8 Model of the arterial system, showing major arteries. After Schaff and Abbrecht [1972].

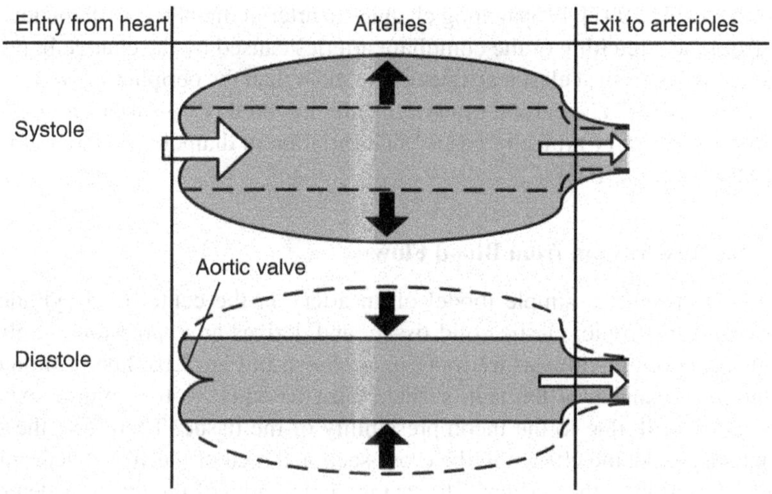

Figure 3.9 Diagram of arterial pressure in systole and diastole. During systole, the artery distends, storing blood; during diastole, the artery contracts so that blood continues flowing into the arterioles after the aortic valve is closed. After Vander et al. [1998].

arteries; the other two-thirds of the blood distends the arteries, raising the arterial pressure. During diastole, when the heart is filling with blood, the stretched arterial walls begin to return to their nonstretched shape, continuing to push blood into the arterioles as the arterial pressure falls. Larger, more central arteries dilate more than peripheral arteries, which are less distensible. Older people have less distension in their arteries than children do, because arteries become more rigid over time [O'Rourke et al., 1992].

Veins have much thinner walls than arteries because the blood in the veins is under lower pressure. Veins have little elastic tissue, and therefore are not distensible like arteries. They can accommodate large volumes of blood with minimal pressure changes, and they lie flat when they are not full and they become cylindrical as they fill with blood. Valves in veins prevent backflow as blood is pumped against gravity by skeletal muscles.

3.3.2 Blood Flow Through Arteries and Veins

With each ventricular contraction, the heart ejects a surge of blood into the aorta, leading to flow, pressure, and diameter waves as the blood propagates through the body. Flow waves are the changes of the velocity of blood flowing through the arteries. The blood velocity varies from 70% to 90% from its mean velocity, which decreases as the arteries get further from the heart [O'Rourke et al., 1992]. Pressure waves are the increase in pressure that propagates from the aorta through other arteries in the body. In the large arteries, the pressure fluctuation is as much as 300% of the mean pressure, while in the peripheral arteries the fluctuation is 40–80% of the mean pressure

[O'Rourke et al., 1992]. Propagating changes in arterial diameter, or diameter waves, result from the stretching of the compliant arteries caused by the change in pressure. The larger, more central elastic arteries dilate more than the peripheral arteries, which are less distensible. The carotid artery typically has an 8–15% variation in diameter, while the radial artery typically has a 1.6% variation in diameter [Buntin and Silver, 1990; Mooser et al., 1988].

3.3.3 Surface Motion from Blood Flow

Lee [1974] presents a simple model of an artery in the center of a cylinder of a homogenous, isotropic, elastic solid tissue, and derives an expression coupling the arterial wall motion with surface motion, as shown in Fig. 3.10. Lee postulates that the volume expansion at the skin surface must be equal to the volume expansion at the vessel wall, due to the incompressibility of the tissue. Therefore, the size of the appendage and the change in the cross-sectional area of the vessel determine the amount of motion at the surface. The change in the area of the artery is determined by its distensibility, which is influenced by several factors, including how close the artery is to the heart, the size of the artery, and the age of the subject. This model is an oversimplification, however; the arteries most frequently palpated are the superficial arteries. When the artery is not in the center of the limb, the arterial expansion is not a bulging of the entire limb, but rather a pulsation of the area of the surface of the limb closest to the artery.

Hong and Fox [1994] used optical interferometry to measure the velocity of skin above superficial arteries and the time delay between the R wave of the electrocardiogram and the pulse. Although the velocity measurements indicate that there is detectable motion due to pulse at the skin surface, they did not quantitatively measure the displacement. In addition to the pulse points measured by Hong and Fox, visible pulsations are available at the aortic artery (in the second right intercostal space at

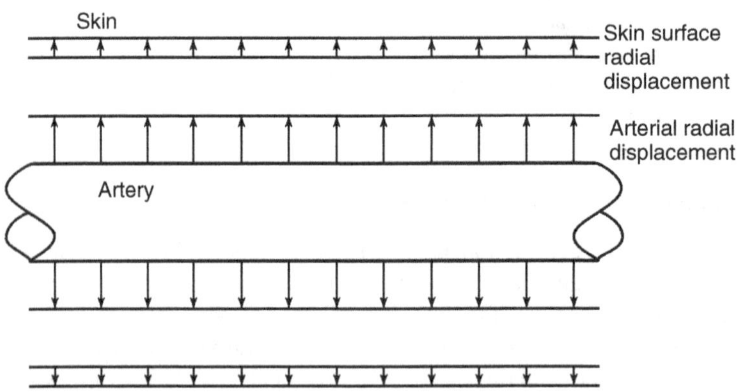

Figure 3.10 Lee's model of an artery in tissue for analyzing surface motion with radial motion of the vessel wall. After Lee [1974].

the suprasternal notch) and at the pulmonic artery (in the third left intercostal space) [Awtry and Loscalzo, 2001b].

Mohri et al. [1987] used a small magnet placed on the chest skin surface and a magnetic field sensor with 2-μm resolution to measure the blood vessel displacement at the skin surface. They found that the carotid artery produced a skin displacement of 0.06 mm, while the jugular vein produced a skin displacement of 0.01 mm. In Mohri et al. [1985], the same authors measured the carotid artery to produce a skin displacement of 0.05 mm, the radial artery to produce a skin displacement of 0.03 mm, the finger pulse to produce a skin displacement of 0.01 mm, and the jugular vein to produce a skin displacement of 0.005 mm. However, this is a contacting measurement, and the measurement may have been affected by the presence of the magnet. The same sensors were used to measure chest wall motion due to the heart, and they measured the chest wall motion to be 20 times less than that measured with noncontact techniques. In Mohri et al. [1985], the arterial motion was similar in amplitude to the heart motion.

The jugular vein is covered by a muscle and is usually not visible as a discrete structure, but its pulsations are transmitted to the skin of the neck, where they are usually visible. The jugular venous pulse has two peaks and two troughs, distinguishing it from the carotid arterial pulse, which has a single upstroke. The venous pulses are typically quite distinct when the patient is at a 45° or greater angle, but are not typically visible in upright healthy subjects [Braunwald and Perlkoff, 2001].

The measurements that have been made indicate that there is measurable motion at the skin surface at superficial pulse sites. This motion may be measurable by Doppler radar, and measurement of the amount of surface motion due to pulse is an interesting area for future research that has not been thoroughly explored. It is expected that motion due to arterial pulsation is less than that due to heartbeat.

3.3.4 Circulatory System Motion: Variation with Age

Although there is no significant change in left ventricular ejection volume with age, the arterial pressure wave varies greatly with age. With age, arterial wall thickness increases, arterial diameter increases, and arterial distensibility decreases [Kawasaki et al., 1987]. In Meinders and Hoeks [2004], the arterial wall rigidity was expressed as

$$\alpha = 0.421 + (0.0602 \times \text{age})$$

This indicates that pulses will be smaller and more difficult to measure in older subjects. The data provided by Meinders and Hoeks [2004] indicate that the change in cross-sectional area of the artery with a pulse is 50% less for 60–70 year olds than it is for 20–30 year olds. A decrease in the amount of change in the diameter of the arteries would decrease the amplitude of skin surface pulsations. A decrease in the amplitude of skin surface motion with age would reduce the signal-to-noise ratio (SNR) in the measurement of pulse with Doppler radar, as the SNR of Doppler radar is proportional to the amount of motion at the skin surface.

3.4 INTERACTION OF RESPIRATORY, HEART, AND CIRCULATORY MOTION AT THE SKIN SURFACE

Many studies have shown that the motion of the lungs and diaphragm due to respiration moves and deforms the heart [McLeish et al., 2002; Nehrke et al., 2001; Shechter, 2001]. However, apparently no studies have been performed on the interaction of respiratory and cardiac motion at the chest's surface. In the studies of chest wall motion due to the heart, the subjects were holding their breath during the measurements; therefore, it is not known how the respiratory motion of the heart affects the chest wall motion due to the heart.

The motion and deformation of the heart due to respiration have been measured with MRI and coronary angiography, in order to create models so that MRIs of the heart can be performed without the patient needing to hold his or her breath. Results indicate that the feet-to-head motion of the heart is roughly linear with the diaphragm motion in the same direction, but some subjects had a good degree of hysteresis in this motion [Nehrke et al., 2001]. In Nehrke et al. [2001], the heart moved from 12 to 24 mm due to respiratory motion in 10 healthy volunteers. In McLeish et al. [2002], combined rotations and translations led to a 22.5 ± 4.5 mm total displacement of the apex of the left ventricle in the eight healthy subjects. The heart also rotated $3.8 \pm 1.9°$, and the left ventricle deformed up to 4 mm due to respiration.

Rosa [1959] studied how the acceleration of the skin surface as measured with a vibrocardiogram was affected by respiration by taking measurements during normal respiration, full inspiration breath-hold, and full expiration breath-hold. They found the acceleration patterns that corresponded to ventricular contraction to be much more reproducible during the breath-hold measurements than with normal respiration, indicating that the rotary heart movements and respiratory displacement of the heart distort the skin surface motion [Rosa, 1959].

The translation, rotation, and deformation of the heart due to respiration certainly affect how the heart's motion interacts with the chest wall at different points in the respiratory cycle. The studies of chest wall motion due to the heart beating all required the subject to hold his/her breath. There are no known studies of skin surface chest motion due to beating of the heart during respiration or at different levels of inspiration. Since the contraction of the left ventricle causes the largest motion at the chest wall in healthy subjects, a 4-mm deformation of the left ventricle likely changes the motion due to heartbeat at the skin surface.

In addition, a 2-cm motion of the heart could affect the part of the heart that causes the largest skin surface motion, changing the relationship of the peak due to the Doppler signal.

3.5 MEASUREMENT OF HEART AND RESPIRATORY SURFACE MOTION

Surface motion measurement is by definition noninvasive, which makes such measurements convenient for home-monitoring and long-term monitoring. Some measurements are noncontact and noninvasive, and these can be used without the knowledge of the subject, with reduced risk of the measurement device affecting the parameter being measured. However, these measurements do not necessarily measure the same parameters as gold-standard measurements. Surface motion measurement of the heart measures physical motion, which is different from the electrical signal provided by the electrocardiogram. Surface motion measurement of respiration measures the motion of the abdominal wall and rib cage, but does not directly measure the airflow. Doppler radar measurement of heart and respiration is also a measurement of surface motion and is discussed in this section followed by other methods.

3.5.1 Radar Measurement of Physiological Motion

According to Doppler theory, as presented in Chapter 2, a continuous wave radar with a stationary person's chest as the target should receive a signal similar to the transmitted signal with its phase modulated by the time-varying chest position and a received power determined by the radar system properties, the environment, and the area of the moving part of the body. When the phase is demodulated, the chest displacement over time can be inferred, from which heart and respiratory rates can be determined. Analog and digital signal processing remove noise and interference, separate the heart and respiratory signals, determine the heart and respiratory rates, and prepare the signal for display or recording. Previous work in microwave monitoring of heart and respiration is described in detail in Chapter 1. This system works from a distance, noncontact, and through clothing. It does not require contact with the subject or that the subject be wired to a monitor. Because it is a motion sensor, it does require that the subject be still to obtain an accurate rate. It will measure any motion within the antenna beam; however, with a single transceiver it cannot distinguish between motions from two different sources.

3.5.2 Surface Motion Measurement of Respiration Rate

Two main categories of surface measurement of respiratory motion exist: measuring the circumference or area of the thorax and/or abdomen, and measuring the linear displacement of the thorax and/or abdomen. Both types of measurements are listed in Table 3.2. Inductance plethysmography uses bands around the chest and abdomen that vary in inductance as the bands are stretched, and piezoelectric strain gauge straps emit a voltage when the circumference of the chest changes. Strain gauges measure the deformation of the chest. When displacement of both the thorax and abdomen are measured and calibrated with a volume measurement, changes in tidal volume can be detected with surface motion measurements.

In Kondo et al. [1997], a laser sensor was used to measure the anteroposterior chest wall motion. This is a noncontact measurement, offering no resistance to respiration and no tactile stimuli, which should ensure a noninvasive measurement of respiration that does not alter the respiratory pattern. The laser monitor measures the distance between the chest wall and the sensor, and obtains a respiratory waveform by plotting the change in the distance over time. The laser monitor can track

TABLE 3.2 Techniques for Surface Measurement of Respiration Rate

Technique	Description	Circumference or Linear Displacement?	Contact or Noncontact?	Pros and Cons?
Doppler radar	Radar signal is directed at subject's chest; linear motion of entire chest is measured	Linear	Noncontact	Subject can remain clothed and unwired. Artifacts due to subject motion
Impedance plethysmography [Gribbin, 1983; Allison et al., 1964; Wolf and Arnold, 2005]	Injects small amounts of electrical current into the subject to measure the subject's impedance changes with respiration	Thoracic impedance (indirectly circumference)	Contact	Can make a cross-sectional "slice" image of subject. May require many electrodes. May suffer from interference
Inductance plethysmography [Gribbin, 1983; Lee-Chiong, 2003; Wolf and Arnold, 2005]	Band worn around chest wire coils stitched into the elastic band change inductance as circumference changes	Circumference and area	Contact	Measures a complex function of circumference and cross-sectional area. Displacement in transducer bands can lead to inaccuracies
Magnetometer [Lee-Chiong, 2003]	Transmitter and receiver coils on chest, abdomen, and back measure changes in displacement	Linear	Contact	Sensor can also be used to monitor changes in body position
Strain gauges [Fadel et al., 2004; Lee-Chiong, 2003]	Strain gauges placed on chest measure changes in circumference	Circumference	Contact	Displacement of body movements influences signal quality. Overstretching or understretching the gauges can affect accuracy
Laser displacement [Kondo et al., 1997, 2000]	Laser pointed at chest measured the change in linear displacement	Linear	Noncontact	Subject cannot be clothed
Linearized magnetometer [Gribbin, 1983]	One coil is driven to make a weak magnetic field; receiving coils determine their position in the field	Linear	Contact	Rotational motion of coils creates and artifact

rapid changes in lung volume with almost no lag. A monitor with multiple laser sensors could simultaneously monitor multiple points on the chest, and better model the volumes of respiration.

In magnetometer measurements, one coil is driven by an oscillator to produce a weak magnetic field, while other coils are attached to the skin on the thorax and abdomen. The coils on the skin pick up the magnetic field and can determine their position in the field. Magnetometers are susceptible to rotational movement, which creates artifacts [Gribbin, 1983]. Magnetometers were used to measure the anteroposterior motion of the rib cage and abdomen [Gribbin, 1983], and to measure displacement between the abdomen and the sternum [McCool et al., 2002].

In McCool et al. [2002], two transmitter coils operating at two different frequencies are placed near the spine at the sternal level and on the abdomen. Two receiving coils are also placed on the body: one tuned to both frequencies is placed on the sternum to measure the sternal–umbilical displacement and the rib cage anteroposterior displacement. The other is tuned only to the frequency of the abdominal transmitter and measures the anteroposterior abdominal displacement. With these three measurements, after calibration, respiratory volume can be estimated by using a three-degree-of-freedom model.

3.5.3 Surface Motion Measurement of Heart/Pulse Rate

Although common clinical methods of measuring pulse rate do not use surface motion, several methods of measuring pulse rate from surface motion have been developed. These techniques for surface measurement of pulse rate are summarized in Table 3.3. They can be broadly grouped into two categories: global measurement of the chest wall and measurement of a small surface on the chest wall or at a pulse point. Some of these methods measure the chest wall motion resulting from the heart beating against the chest wall, and others measure the surface motion resulting from arterial and venous pulses. Mechanocardiography is an all-encompassing term for the measurement of the motion or vibration of the chest wall due to the heart.

Apexcardiography is the measurement of chest motion at the apex relative to the rest of the chest wall. Positioning the apexcardiograph transducer for maximum signal typically requires repeated exploration of the apical area, and sensitivity to position results in poor reproducibility. Only the apex impulse measurements can be made with regularity, and this requires experience [Bancroft and Eddleman, 1967]. Apexcardiography represents the displacements of the precordium overlying the apex of the heart, caused by left ventricular movement. The apexcardiogram measures the movement of the chest wall and is indicative of the entire left ventricle. Its contour differs from that perceived in palpation [Braunwald and Perlkoff, 2001]. The device often has a funnel-shaped applicator connected to a microphone [Tafur et al., 1964] or a piezoelectric transducer [Bancroft and Eddleman, 1967].

The kinetocardiogram measures chest wall displacement at a single point (typically the apex) similar to the apexcardiogram. The difference is that the kinetocardiogram measures relative to an external fixed laboratory coordinate system, not relative to the rest of the chest. These readings most closely resemble the movements

TABLE 3.3 Techniques for Surface Measurement of Pulse Rate

Technique	Description	Contact or Noncontact?	Global or Small Area?	Pros and Cons
Doppler radar	Radar signal is directed at subject's chest; linear motion of entire chest is measured	Noncontact	Global	Subject can remain clothed and unwired Artifacts due to subject motion
Apexcardiography [Bancroft and Eddleman, 1967; Braunwald and Perlkoff, 2001]	Measures chest wall displacement at the apex relative to the rest of the chest	Contact	Small area	Difficult to place correctly
Kinetocardiogram/impulse cardiogram [Bancroft and Eddleman, 1967; Braunwald and Perlkoff, 2001]	Measures chest wall displacement relative to the laboratory coordinate system	Contact	Small area	Reading closely resembles palpation
Cardiokymogram (displacement cardiograph) [Fenton and Vas, 1973]	A coil that is part of a tuned circuit is placed near the chest of the patient; as the coil's environment changes, the output frequency changes	Noncontact	Global (depends on the size of the coil)	Measures superposition of heart motion and chest wall motion
Laser displacement system [Aubert et al., 1984; Ronaszeki et al., 1990]	Laser beam is pointed at the chest; displacement is measured	Noncontact	Small area	Not affected by loading Clothing must be removed
Ballistocardiogram/seismocardiogram [Mounsey, 1957; Crow et al., 1994]	Accelerometer is attached to the chest	Contact	Small area	Measures acceleration rather than displacement

detected by palpation [Bancroft and Eddleman, 1967]. The kinetocardiogram records the motion of specific points on the chest wall relative to a fixed point in space, and its contour is similar to that perceived by palpation [Braunwald and Perlkoff, 2001]. The device may be similar to the apexcardiogram, but has a flat applicator rather than a funnel [Eddleman et al., 1953].

The displacement cardiograph, also known as the cardiokymograph, consists of a coil that is part of a tuned circuit oscillator and is placed between 5 and 15 mm from the subject's chest wall. Changes in the location and volume of the chest due to the heart beating and respiration alter the loading of the coil, and therefore the frequency of oscillation. This frequency is compared with that of a reference frequency, and the difference is converted into an output voltage. The field created by the coil penetrates tissues so that the motion of the heart itself is sensed as well as chest wall motion. The system is much more sensitive to chest wall motion than the heart motion, but the heart motion is greater. The output of this system is qualitative because of the superposition of the heart motion and chest motion, and because of the complicated and irreproducible conversion between chest position and oscillation frequency. The cardiokymograph can detect heart motion in patients for whom the apexcardiograph is not detectable, including patients with emphysema, and others in whom no impulse can be palpated at the apex [Fenton and Vas, 1973].

The laser displacement system points a laser beam at the chest wall and measures the displacement [Aubert et al., 1984]. This requires the subject to be unclothed since the light at optical frequencies cannot penetrate clothing. The laser beam is focused at a small point on the chest, typically the apex [Aubert et al., 1984].

The ballistocardiograph, also known as the seismocardiograph, consists of an accelerometer strapped to the subject's chest, but is not a quantitative measurement of displacement [Mounsey, 1957; Crow et al., 1994; Salerno and Zanetti, 1991]. It can be used to sense the heart rate but needs to be placed on the area of the chest that is moving, such as the apex or sternum. The term "ballistocardiograph" can also refer to the measurement of the recoil of the human body due to the momentum of the blood being pumped by the heart; this type of ballistocardiograph is typically measured by a scale [Inan et al., 2009].

Doppler radar systems can simultaneously measure chest motion caused by heart and respiratory movements. In order to analyze the heart signals when the measurement subject is not holding his/her breath, it is necessary to use filters to attenuate the respiration signal relative to the heart signal so that the heart signal is dominant. This measurement is noncontact and works through clothing. No other respiratory measurements are noncontact and operate through clothing; the cardiokymograph could theoretically operate through clothing but it needs to be placed very close to the chest.

REFERENCES

Allison RD, Holmes EL, Nyboer J. Volumetric dynamics of respiration as measured by electrical impedance plethysmography. J Appl Physiol 1964;19:166–173.

American Thoracic Society/European Respiratory Society Task Force. 2004. Standards for the diagnosis and management of patients with COPD [Internet]. Version 1.2. New York:

American Thoracic Society. Available from: http://www.thoracic.org/go/copd. Updated 2005 Sept 8.

Arcem OX, Knudson OA, Ellison MC, Baselga P, Ivy DD, DeGroff C, Valdes-Cruz L. Longitudinal motion of the atrioventricular annuli in children: reference values, growth-related changes, and effects of right ventricular volume and pressure overload. J Am Soc Echocardiogr 2002;15(9):906–916.

Aubert AE, Welkenhuysen L, Montald J, de Wolf L, Geivers H, Minten J, Kesteloot H, Geest H. Laser method for recording displacement of the heart and chest wall. J Biomed Eng 1984;6(2):134–140.

Awtry EH, Loscalzo J. Structure and function of the normal heart and blood vessels. In: Carpenter CCJ, Griggs RC, Loscalzo J, editors. Cecil Essentials of Medicine. 5th ed. New York: W. B. Saunders Company; 2001a. p 21–29.

Awtry EH, Loscalzo J. Evaluation of the patient with cardiovascular disease. In: Carpenter CCJ, Griggs RC, n J, editors. Cecil Essentials of Medicine. 5th ed. New York: W. B. Saunders Company; 2001b. p 30–42.

Bancroft WH Jr, Eddleman EE Jr. Methods and physical characteristics of the kinetocardiographic and apexcardiographic systems for recording low-frequency precordial motion. Am Heart J 1967;73(6):756–764.

Berson AS, Pipberger HV. Measurement of chest wall vibrations due to the activity of the heart. J Appl Physiol 1966;21(2):370–374.

Brandt CM, Annoni H, Harthong J, Reiner JM, Krauskhopff R. Evaluation of chest wall distortion related to cardiac activity by structured lights: a study of the apical impulse by the Moiré Technique. Acta Cardiol 1986;41(3):207–213.

Braunwald E, Perlkoff JK. Physical examination of the heart and circulation. In: Braunwald E, Zipes DP, Libby P, editors. Heart Disease: A Textbook of Cardiovascular Medicine. New York: W. B. Saunders Company; 2001. p 45–81.

Buntin CM, Silver FH. Noninvasive assessment of mechanical properties of peripheral arteries. Ann Biomed Eng 1990;18(5):549–566.

Crow RS, Hannan P, Jacobs D, Hedquist L, Salerno DM. Relationship between seismocardiogram and echocardiogram for events in the cardiac cycle. Am J Noninvasive Cardiol 1994;8(1):39–46.

DeGroote A, Wantier M, Cheron G, Estenne M, Pavia M. Chest wall motion during tidal breathing. J Appl Physiol 1997;83(5):1531–1537.

Deliyannis AA, Gillam PMS, Mounsey JPD, Steiner RE. The cardiac impulse and the motion of the heart. Br Heart J 1964;26:396–411.

Dressler W. Pulsations of the chest wall. Arch Intern Med 1937;60:225–239.

Eddleman EE Jr, Willis K, Reeves TJ, Harrison TR. The kinetocardiogram. I. Method of recording precordial movements. Circulation 1953;8:269–275.

Fadel PJ, Barman SM, Phillips SW, Gebber GL. Fractal fluctuations in human respiration. J Appl Physiol 2004;97:2056–2064.

Fenton TR, Vas R. Measuring characteristics of the displacement cardiograph. Med Biol Eng 1973;11(5):552–558.

Flint A. Practical Treatise on the Diagnosis, Pathology, and Treatment of Diseases of the Heart. Philadelphia: Blanchard and Lea; 1859.

Gillam PMS, Deliyannis AA, Mounsey JPD. The left parasternal impulse. Br Heart J 1964;26:726–736.

Global Initiative for Chronic Obstructive Lung Disease (GOLD). 2008. From the global strategy for the diagnosis, management and prevention of COPD. Available from: http://www.goldcopd.org/.

Gribbin HR. Using body surface movements to study breathing. J Med Eng Technol 1983;7(5):217–223.

Hong H-D, Fox MD. No touch pulse measurement by optical interferometry. IEEE Trans Biomed Eng 1994;41(11):1096–1099.

Ikegaya K, Suzumura N, Funada T. Absolute calibration of phonocardiographic microphones and measurements of chest wall vibration. Med Biol Eng Comput 1971;9(6):683–692.

Inan OT, Etemadi M, Paloma A, Giovangrandi L, Kovacs GTA. Non-invasive cardiac output trending during exercise recovery on a bathroom-scale-based ballistocardiograph. Physiol Meas 2009;30(3):261.

Kawasaki T, Sasayama S, Yagi S, Asakawa T, Hirai T. Non-invasive assessment of the age related changes in stiffness of major branches of the human arteries. Cardiovasc Res 1987;21(9):678–687.

Ko K, Erickson R, Schmidt T, Webster J. Holographic interferometry of cerebral pulsations. Surg Neurol 2005;63(2):118–122.

Kondo T, Kobayashi I, Taguchi Y, Ohta Y, Yanagimachi N. A dynamic analysis of chest wall motions with MRI in healthy young subjects. Respirology 2000;5(1):19–25.

Kondo T, Uhlig T, Pemberton P, Sly PD. Laser monitoring of chest wall displacement. Eur Respir J 1997;10:1865–1869.

Lee J-S. On the coupling and detection of motion between an artery with a localized lesion and its surrounding tissue. J Biomech 1974;7(5):403–409.

Lee-Chiong TL Jr. Monitoring respiration during sleep. Clin Chest Med 2003;24(2):297–306.

Macklem PT. The act of breathing. In: Russo C, editor. The Thorax. New York: Marcel Dekker, Inc.; 1995. p 445–456.

Macklem PT, Gross D, Grassino A, Roussos C. Partitioning of inspiratory pressure swings between diaphragm and intercostal/accessory muscles. J Appl Physiol 1978;44(2): 200–208.

McCool FD, Wang J, Ebi KL. Tidal volume and respiratory timing derived from a portable ventilation monitor. Chest 2002;122(2):684–691.

McLeish K, Hill DLG, Atkinson D, Blackall JM, Razavi R. A study of the motion and deformation of the heart due to respiration. IEEE Trans Med Imag 2002;21(9):1142–1150.

Mead J. Control of respiratory frequency. J Appl Physiol 1960;15(3):325–336.

Meinders JM, Hoeks APG. Simultaneous assessment of diameter and pressure waveforms in the carotid artery. Ultrasound Med Biol 2004;30(2):147–154.

Mohri K, Jinnouchi T, Kawano K. Accurate mechanocardiogram sensors using amorphous star-shaped core multivibrator combined with a magnet. IEEE Trans Magnet 1987;23(5): 2212–2214.

Mohri K, Kondo T, Sugino H, Yamasaki J, Yoshino K. Non-contact linear displacement sensors using amorphous-core multivibrators for mechanocardiography. IEEE Trans Magnet 1985;21(5):2071–2073.

Mooser V, Etienne J-D, Farine P-A, Monney P, Perret F, Cecchini M, Gagnebin E, Bornoz C, Tardy Y, Arditi M, Meister J-J, Leuenberger C-E, Saurer E, Mooser E, Waeber B, Brunner HR. Non-invasive measurement of internal diameter of peripheral arteries during the cardiac cycle. J Hypertens 1988;6(4 Suppl):S179–S181.

Mounsey P. Precordial ballistocardiography. Br Heart J 1957;19:259–271.

Nehrke K, Bornert P, Manke D, Bock JC. Free-breathing cardiac MR imaging: study of implications of respiratory motion – initial results. Radiology 2001;220(3):810–815.

O'Rourke MF, Kelly R, Avolio A. The Arterial Pulse. Philadelphia: Lea and Febiger; 1992.

Opie LH. Mechanisms of cardiac contraction and relaxation. In: Braunwald E, Zipes DP, Libby P, editors. Heart Disease: A Textbook of Cardiovascular Medicine. New York: W. B. Saunders Company; 2001. p 443–478.

Osmond DG. Functional anatomy of the chest wall. In: Russo C, editor. The Thorax. New York: Marcel Dekker, Inc.; 1995. p 413–444.

Owen A. Effect of increasing age on diastolic motion of the left ventricular atrioventricular plane in normal subjects. Int J Cardiol 1999;69:127–132.

Ramachandran G, Singh M. Three-dimensional reconstruction of cardiac displacement patterns on the chest wall during the P, QRS, and T-segments of the ECG by laser speckle interferometry. Med Biol Eng Comput 1989;27(5):525–530.

Ramachandran G, Swarnamani S, Singh M. Reconstruction of out-of-plane cardiac displacement patterns as observed on the chest wall during various phases of ECG by capacitance transducer. IEEE Trans Biomed Eng 1991;38(4):383–385.

Rodarte JR, Shardonofsky FR. Respiratory system mechanics. In: Murray JF, Nadell JA, editors. Textbook of Respiratory Medicine. 3rd ed. New York: W. B. Saunders Company; 2000. p 91–117.

Ronaszeki A, Aubert AE, de Geest H. Laser apexcardiogram in healthy young men: a comparative study with the conventional method. Acta Cardiol 1990;45(3):203–210.

Rosa LM. The 'displacement' vibrocardiogram of the precordium in the low frequency range. Am J Cardiol 1959;4(2):191–199.

Salerno DM, Zanetti J. Seismocardiography for monitoring changes in left ventricular function during ischemia. Chest 1991;100(4):991–993.

Schaff BW, Abbrecht PH. Digital computer simulation of human systemic arterial pulse wave transmission: a nonlinear model. J Biomech 1972;5(4):345–364.

Schlant RC, Silverman ME, Roberts WC. Anatomy of the heart. In: Hurst JW, Schlant RC, Rackley CE, Sonnenblick EH, Wenger NK, editors. The Heart, Arteries, and Veins. San Francisco: McGraw Hill; 1990. p 14–35.

Singh M, Ramachandran G. Reconstruction of sequential cardiac in-plane displacement patterns on the chest wall by laser speckle interferometry. IEEE Trans Biomed Eng 1991;38(5):483–489.

Tafur E, Cohen LS, Levine HD. The normal apex cardiogram: its temporal relationship to electrical, acoustic, and mechanical cardiac events. Circulation 1964;30:381–391.

Tobin MJ, Chadha TS, Jenouri G, Birch SJ, Gazeroglu HB, Sackner MA. Breathing patterns: 1. Normal subjects. Chest 1983;84(2):202–205. DOI: 10.1378/chest.84.2.202.

Vander A, Sherman J, Luciano D. Human Physiology: The Mechanisms of Body Function. 7th ed. San Francisco: McGraw Hill; 1998.

Vermarien H, van Vollenhoven E. The recording of heart vibrations: a problem of vibration measurement on soft tissue. Med Biol Eng Comput 1984;22(2):168–178.

Wilson TA, Rehder K, Krayer S, Hoffman A, Whitney CG, Rodarte JR. Geometry and respiratory displacement of human ribs. J Appl Physiol 1987;62(5):1872–1877.

Wolf GK, Arnold JH. Noninvasive assessment of lung volume: respiratory inductance plethysmography and electrical impedance tomography. Crit Care Med 2005; 33(3 (Suppl)):S163–S169.

Yip GW, Zhang Y, Tan PY, Wang M, Ho P-Y, Brodin L-A, Sanderson JE. Left ventricular long-axis changes in early diastole: impact of systolic function on diastole. Clin Sci 2002;102(5):515–522.

4

PHYSIOLOGICAL DOPPLER RADAR OVERVIEW

ADITYA SINGH[1], BYUNG-KWON PARK[2], OLGA BORIC-LUBECKE[3], ISAR MOSTAFANEZHAD[4], AND VICTOR M. LUBECKE[3]

[1]*University of Hawaii Neuro-science and MRI research Program, John A. Burns School of Medicine, Honolulu, Hawaii, United States*
[2]*DAS Sensor SW Engineering Team, Hyundai Mobis Mechatronics R&D Center, Gyeonggi-Do, South Korea*
[3]*Department of Electrical Engineering, University of Hawaii at Manoa, Honolulu, Hawaii, United States*
[4]*Department of Physics and Astronomy, University of Hawaii at Manoa, Honolulu, Hawaii, United States*

A simplified block diagram of a Doppler radar system for physiological monitoring is shown in Fig. 4.1. It consists of three modules: radio frequency (RF) front-end module, baseband module, and signal-processing module. In this chapter, design considerations for each of these modules are discussed.

Microwave signal is generated in the radar front end and transmitted through the transmit antenna toward the subject. Reflected signal from the subject, which is phase-modulated due to patient's physiological motions, is received using the receive antenna. The received signal is down-converted from microwave region to DC (or near DC) and passes through amplifiers, filters, and data acquisition (DAQ). Digital-to-analog conversion within the DAQ creates a digital version of the baseband radar signal, which can be processed further using numerous digital signal-processing (DSP) techniques.

We can define transmit and receive signals as follows. The transmit signal can be expressed in time domain as

$$S_t(t) = \cos(\omega_0 t) \tag{4.1}$$

Doppler Radar Physiological Sensing, First Edition.
Edited by Olga Boric-Lubecke, Victor M. Lubecke, Amy D. Droitcour, Byung-Kwon Park, and Aditya Singh.

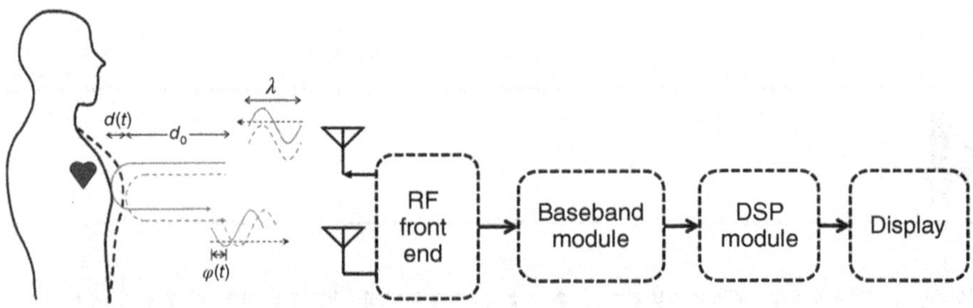

Figure 4.1 Simplified block diagram of a physiological Doppler radar system.

where ω_0 is the radian frequency of the transmitted microwave signal (radar's RF). This transmit signal will travel until it is incident on the human subject's body where it is reflected back to the receive antenna. Signal received at the antenna has undergone phase delay and amplitude change, which can be expressed as

$$S_r(t) = A\cos\left(\omega_0 t + \frac{2\pi}{\lambda}\left(2d_0 + 2d\left(t\right)\right)\right) \tag{4.2}$$

where d_0 is the static distance of radar antenna to the human subject and $d(t)$ represents chest displacement due to physiological motion (heart beat and respiration), and A is the amplitude of the received signal. Wavelength is related to the radian frequency through

$$\lambda = \frac{c}{f} = 2\pi\frac{c}{\omega} \tag{4.3}$$

where c is the speed of electromagnetic waves (same as the speed of light) in the surrounding medium of (air). For example, for $f = 2.4$ GHz, λ (wavelength) is calculated to be 12.5 cm. The next section describes each of the system components in more detail and then discusses the problems, issues, and trade-offs in the design.

4.1 RF FRONT END

A simple but robust approach for a Doppler transceiver system is depicted in Fig. 4.2. Transmit and receive antennas, signal source, splitter, and a frequency mixer are key components. Output of the system is a comparison of the transmitted and received signals.

A simple way to compare phase of two sinusoids is to multiply them together:

$$S_r(t)S_t(t) = A\cos\left(\omega_0 t + \frac{2\pi}{\lambda}\left(2d_0 + 2d\left(t\right)\right)\right)\cos(\omega_0 t) \tag{4.4}$$

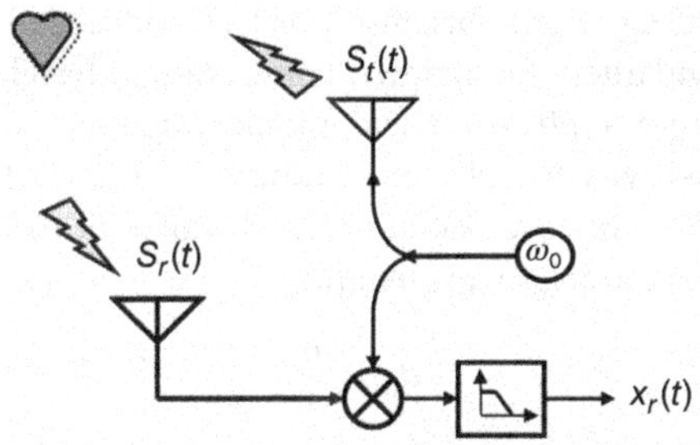

Figure 4.2 A simple Doppler transceiver architecture denoting transmit and receive antennas, signal source, and a frequency mixer.

Applying the following trigonometric identity,

$$\cos a \cos b = \frac{\cos(a-b) + \cos(a+b)}{2} \tag{4.5}$$

we can obtain

$$S_r(t)S_t(t) = \frac{A}{2}\cos\left(\frac{2\pi}{\lambda}\left(2d_0 + 2d\,(t)\right)\right) + \frac{A}{2}\cos\left(2\omega_0 t + \frac{2\pi}{\lambda}\left(2d_0 + 2d\,(t)\right)\right) \tag{4.6}$$

In this equation, the second term on the right has twice the frequency of the transmitted signal. Mixer output will be passed through a low-pass filter, which will filter out the higher frequency component. This will result in the received baseband signal as

$$x_r(t) = \frac{A}{2}\cos\left(\frac{2\pi}{\lambda}\left(2d_0 + 2d\,(t)\right)\right) \tag{4.7}$$

which has a variable phase, in the form

$$\phi(t) = \frac{2\pi}{\lambda}2d(t) \tag{4.8}$$

This relates the displacement of the subject to a phase that is detected by the radar. As an example, at 2.4 GHz, given a wavelength of 12.5 cm, a 1-cm displacement will result in a roundtrip phase change of roughly 1 rad or 57.6°. Equation 4.7 states that baseband output of the radar is actually cosine of the phase. This type of receiver is also known as a homodyne or direct conversion receiver, and since this receiver has only one output it is a single-channel receiver. A single-channel receiver is very simple and is capable of producing a baseband signal with sufficient accuracy for extracting vital signs [Droitcour et al., 2001]. In traditional CW Doppler radar, a single-ended receiver similar to that illustrated in Fig 4.2 cannot distinguish approaching and receding targets because positive and negative Doppler shifts fold into one frequency band after signal down-conversion to baseband [Saunders, 1990]. In Doppler radar used for physiological monitoring, this spectrum folding issue results in receiver sensitivity to null and optimum locations of the subject under test [Park et al., 2006]. The null case occurs if the distance of the subject to the radar is an integer multiple of quarter of wavelength:

$$d_0 = \frac{n}{4}\lambda \tag{4.9}$$

In that case, the argument in cosine term in Equation 4.7 becomes

$$\frac{2\pi}{\lambda}(2d_0 + 2d(t)) = n\pi + \phi(t) \tag{4.10}$$

resulting in

$$x_r(t) = \frac{A}{2}\cos(n\pi + \phi(t)) = \pm\frac{A}{2}\cos(\phi(t)) \tag{4.11}$$

As it was discussed earlier, $\phi(t)$ is a small time-varying phase, thus a Taylor series expansion results in

$$x_r(t) = \pm\frac{A}{2}\cos(\phi(t)) \approx \pm\frac{A}{2} \mp \frac{A}{2}\frac{\phi(t)^2}{2} \approx \pm\frac{A}{2} \tag{4.12}$$

This indicates that the received signal is almost constant and will not be very sensitive to physiological motion.

When distance from the subject to the antenna is an odd multiple of eighth of wavelength or

$$d_0 = \frac{2n+1}{8}\lambda \tag{4.13}$$

the receiver is in the "optimum" point, and the received signal will most closely follow physiological motion:

$$x_r(t) = \pm\frac{A}{2}\sin(\varphi(t)) \approx \pm\frac{A}{2}\varphi(t) \tag{4.14}$$

Equations 4.13 and 4.14 indicate that null and optimum points are eighth of a wavelength apart, as illustrated in Fig. 4.3.

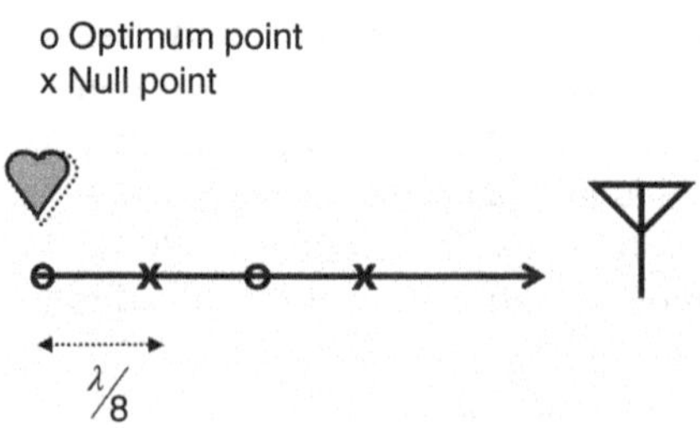

Figure 4.3 Illustration of null and optimum points in a single-channel receiver system.

4.1.1 Quadrature Receiver

In Doppler radar systems, a quadrature receiver is used to develop a phase-coherent receiver. This enables obtaining a velocity vector rather than just a speed with the Doppler shift [Saunders, 1990]. While real signals' positive and negative frequency components are mirror images of each other, complex exponentials can have positive and negative frequencies that do not have the same frequency spectra. Euler's equations define complex exponential phasors as

$$e^{j\omega t} = \cos(\omega t) + j\sin(\omega t) \tag{4.15}$$

and

$$e^{-j\omega t} = \cos(\omega t) - j\sin(\omega t) \tag{4.16}$$

where $j = \sqrt{-1}$. The $\cos(\omega t)$ term describes the phasor's real component, while the $\sin(\omega t)$ term describes the phasor's component along the imaginary, or j axis. Euler's equations can be manipulated to show that

$$\sin(\omega t) = \frac{1}{2j}(e^{j\omega t} - e^{-j\omega t}) \tag{4.17}$$

and

$$\cos(\omega t) = \frac{1}{2}(e^{j\omega t} + e^{-j\omega t}) \tag{4.18}$$

These equations indicate that the cosine has equal positive components at $+\omega$ and $-\omega$, while the sine has a positive component at $+\omega$ and a negative component of equal magnitude at $-\omega$, as is shown in Fig. 4.4.

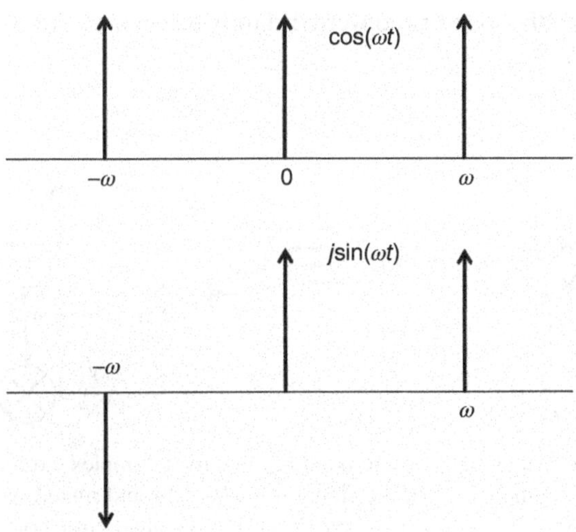

Figure 4.4 Spectra of real cosine signal and an imaginary sine signal when represented in complex notation.

In addition,

$$\sin(\theta) = \cos\left(\theta - \frac{\pi}{2}\right)$$
$$\cos(\theta) = -\sin\left(\theta - \frac{\pi}{2}\right) \tag{4.19}$$

Therefore, a delay in the time domain manifests itself as a phase shift in the frequency domain, and this delay can shift a sine wave to a cosine wave.

In quadrature processing, by convention, the real part of the spectrum is called the in-phase component, and the imaginary part of the spectrum is called the quadrature-phase component. Real signals, those signals that are real in the time domain, have positive and negative frequency components. The positive and negative frequency components of a real spectrum are symmetric about the zero-frequency point. However, the positive and negative frequency components of a quadrature, or imaginary, spectrum are complex conjugates of each other. Complex signals are a combination of in-phase and quadrature, or real and imaginary components.

A complex exponential, for example, $e^{-j2\pi f_{LO}t}$ has only a single frequency component, in this case at a negative frequency, $-f_{LO}$. Although the complex exponential is not real, it can be realized by multiplying the signal by both a sine and a cosine at the local oscillator (LO) frequency, and then the two signals can be combined as in Euler's equation. This is illustrated in Fig. 4.5.

The quadrature receiver architecture illustrated in Fig. 4.5 is commonly used in communication systems to avoid the issue of spectrum folding when the received signal is down-converted to baseband. As illustrated in Fig. 4.6, when an RF signal with an asymmetric spectrum is mixed with a simple cosine signal, which contains positive and negative frequencies, the resulting baseband spectra overlap and cause signal distortion. However, if an RF signal is mixed with a complex exponential, only the positive or negative band is converted into baseband, avoiding the interference problem (Fig. 4.7).

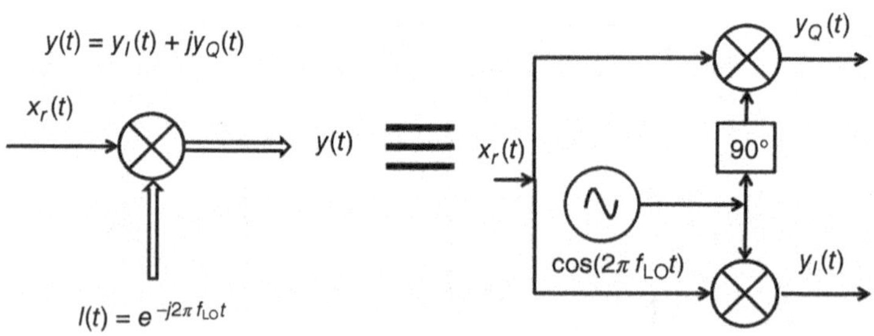

Figure 4.5 For the real signal, $x_r(t)$, to be multiplied by a complex exponential with only a negative frequency component, $l(t)$, the signal must be split and mixed with local oscillator signals to determine the in-phase component, $y_I(t)$, and the quadrature component, $y_Q(t)$. The LO signal on the Q channel is delayed by 90° before mixing. The two components can be summed to create the output: $y(t) = y_I(t) + jy_Q(t)$.

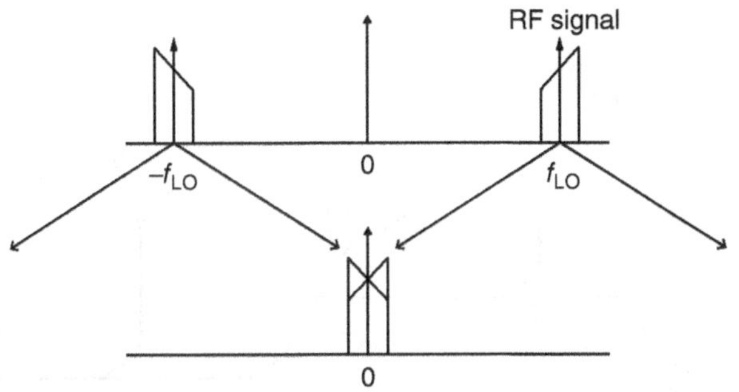

Figure 4.6 Self-image problem with a direct-conversion receiver. If a quadrature receiver is not used, both the positive and negative frequency components are down-converted to baseband, where they can interfere with each other.

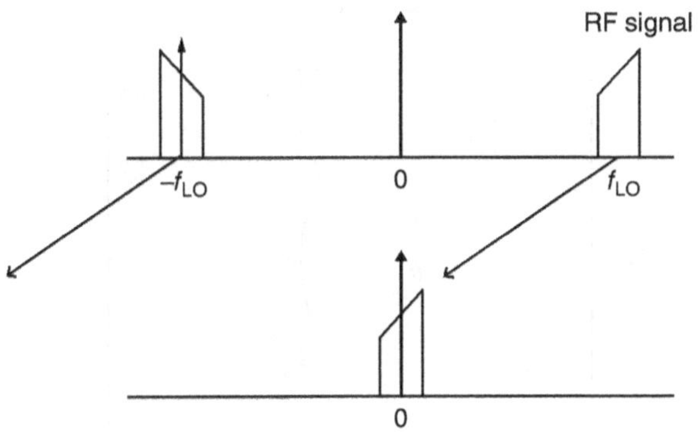

Figure 4.7 Avoiding the self-image problem with a quadrature direct-conversion. When the RF signal is mixed with a complex exponential, only the positive or negative band is converted into baseband, avoiding the interference problem.

Figure 4.8(a) illustrates a Doppler radar system with quadrature receiver. The local oscillator (LO) signal, derived from the transmit signal, is split to feed the two mixers. For the Q channel, it goes through a 90° phase-shifter and is then mixed with the received signal to generate the quadrature-channel output. The I channel output is the same as what was derived before:

$$x_{rI}(t) = \frac{A}{2} \cos\left(\frac{2\pi}{\lambda}(2d_0 + 2d(t))\right) \tag{4.20}$$

Because of the 90° phase shift, the Q channel output will be

$$x_{rQ}(t) = \frac{A}{2} \cos\left(\frac{2\pi}{\lambda}(2d_0 + 2d(t)) - \frac{\pi}{2}\right) = \frac{A}{2} \sin\left(\frac{2\pi}{\lambda}(2d_0 + 2d(t))\right) \tag{4.21}$$

Equations 4.20 and 4.21 indicate that if the subject is in the null point for one of the channels, the other channel will be in optimum point. A simple way to represent

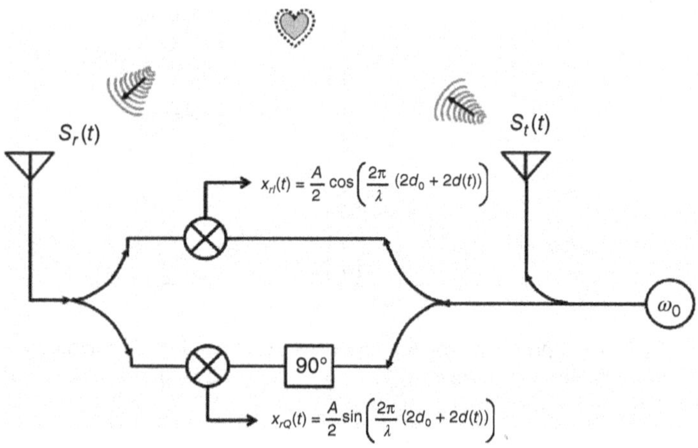

(a)

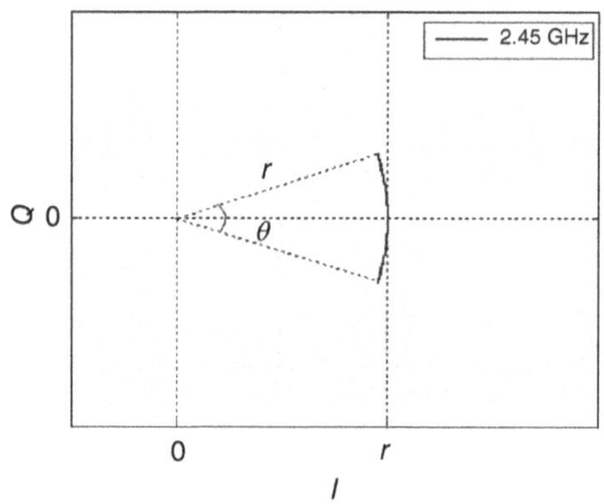

(b)

Figure 4.8 A quadrature receiver system (a) and a sample of received I and Q signals (b).

I and Q channels is to use complex notation:

$$x_{rC}(t) = x_{rI}(t) + jx_{rQ}(t) = \frac{A}{2}\exp\left(j\frac{2\pi}{\lambda}\left(2d_0 + 2d\left(t\right)\right)\right) \tag{4.22}$$

RF transmit and receive signals can be represented in complex form as

$$S_t(t) = \text{Re}[S_{tC}(t)] = \text{Re}[\exp(j\omega_0 t)] \tag{4.23}$$

$$S_r(t) = \text{Re}[S_{rC}(t)] = \text{Re}\left[A\exp\left(j\omega_0 t + j\frac{2\pi}{\lambda}\left(2d_0 + 2d\left(t\right)\right)\right)\right] \tag{4.24}$$

When the received signal (Eq. 4.24) is mixed with a complex exponential, it results in the complex baseband signal of Equation 4.22. Graphical representation of the complex baseband signal in the complex plane, as shown in Fig. 4.8(b), represents an arc with radius r. The radius can yield some information related to target's position with respect to the radar. For a given frequency, the arc length is an indication of the amount of movement in front of the radar with larger arc length representing a larger motion.

It is important to have balanced I and Q channels for accurate signal recovery. Ideally, both channels should have same gain/loss and phase shift in RF and baseband components. Phase and amplitude imbalance between the two receiver chains are induced by the hybrid coupler and mismatches between RF components, baseband components, and the ADC in each receiver chain. Effect of balance imperfections is discussed in Droitcour [2006]. In the simplest form, if the in-phase signal is represented by

$$x_{rI}(t) = \frac{A}{2}\cos\left(\frac{2\pi}{\lambda}\left(2d_0 + 2d\left(t\right)\right)\right) \tag{4.25}$$

Then the quadrature component could be represented by

$$x_{rQ}(t) = A_e \cdot \frac{A}{2}\sin\left(\frac{2\pi}{\lambda}\left(2d_0 + 2d\left(t\right)\right) + \phi_e\right) \tag{4.26}$$

where A_e and ϕ_e are the amplitude and phase imbalance factors, respectively. When there is no imbalance, the arc transcribed in the IQ plane fits that of a circle. Presence of any imbalance causes the arc to resemble an ellipse. Phase imbalance also results in rotation of the arc/ellipse. The effect of imbalances is shown in Fig. 4.9.

4.1.2 Phase Coherence and Range Correlation

An important advantage of the homodyne receiver system is a range correlation effect. This effect helps in reducing the baseband noise significantly, thus enabling detection of very small motions (such as motion associated with heart signals) at distances of several meters. Phase noise is the characteristic of any signal source and is exhibited as signal frequency deviation from an ideal single-line frequency spectrum. Some RF sources have a very low-phase noise and others have a larger phase noise, depending on the type of the active device, and oscillator configuration. Phase noise is generated because of random phase fluctuations within the oscillator. Regardless of its origin, RF phase noise translates to baseband noise after frequency down-conversion and can become a determining factor for baseband signal-to-noise ratio (SNR) [Droitcour, 2006; Razavi, 2001].

Assuming the complex representation for the LO and RF signals with phase noise, we have

$$S_{tC}(t) = \exp(j\omega_0 t + j\theta(t)) \tag{4.27}$$

$$S_{rC}(t) = A\exp\left(j\omega_0 t + j\frac{2\pi}{\lambda}\left(2d_0 + 2d\left(t\right)\right) + j\theta\left(t - \frac{2d_0}{c}\right)\right) \tag{4.28}$$

In these expressions, $\theta(t)$ represents the signal generator phase noise. This phase noise is present in the transmitted signal and a delayed version of it will be received at

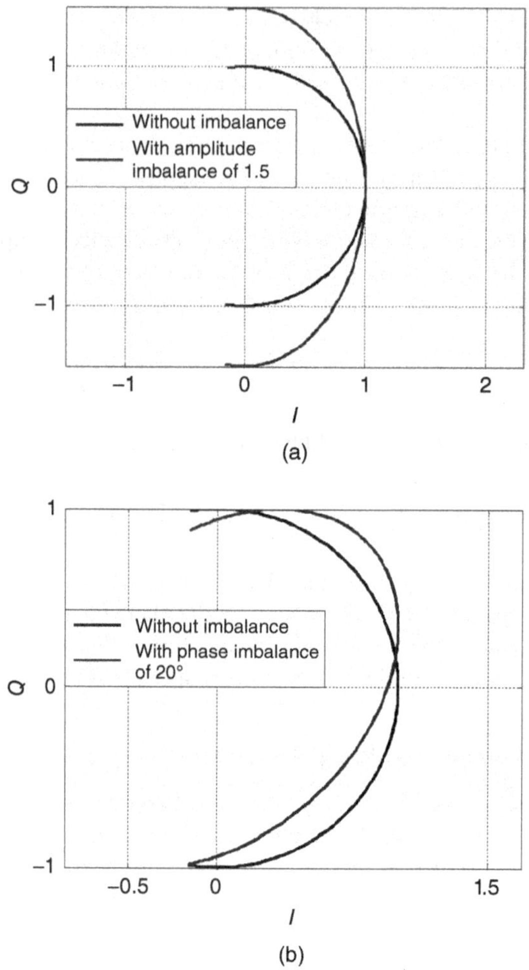

Figure 4.9 Simulated plots showing the effect of amplitude imbalance between the *I* and *Q* channels (a), and the effect of phase imbalance between the *I* and *Q* channels (b).

the receive antenna as in Equation 4.28. As discussed in the previous section, the transmit and receive signals are mixed together (multiplied) to yield the baseband signal, which in this case will be

$$x_{rC}(t) = \frac{A}{2}\exp\left(j\frac{2\pi}{\lambda}\left(2d_0 + 2d\left(t\right)\right) + j\theta(t) - j\theta\left(t - \frac{2d_0}{c}\right)\right) \tag{4.29}$$

Equation 4.29 shows the effect of phase noise on the baseband signal. If the generator is perfect ($\theta(t) = 0$), the phase noise will not appear in baseband. Another condition is when the delay, $2d_0/c$, is small causing the transmit and receive phase noise to be

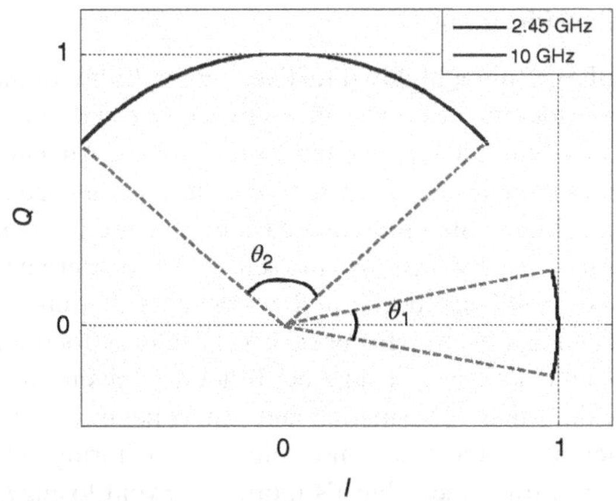

Figure 4.10 Simulated *IQ* plot of a motion as seen by two different frequencies. The arc length is about four times longer at 10 GHz compared with 2.45 GHz ($\theta_2 > \theta_1$).

very similar (correlated) and cancel each other in the baseband signal. This is known as the range correlation effect and is discussed in Budge and Burt [1993]. In order to benefit from this phenomena, same signal source has to be used to generate the transmit signal and to provide LO to down-convert the received signal.

4.1.3 Frequency Choice

In principle, Doppler radar can operate at any frequency. However, physical constraints such as antenna size, measurement environment (detection through air or through some barrier), and distance to target, will determine optimum operation frequency for a particular application. An important feature in frequency choice is the resolution. The amount of phase modulation in radians is $4\pi x(t)/\lambda$, where $x(t)$ is the chest motion. The higher the frequency, the shorter the wavelength, and therefore the greater the phase modulation. On the complex *IQ* plot, greater phase modulation will result in a longer arc as shown in Fig. 4.10. The arc length is about four times longer at 10 GHz as compared with 2.45 GHz.

The choice of an unlicensed band is important for Federal Communications Commission (FCC) compliance, and also in order to have a range of commercially available antennas, and radio frequency components to choose from. The FCC unlicensed bands at RF and microwave frequencies are 902–928 MHz, 2.4–2.4835 GHz, 5.725–5.875 GHz, 24.0–24.25 GHz, and 57–64 GHz.

Frequency choice is also related to the antenna size, and far-field considerations. A higher frequency means that the same antenna gain and directivity can be obtained with a physically smaller antenna. For a given directivity, the distance to the far field region is directly proportional to the wavelength, thus as the frequency increases, far-field limit moves closer to the antenna.

4.1.4 Antenna Considerations

The choice of whether to use a highly directive, a wide-beam, or an omnidirectional antenna requires consideration of trade-offs between size and directivity. In general, the higher the directivity of an antenna, the larger its area. An antenna with a larger size has a larger region that is near-field, in which the antenna pattern is not constant over varying range. A highly directive antenna could focus on only the desired target. This would enable increased selectivity (the sensitivity to alternate targets would be greatly reduced) and would also decrease the sensitivity to clutter, since less clutter would be in the antenna's beam. However, if the beam is focused to cover only a small area on the subject's chest, it may be difficult to ensure that beam is on the appropriate part of the subject [Vermarien and van Vollenhoven, 1984]. In addition, in applications where the subject may move during monitoring, a highly directional antenna would need to track the subject's motion to avoid losing contact. It is possible to increase selectivity with broad-beam antennas by using several transceivers [Samardzija et al., 2005]. Finally, a large antenna has some drawbacks of its own; it makes the entire system less portable and it may intimidate subjects.

Another important question is whether to use separate antennas for transmitting and receiving, or to use a single antenna for both transmitting and receiving with a ferrite circulator to provide isolation between the transmitted signal and the received signal. Antennas are generally larger and more massive than drop-in circulators. The price of an additional commercially made antenna is similar to the price of an on-board circulator. However, patch antennas developed on printed circuit boards could decrease the cost of mass-produced items. Drop-in circulators are specified to provide between 20 and 26 dB of isolation, and antenna spacing and design affects the isolation between the antennas.

Using two antennas leads to a bistatic radar system, which may affect the radar cross section of the target. However, as long as the two antennas are kept near each other this effect will be minimal [Skolnik, 1961]. If the antennas are near each other, care must be taken to minimize leakage between the two antennas. When the antennas are spaced and angled appropriately, the dominant source of leakage generally is backscatter from nearby clutter, which is unavoidable [Banks, 1975].

4.1.5 Power Budget

The distinct electrical properties of human tissue determine how a transmitted signal penetrates or reflects at all tissue layer boundaries. The tissue dielectric properties determine how much of the power is attenuated per unit distance due to tissue conductivity (σ), how much is transmitted to the next layer due to permittivity (ε) difference between adjacent layers, and how much is reflected back toward the skin surface. Biological tissue is nonmagnetic; therefore, its permeability (μ) can be assumed as 1. If the dielectric constants between adjacent layers have a large difference, reflection may be larger than transmission or vice versa, and if the tissue's conductivity is higher, then dissipation rate of the penetrating signal per unit length becomes bigger.

The intrinsic impedance of the materials, η, is calculated as

$$\eta = \frac{j\omega\mu}{\gamma} \tag{4.30}$$

where ω is the radial velocity, $2\pi f$, and γ is the propagation constant,

$$\gamma = j\omega \cdot \sqrt{\mu\varepsilon} \cdot \sqrt{1 - \frac{j\sigma}{\omega\varepsilon}} \tag{4.31}$$

The reflection coefficient, Γ, at the interface between free space and the material with intrinsic impedance is

$$\Gamma = \frac{\eta - \eta_0}{\eta + \eta_0} \tag{4.32}$$

where η_0 is the impedance of free space.

The transmission coefficient, T, at the same interface is

$$T = 1 + \Gamma = \frac{2\eta}{\eta + \eta_0} \tag{4.33}$$

Using (4.31) and (4.32), at the frequency of interest or at 2.4 GHz, the reflection coefficient at the air–skin interface is approximately −0.1, while the transmission coefficient is 0.29. The portion of the power that is reflected is equal to the square of the reflection coefficient and can be expressed as

$$\frac{P_{\text{refl}}}{P_{\text{inc}}} = \Gamma^2 \tag{4.34}$$

Thus, the reflection power at the surface of the skin is 51% of the incident power.

An estimate of the amount of reflected power at each internal skin layer boundary is presented in Fig. 4.10. Values of the intrinsic impedance and attenuation coefficient at 2.4 GHz were taken from Gabriel [1996], with an estimated assumption of each layer type and thickness encountered. As shown in Fig. 4.11, most of reflected signal comes from air/skin interface which is about 51% of total incident power. Also, it is clear that most of penetrated power is absorbed and only about 9% of it can be re-radiated. Therefore, 51% of incident power is reflected to the radar system, and 91% of it comes from the air/skin interface. From this result, it is clear that noncontact Doppler radar systems detect primarily skin surface motion. Changes in the shape and volume of the heart during systole move the ribs and soft tissue near the heart, causing the chest to pulse with each heartbeat. The contraction and relaxation of the left ventricle causes a larger chest motion than other heart actions in a healthy human. During isovolumetric contraction, the heart normally undergoes a partial rotation in a counter-clockwise direction (when facing a subject), causing the lower front part of the left ventricle to strike the front of the chest wall [Braunwald and Perlkoff, 2001]. The left ventricle also shortens as it contracts, making the heart more spherical, increasing its diameter and further adding to the

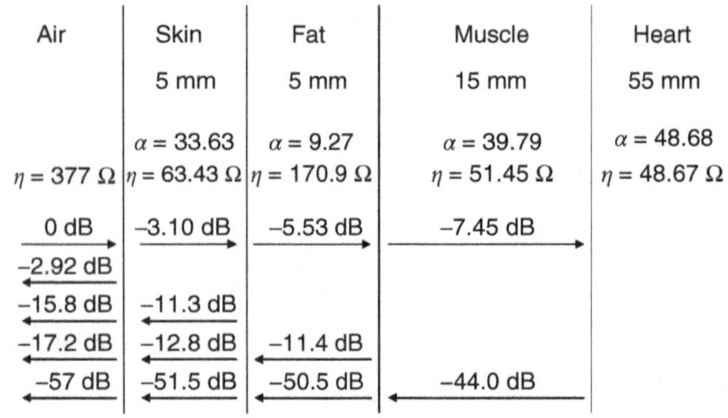

Figure 4.11 Power budget of 2.4 GHz radio wave at each internal layer in the body. Most of reflected signal comes from air/skin interface, which is about 51% of the incident power.

impulse on the chest wall [Dressler, 1937]. The peak outward motion of the left ventricular impulse occurs either simultaneously with or just after the opening of the aortic valve then the left ventricular apex moves inward [Braunwald and Perlkoff, 2001; Deliyannis et al., 1964]. The left ventricular motion causes the chest to pulse outward briefly, and the adjacent chest retracts during ventricular ejection [Gillam et al., 1964]. This impulse occurs at the lowest point on the chest where the cardiac beat can be seen, and it is normally above the anatomical apex, in the fourth and fifth intercostals spaces in the left midclavicular line [Awtry and Loscalzo, 2001]. For gas exchange to occur in the lungs, air with carbon dioxide needs to be removed from the lungs and air with oxygen needs to be inspired. In respiration, muscles contract to generate changes in thorax volume, which create pressure differences between the thorax and the external environment, causing air to move in and out of the lungs, from areas of high pressure to areas of low pressure. The motions of the thorax and the abdomen cause significant displacements at the skin surface that are measurable with Doppler radar, allowing noncontact measurement of respiratory effort.

The FCC Code of Federal Regulations (CFR), Section 15.427 [FCC, 2000] states that the maximum output power in the 2400–2435 MHz unlicensed band is 1 W. For antenna gain greater than 6 dBi, the output power must be reduced by 1 dB for every 3 dB that the antenna gain exceeds 6 dBi. Most consumer wireless devices that operate in this band have radio output powers between 10 and 300 mW, with some Bluetooth devices transmitting as little as 1 mW. These devices typically have antenna gains near 3 dBi. The radar systems developed for physiological monitoring typically operate with an output power between 1 and 10 mW, at the low end of consumer wireless products, and have an antenna gain of about 6 dBi, about double that of consumer electronics. Given the similarity in the levels of radiation, this monitor poses no greater risk to humans than do 2.4-GHz infant monitors, wireless LAN, or cordless telephones.

4.2 BASEBAND MODULE

Once a relatively noise free signal is available at the mixer output in baseband, it needs to be amplified and digitized to be ready for baseband processing. Low-noise baseband amplifiers will amplify I and Q channel outputs, in addition to low-pass filtering before sampling. A DAQ device is used to sample baseband channels and send them to a computer. Figure 4.12 shows a more detailed connection of the DAQ to low-noise amplifiers (LNAs) and the computer.

Since the signals are weak (order of millivolts), a low-noise baseband amplifier is needed to handle I and Q signals to bring the signal amplitude to the appropriate level for the DAQ system. Usually, a voltage gain of 100–1000 is applied depending on the application [Droitcour, 2006; Park, 2007; Massagram, 2008]. The signals also need to be low-pass filtered to prevent aliasing when they are being sampled in the DAQ. According to Nyquist's sampling theorem, a signal with a bandwidth of W Hertz needs to be sampled at 2 W samples per second (SPS) so that it can be ideally reconstructed [Proakis and Manolakis, 2006]. Physiological signals have a bandwidth of dc to about 8 Hz [Droitcour, 2006; Massagram, 2008]. Ideally, a sampling frequency of 16 Hz should be enough for capturing the radar outputs, but usually sampling rates higher than 100 Hz are used to minimize the effect of any out of band interference. Also, a higher sampling rate lowers the noise floors of the baseband signals [Proakis and Manolakis, 2006].

4.2.1 Analog Signal Conditioning and Coupling Methods

A major issue in a homodyne (direct conversion) receiver is the DC offset at the receiver output [Razavi, 2001]. There are two main sources of the DC offset: (1) RF signal that is returned from other objects (background clutter) has the same frequency as the LO and mixes with the LO in the mixer to yield a DC signal, and (2) LO-RF leakage within the mixer circuit and other internal reflections that cause DC output. When such a DC offset is present in the system, the baseband signals can be represented as

$$\begin{aligned} x_{rI}(t) &= V_I + \frac{A}{2}\cos\left(\frac{2\pi}{\lambda}\left(2d_0 + 2d\left(t\right)\right)\right) \\ x_{rQ}(t) &= V_Q + \frac{A}{2}\sin\left(\frac{2\pi}{\lambda}\left(2d_0 + 2d\left(t\right)\right)\right) \end{aligned} \tag{4.35}$$

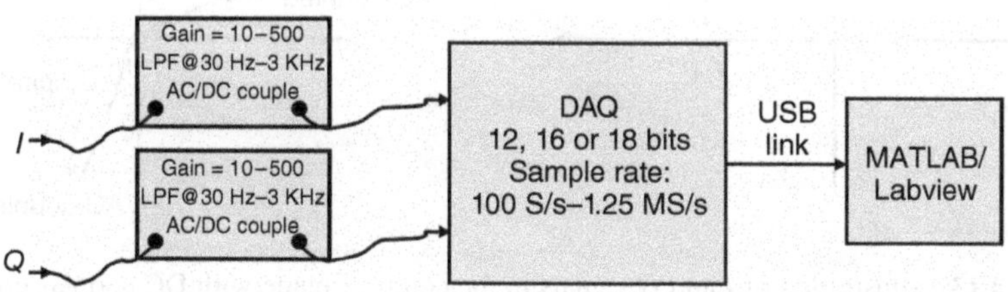

Figure 4.12 Baseband signal amplifiers, low-pass filtering, and data acquisition for the radar system.

In most communication systems, this DC offset may not be an issue since the message signal (information) is shaped in such a way that it carries little or no content close to DC. However, respiratory rate of a human subject is usually around 0.1–0.3 Hz, which is extremely close to DC. This constraint mandates that any component that manipulates baseband signals needs to have a very good low-frequency response in order to preserve the respiration and heart rate traces for further processing. One way to deal with this problem is to use DC coupling at the baseband-level amplifiers. DC coupling preserves the respiration and heart traces, but because of relatively large amount of gain needed to amplify the baseband signal to a desired level for DAQ, dynamic range will be limited since the signal might saturate the amplifiers. AC coupling, on the contrary, is a simple solution; however, the frequency response of AC coupling, which is essentially a high-pass filter, will adversely affect the shape of respiration signal and the system time response, and ultimately accuracy of measurements [Massagram et al., 2009]. A more sophisticated DC cancellation technique might be used as in Vergara and Lubecke [2007] and Vergara et al. [2008], where the baseband DC is monitored and cancelled using two stages comprised of variable and fixed gain amplifiers and feedback from the software. In Mostafanezhad and Boric-Lubecke [2011] and Chapter 8, an RF-based approach to cancel the baseband DC is discussed. Figure 4.13(a) depicts how the complex *IQ* signal will look like when it is DC coupled. This is essentially a plot of the baseband received signal as in Equation 4.22. The arc on the dotted circle is the actual received data, while the circle is estimated for demodulation purposes discussed in the next section.

AC coupling will result in a distortion (the ribbon shape in the *IQ*) in the received signal an exaggerated effect of which is plotted in Fig. 4.13(b). AC coupling will also cause a longer response time. For example, assuming a high-pass filter with a single pole at 0.03 Hz, the settling time will be 12 s. That is, if there are any changes in the measurement environment, it will take 12 s before the system can produce a meaningful output.

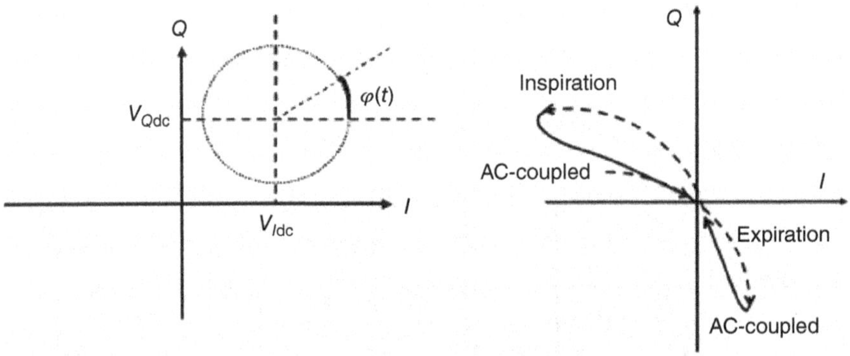

Figure 4.13 Illustration of *I* and *Q* signals in the complex plane with DC and AC coupling. The dotted circle is a circle fitted on the actual data (arc). DC coupling (a) and exaggerated effects of AC coupling (b) is shown.

Problem with low-frequency content of the baseband channels is not limited to issues with AC and DC coupling. Baseband amplifier noise is a major contributor to baseband noise. The low-frequency noise known as *1/f* noise or Flicker noise is generated and present in all low-frequency instruments. This noise can bury important respiration and heart beat spectrum to a degree that rate estimation is not possible. In Chapter 8, a coherent low-IF radar architecture that reduces the baseband noise problem and overcomes DC coupling issues is discussed. Reducing noise level of the system is critical in system performance, since it can help achieve a better SNR using the same transmitter power resulting in longer range, which can enable many applications.

4.2.2 Data Acquisition

Once the baseband analog signal is conditioned and preprocessed (filtered and coupled), it will be sampled using an analog-to-digital converter clocked at sampling frequency of 100–10,000 SPS. Usually DAQ devices have variable input voltage ranges. This voltage range has to be set given the range of the baseband signal to be acquired. Bit depth (or resolution) is another factor determining the baseband noise. Quantization noise is illustrated in Fig. 4.14. In order to clearly show the effect of quantization noise, a random signal is quantized using only four levels.

As an example, a 16-bit DAQ quantizes the input signal into $2^{16} = 65,536$ levels. Thus, assuming an input range of $\pm 10\,\text{V}$, we have

$$\Delta V = \frac{20V}{2^{16}} = 0.3\,\text{mV} \tag{4.36}$$

where ΔV is the smallest change in voltage level measurable by DAQ in an ideal case. In practice, though, this number is larger due to imperfections and noise in the ADC process.

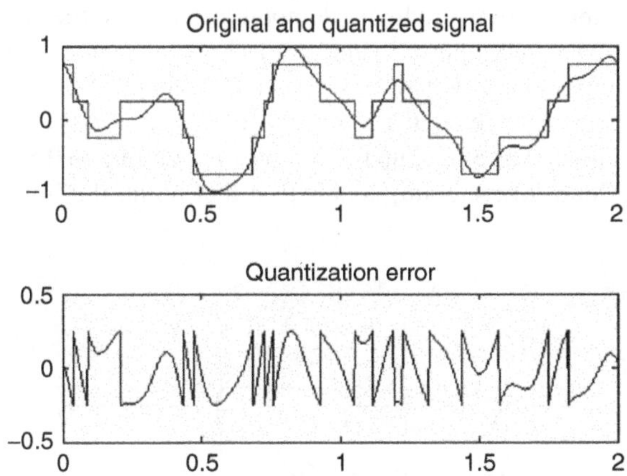

Figure 4.14 Illustration of the quantization noise in the data acquisition system.

4.3 SIGNAL PROCESSING

4.3.1 Phase Demodulation

For a single-channel receiver, use of baseband output is straightforward. The single-channel baseband output is essentially proportional to phase change, and can be directly used for estimating heart and respiration rates. No further phase demodulation in required, nor possible in this case. To overcome the limitations of the single-channel receiver due to null and optimum points, a quadrature receiver is commonly used. Thus, a proper method is needed to combine *I* and *Q* channels to obtain an optimum signal for further processing. This process is called demodulation throughout this work. There are several methods available to demodulate the *I* and *Q* channels [Droitcour, 2006; Park et al., 2007].

A simple method is to monitor both channels, if one of the channels is in null point choose the other channel as demodulation output since, it will be in optimum point. On the other hand, if both channels have comparably strong enough signals, simply add them together. If *I* and *Q* channels have opposite signs, then subtract them and call the result demodulation output. This linear operation of combining the two signals is also called linear demodulation. Basically, multidimensional data can be projected into a single dimension. This method emphasizes on the individual data's similarities and suppresses redundancy. Steps for linear demodulation are as follows:

1. Remove the mean value of both *I* and *Q* signals (DC cancellation).
2. Calculate the covariance matrix of the *I* and *Q* signals.
3. Calculate eigenvectors and eigenvalues of the covariance matrix.
4. Multiply the eigenvectors matrix with the data and pick the output with largest variance (information).

Graphically, this can be explained by rotating the arc of *IQ* data such that it will be in line with one of the axes resulting in maximum demodulated signal (Fig. 4.15).

The main assumption for using linear demodulation is that the arc of *IQ* signals is very small such that it can be approximated with a line (hence the name linear demodulation). This assumption may not always be valid. Larger physiological displacement or a higher RF (smaller wavelength) can cause a larger arc in the *IQ* plane. The arc can still be approximated as a line, up to 46.8° corresponding to 0.81 cm of displacement [Droitcour, 2006; Massagram, 2008]. Past that point, the linear demodulation

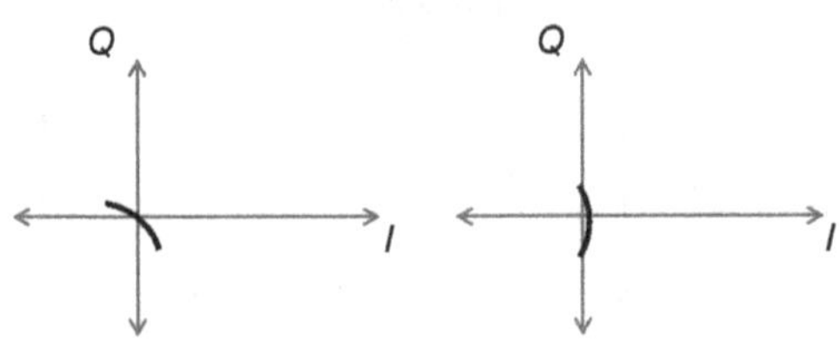

Figure 4.15 Plots illustrating the principle for linear demodulation.

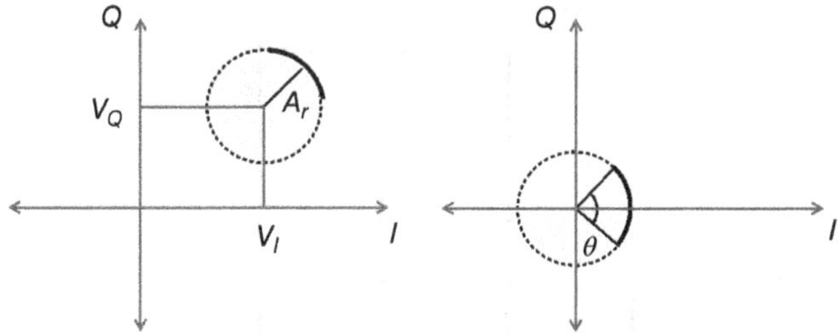

Figure 4.16 Plots illustrating the principle for nonlinear (arctangent) demodulation technique.

will not yield optimum result and a nonlinear method has to be used. Arctangent demodulation has been introduced in Park et al. [2007] as a nonlinear alternative. The goal is to estimate the phase fluctuations in the argument of Equation 4.22, which will lead to physiological displacement. This method is based on approximating a circle on the arc as depicted in Fig. 4.16. The circle is identified by its center and radius. Once parameters of the circle are known, the arc can be projected on the circle and the phase change can be calculated using

$$\phi(t) = \arctan\left(\frac{V_{Q\mathrm{dc}} + x_{rI}}{V_{I\mathrm{dc}} + x_{rQ}}\right) \tag{4.37}$$

hence the name arctangent demodulation. In order for arctangent demodulation to yield desirable results, it is important to estimate the correct value of the signal component. As discussed earlier, LO leakage can cause offset, which has to be discarded from the calculations before arctangent demodulation. For a small arc, both methods yield similar results in terms of obtaining respiratory and heart rates. However, while the output of linear demodulation yields relative displacement, nonlinear demodulation provides absolute displacement. Also as the arc becomes larger (no longer like a line), arctangent demodulation provides more accurate output. It should also be mentioned that under ideal conditions (DC coupling, no noise) linear demodulation results in AM modulation of the heart signals decreasing the accuracy [Boric-Lubecke et al., 2011].

4.3.2 Demodulated Phase Processing

Various types of information can be extracted from respiratory or heart trace signals. For example, visual inspection of ECG can unveil abnormalities. As discussed in Chapter 1, heart rate and respiration rate monitoring can help diagnose many health problems. Both heart and respiration rates are time-varying quantities. Heart rate variability is discussed in Droitcour [2006] and Massagram et al. [2009]. Heart rate of a healthy individual is a time-dependent measure even if the subject is not going through physical activity. Thus, a time-dependent rate analysis tool is needed to address this need. Estimation theory provides a background for development of tools

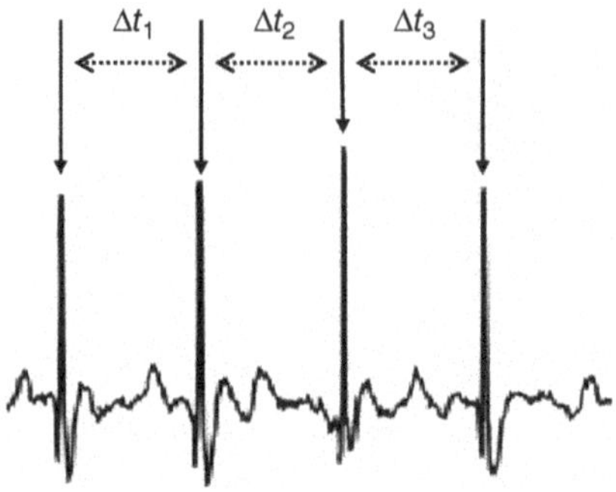

Figure 4.17 Peak detection for a sample ECG signal. Δt_i is the peak to peak distance that can be used to estimate the heart rate.

to estimate heart and respiration rates in beats per minute (BPM). Most important of these tools are: (1) peak detection, (2) short-time Fourier transform (FT), and (3) autocorrelation.

Peak detection is a method that yields the time between two consecutive peaks. This method works well when SNR is high and with the type of signal that has a sharp peak similar to that of ECG or finger pulse. Figure 4.17 illustrates how peak detection works for an ECG signal. The peaks are detected and the time distance between two consecutive peaks is used to yield the heart rate. In this case, the sharp peaks of the ECG QPRS complex make the method feasible. Peak detection can be very accurate making it suitable for HRV analysis, but this method is very susceptible to noise as any small background noise can locally move the location of the peak and add error. Because of these issues, peak detection may not be the optimum rate estimation algorithm for radar signal.

Frequency domain analysis can be used to estimate rate of periodic phenomena. Due to the inherent variability in the rate, a windowed method is required. The short-time Fourier transform method divides the data into chunks of proper length (windows), calculates the fast Fourier transform (FFT) of each window of data, average the FFT over multiple windows to yield a representation of power spectral density (PSD) of a segment of data and ultimately find the peak in the PSD. The window length is determined depending on the application. Rule of thumb is to have a window long enough to accommodate 5–10 periods of the signal. This results in a 6–10 s window for heart beat rate estimation. For respiration, due to the low-frequency nature of the signal, a longer window length is required, usually 10–18 s [Droitcour et al., 2009]. Figure 4.18 depicts how short-time FT operates on a finger pulse data.

Another method is autocorrelation, which is used to find repeating patterns in semi-periodic signals [Bracewell, 1965]. This method looks for similarity between observations (windows of signal) by calculating cross correlation of a signal with itself. Autocorrelation is the correlation of a signal with its delayed version. For a windowed signal with a window length of N:

$$a_{xx}[m] = \frac{1}{N} \sum_{n=0}^{N-m-1} x[n]w[n] \cdot x[n+m]w[n+m] \tag{4.38}$$

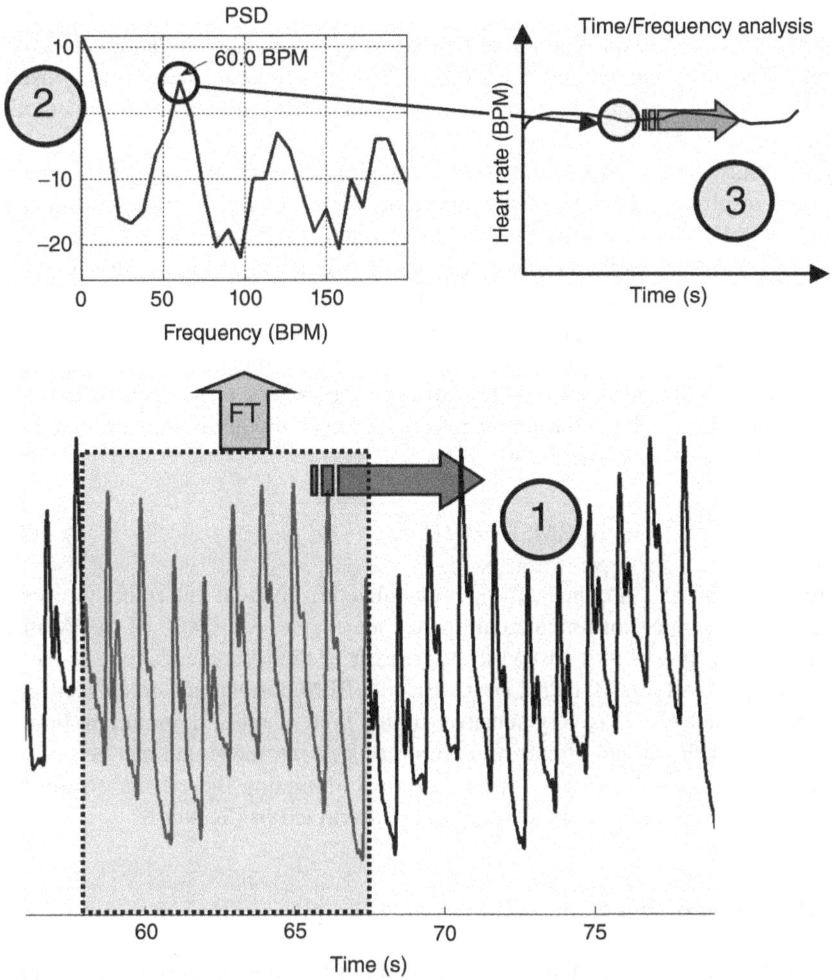

Figure 4.18 Illustration of the short-time FT frequency analysis: (1) a short-time moving window FFT slides through the time domain data, (2) spectrum of the window of the time domain data is calculated, (3) peak of interest in the frequency spectrum is mapped to a time plot, which shows frequency versus time analysis.

a_{xx} is the autocorrelation of x where m is the delay and $w[n]$ is the windowing function usually used to taper the ends of a segment of data to create a smoother autocorrelation. Peaks in the autocorrelation sequence are detected and heart rate as a function of time is calculated.

All window-based processing methods in general increase rate estimation accuracy by averaging information in time or frequency domain from several sliding windows of data, which also increases SNR. However, a major drawback is that it limits variability of the estimated rate. For example, a window length of 10 s will limit any variation in the estimated rate to be below 1/10 Hz, that is, estimated heart rate cannot track any changes faster than 0.1 Hz (6 BPM) in 10 s of data.

In addition to the classical methods described earlier, there are other methods devised for rate estimation or signal conditioning before rate estimation. A combination of filtering and Hilbert transform (HT) for estimating heart rate is used in Mostafanezhad et al. [2007]. A brief comparison of different estimation methods is also discussed in Mostafanezhad et al. [2007].

The window-based short-time Fourier transform is typically used for extracting heart and respiratory rates from reference and radar signals. In order to benchmark the system performance during an experiment, physiological signals are also recorded using contact-based methods as a reference. A piezoelectric finger pulse sensor, or ECG may be used as a reference for heart activity. A chest belt, commonly using a piezoelectric sensor, is strapped around the subject's chest to monitor respiration. In order to minimize interference, these contact-based references are acquired through a DAQ device other than the DAQ used to record radar's I and Q channels. Heart rate obtained from radar and reference is compared using the root mean square error (RMSE) criteria. The RMSE is defined as

$$\text{RMSE} = \sqrt{(hr_{\text{radar}} - hr_{\text{ref}})^2} \tag{4.39}$$

where hr_{radar} and hr_{ref} are the heart rates calculated from radar and reference, respectively, and is the operator for calculating the mean value. An RMSE of 1 BPM means the radar signal is within a 1 BPM average error of the reference signal. If the subject's average heart rate 70 BPM, an RMSE of 1 BPM corresponds to an accuracy of 99%. Since radar signal is very sensitive to any kind of motion, including fidgeting motion of the subject and clutter motion, preprocessing the signal can help the rate estimation algorithms. Several approaches for minimizing the effects of unwanted motion on physiological signal extraction are discussed in Chapter 8.

4.4 NOISE SOURCES

Noise and interference-related issues are some of the most important challenges in Doppler radar physiological monitoring. Noise and interference can be categorized into electrical and mechanical noise. Sources of electrical noise include thermal noise, RF generator phase noise, and $1/f$ noise. Sources of mechanical noise include mechanical movements of the antenna, fidgeting of the human subject, other human subject, and other unwanted background motion. Note that the stationary background clutter is not an issue for CW Doppler radar, since all reflections from stationary objects result in a DC component that can be isolated from the desired signals.

4.4.1 Electrical Noise

Thermal noise is caused by random movements of electrons within an electric conductor. A resistance can be modeled as a noise source and a pure resistance in an electric circuit. A popular equivalent circuit for resistor noise is depicted in Fig. 4.19. Thermal noise falls in the category of white noise, that is, it has a flat PSD.

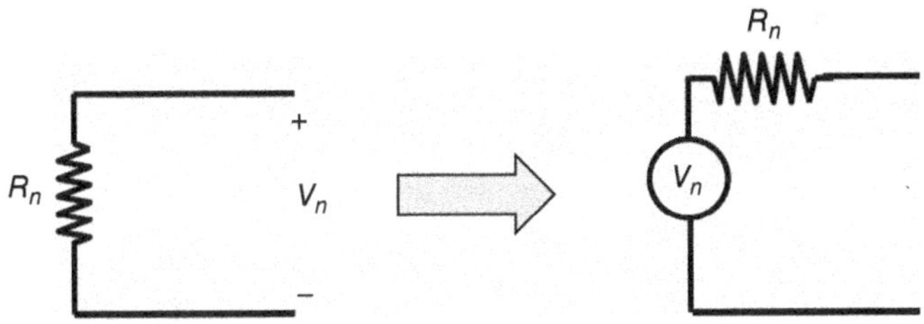

Figure 4.19 Equivalent circuit for thermal noise in conductors.

PSD of thermal noise is calculated using

$$\overline{V_n^2} = 4KTR\left(\frac{\mathrm{V}^2}{\mathrm{Hz}}\right) \tag{4.40}$$

where K is the Boltzmann's constant in joules per Kelvin, T is temperature in Kelvin, R is the resistance value in Ohms. Equation 4.40 yields power of noise per bandwidth. Usually, square root of Equation 4.40 is used for identifying the noise level. For example, a 1 kΩ resistor at room temperature generates $\sqrt{\overline{V_n^2}} = 4.07\,\mathrm{nV}/\sqrt{\mathrm{Hz}}$ or $-167\,\mathrm{dBV_{rms}}/\sqrt{\mathrm{Hz}}$ of noise voltage.

In Doppler radar physiological monitoring, noise levels are typically larger than this value caused by other sources of noise and interference. Thus, thermal noise will not be considered to be a major issue. As discussed earlier in this chapter, depending on the radar configuration, RF generator phase noise plays an important role in overall SNR of the system.

$1/f$ noise contributes to the baseband noise of mixers and amplifiers. In electronic devices, it is known as Flicker noise and results from a variety of effects such as impurities in a conductive channel and generation and recombination noise in a transistor. It is related to a direct current and it usually dominates at low frequencies [Razavi, 2001]. Since Flicker noise decays with frequency, at some point Flicker noise is shadowed by white noise. This frequency is known as corner frequency and is illustrated in Fig. 4.20.

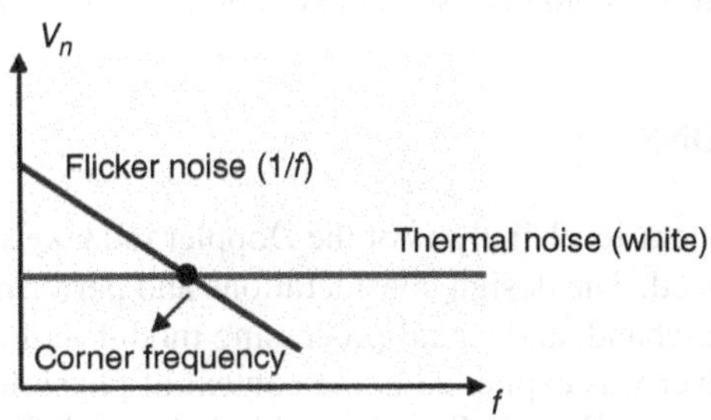

Figure 4.20 Flicker noise versus thermal noise and the concept of corner frequency.

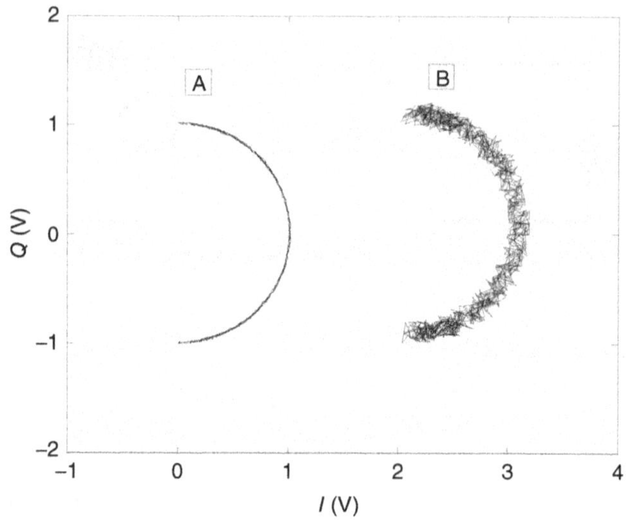

Figure 4.21 Simulated baseband plots showing the effect of distributed random noise on data. Arc "A" shows the *IQ* plot with very little noise while arc "B" shows the *IQ* plot with a lot of noise.

Corner frequency of a device depends on the technology used in the device and it is usually in the kilohertz range for bipolar devices. As it is shown in Fig. 4.20, due to the $1/f$ nature of the noise it can become substantial at low frequencies reducing the SNR of the system especially since physiological motion spectrum contains significant low frequency content. Figure 4.21 shows the effect of electrical noise on simulated baseband data.

4.4.2 Mechanical Noise

Unwanted antenna motions create unwanted phase shift in the radar data that can obscure physiological motion and make physiological signal extraction very challenging. Similar challenges result from human subject fidgeting motion and background clutter motion. In Chapter 8, several techniques for reducing the effect of unwanted motion on physiological signal extraction are discussed.

4.5 CONCLUSIONS

In this chapter, a system-level analysis of the Doppler radar system for physiological sensing was introduced. The design considerations and performance trade-offs were discussed for RF, baseband, and signal-processing modules of the system. The need for quadrature receiver was explained in the context of physiological monitoring, as well as associated trade-offs including channel imbalance, DC offset, baseband coupling, phase demodulation, and noise issues. Graphical representation of quadrature outputs was introduced to explain trade-off of frequency operation, effects of channel imbalance and AC coupling, and demodulations methods. The operation frequency and power requirements were analyzed in terms of component availability, safety, and penetration into human tissue. The power used for typical physiological Doppler radar is similar to that of common wireless devices such as cordless phones, cellular phones, baby monitors, and wireless LAN, and thus does not introduce a significant health risk. At 2.4 GHz, if the radar is used at same distance from the subject, it was determined that more than 90% of the reflected power comes from the body surface, and thus internal organ motion will not be considered for noncontact applications. Signal-processing methods commonly used for rate extraction, including peak detection, FFT, and autocorrelation were introduced. Finally, noise sources that cause SNR were discussed. Design concepts and performance issues introduced in this chapter are examined in more detail in the following chapters, with proposed solutions and associated experimental data.

REFERENCES

Awtry EH, Loscalzo J. Evaluation of the patient with cardiovascular disease. In: Carpenter CCJ, Griggs RC, Loscalzo J, editors. Cecil Essentials of Medicine. 5th ed. New York: W. B. Saunders Company; 2001. p 30–42.

Banks DS. Continuous wave (CW) radar. Electron Prog 1975;17(2):34–41.

Boric-Lubecke O, Lubecke V, Mostafanezhad I, Amplitude modulation issues in Doppler radar heart signal extraction. IEEE RWS 2011; January 2011; Phoenix, AZ.

Bracewell R. The Fourier Transform and Its Applications. New York: McGraw-Hill; 1965.

Braunwald E, Perlkoff JK. Physical examination of the heart and circulation. In: Braunwald E, Zipes DP, Libby P, editors. Heart Disease: A Textbook of Cardiovascular Medicine. New York: W. B. Saunders Company; 2001. p 45–81.

Budge MC, Jr., Burt MP. Range correlation effects in radars. Record of the 1993 IEEE National Radar Conference; 1993; p 212–216.

Deliyannis AA, Gillam PMS, Mounsey JPD, Steiner RE. The cardiac impulse and the motion of the heart. Br Heart J 1964;26:396–411.

Dressler W. Pulsations of the chest wall. Arch Intern Med 1937;60:225–239.

Droitcour AD. Non-contact measurement of heart and respiration rates with a single-chip microwave Doppler radar [PhD dissertation]. Stanford University; 2006.

Droitcour A, Boric-Lubecke O, Kovacs G. Signal-to-noise ratio in Doppler radar system for heart and respiratory rate measurements. IEEE Trans Microwave Theory Tech 2009;57(10):2498–2507.

Droitcour A, Lubecke VM, Lin J, Boric-Lubecke O. A microwave radio for Doppler radar sensing of vital signs. IEEE MTT-S International Microwave Symposium Digest; 2001 May; Phoenix, AZ, USA. Vol. 1, p 175–178.

Federal Communication Commission. Code of Federal Regulations. Title 47: Telecommunications, Part 15: Radio Frequency Devices; 2000.

Gabriel C. Compilation of the dielectric properties of body tissues at RF and microwave frequencies. Radiofrequency Radiation Division, Brooks AFB, San Antonio, TX, Contract AL/OE-TR-1996-0037; 1996.

Gillam PMS, Deliyannis AA, Mounsey JPD. The left parasternal impulse. Br Heart J 1964;26:726–736.

Massagram WJ. A study of feasibility in long-term cardiopulmonary monitoring via Doppler radar [PhD dissertation]. University of Hawaii at Manoa; 2008.

Massagram W, Lubecke VM, Host-Madsten A, Boric-Lubecke O. Assessment of heart rate variability and respiratory sinus arrhythmia via Doppler radar. IEEE Trans Microwave Theory Tech 2009;57(10):2542–2549.

Mostafanezhad I, Boric-Lubecke O. An RF-based analog linear demodulator. IEEE Microwave Wireless Compon Lett 2011;21(7):392–394.

Mostafanezhad I, Massagram W, Hafner N, Petrochilos N, Host-Madsen A, Lubecke V, Boric-Lubecke O. Comparison of heart rate estimators for Doppler radar monitoring. IASTED-07 (SIP-07, Hawai'i); 2007 August.

Park B. Cardiopulmonary monitoring using Doppler radar [PhD dissertation]. University of Hawaii at Manoa; 2007.

Park B-K, Boric-Lubecke O, Lubecke VM. Arctangent demodulation with DC offset compensation in quadrature Doppler radar receiver systems. IEEE Trans Microwave Theory Tech 2007;55(5):1073–1079.

Park B-K, Yamada S, Boric-Lubecke O, Lubecke V. Single-channel receiver limitations in Doppler radar measurements of periodic motion. IEEE Radio Wireless Symp 2006;1:17–19.

Proakis JG, Manolakis DM. Digital Signal Processing. Prentice Hall; 2006.

Razavi B. Design of Analog CMOS Integrated Circuits. McGraw-Hill; 2001.

Samardzija D, Boric-Lubecke O, Host-Madsen A, Lubecke VM, Droitcour AD, Kovacs GTA. Applications of MIMO techniques to sensing of cardiopulmonary activity. Proceeding of IEEE/ACES Conference on Wireless Communications and Applied Computational Electromagnetics; 2005, p 618–621.

Saunders K. CW and FM radar. In: Skolnik MI, editor. Radar Handbook. 2nd ed. San Francisco: McGraw-Hill, Inc.; 1990. p 14.1–14.45.

Skolnik MI. An analysis of bistatic radar. IRE Trans Aerosp Navig Electron 1961;8:19–27.

Vergara AM, Boric-Lubecke O, Lubecke V. DC information preservation for cardiopulmonary monitor utilizing CW Doppler radar, IEEE Eng Med Biol Soc; August 2008. p 1246–1249, 2008.

Vergara AM, Lubecke V. Data acquisition system for Doppler radar vital-sign monitor. IEEE Eng Med Biol Soc; 2007 August. p 2269–2272; 2007.

Vermarien H, van Vollenhoven E. The recording of heart vibrations: a problem of vibration measurement on soft tissue. Med Biol Eng Comput 1984;22:168–178.

5

CW HOMODYNE TRANSCEIVER CHALLENGES

ADITYA SINGH[1], ALEX VERGARA[2], AMY D. DROITCOUR[3], BYUNG-KWON PARK[4], OLGA BORIC-LUBECKE[5], SHUHEI YAMADA[5], AND VICTOR M. LUBECKE[5]

[1]*University of Hawaii Neuro-science and MRI research Program, John A. Burns School of Medicine, Honolulu, Hawaii, United States*
[2]*R & D Department, Theranova LLC, San Francisco, California, United States*
[3]*Wave 80 Biosciences, Inc., San Francisco, California, United States*
[4]*DAS Sensor SW Engineering Team, Hyundai Mobis Mechatronics R&D Center, Gyeonggi-Do, South Korea*
[5]*Department of Electrical Engineering, University of Hawaii at Manoa, Honolulu, Hawaii, United States*

As discussed in the previous chapter, continuous-wave (CW) homodyne radar has been the most commonly used architecture for physiological monitoring, due to its ease of implementation. However, there are several challenges associated with this architecture that limit the performance of physiological radar. Figure 5.1 shows the simplified block diagram of a physiological radar, indicating radio frequency (RF) front-end, baseband, and signal processing modules. In this chapter, specific challenges related to each of those modules are examined.

5.1 RF FRONT END

Challenges associated with CW homodyne RF front end due to direct down-conversion to baseband are common to any homodyne radio used in communications or radar. However, specific issues for radar application in physiological monitoring arise from the facts that physiological signals occupy a very-low-frequency spectrum

Doppler Radar Physiological Sensing, First Edition.
Edited by Olga Boric-Lubecke, Victor M. Lubecke, Amy D. Droitcour, Byung-Kwon Park, and Aditya Singh.

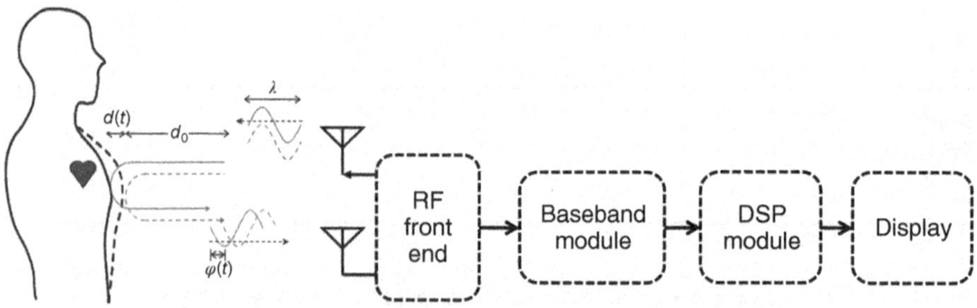

Figure 5.1 Physiological radar system block diagram.

near DC and that two physiological signals associated with cardiac activity and respiratory efforts coexist and may cross-couple.

5.1.1 Single-Channel Limitations

In a typical Doppler radar used to measure target velocity, if the signal spectrum is asymmetrical, direct down-conversion to baseband using a single-channel receiver will results in spectrum folding that corrupts the data. This will result in the ambiguity in speed direction, that is, it will not be possible to distinguish approaching and receding targets. In a Doppler radar for physiological monitoring, these effects will be seen as detection sensitivity to target position. In Chapter 4, "null" and "optimum" points concept was introduced, assuming a single source of periodic motion. In this section, the effect of "null" and "optimum" point are examined in the presence of two sources of periodic motion, describing respiratory effort and cardiac activity [Park et al., 2006].

Typically, a Doppler radar motion-sensing transceiver transmits a CW signal and demodulates the signal reflected from a target. According to Doppler theory, when the target has time-varying movement with zero net velocity, the reflected signal is phase-modulated in proportion to the position of the target rather than the velocity. A stationary human body presents two independent time-varying movements with zero net velocity based on respiration and cardiac activity, and the largest reflection of incident RF power occurs at the body surface. Thus, the phase of a reflected signal will be proportional to positional variations across the body surface corresponding to the motion of the heart and lungs. In terms of demodulation, there are two extreme cases, with respect to the nominal distance of the target, which are called the *optimum* case and the *null* case [Droitcour et al., 2004]. For the optimum case, the demodulated phase variation is large and linearly proportional to chest displacement; thus, it is possible to get highly accurate data. In the null case, on the other hand, the phase variation is smaller and proportional to the square of the displacement. As a result, the demodulated heart- and respiration-related phase data are self- or mutual-coupled, causing distortion of the actual displacement data.

A CW radar typically transmits a single-tone signal and is expressed as

$$T(t) = \cos(2\pi f_0 t + \phi(t)) \tag{5.1}$$

where f_0 is the oscillation frequency, t is the elapsed time, and $\phi(t)$ is the phase noise of the oscillator. If the subject is located at a nominal distance d_0 with two independent time-varying displacement given by $x(t)$ and $y(t)$, then the round trip distance of radar signal is $2d(t) = 2d_0 + 2x(t) + 2y(t)$. Thus, when chest movement period $T \gg d_0/c$ and $x(t) \ll d_0$, the received signal can be approximated as

$$R(t) = \cos\left[2\pi f_0 t - \frac{4\pi d_0}{\lambda} - \frac{4\pi x(t)}{\lambda} - \frac{4\pi y(t)}{\lambda} + \phi\left(t - \frac{2d_0}{c}\right)\right] \tag{5.2}$$

Since the same oscillator is used for transmitted and local oscillator (LO) signals, the resulting low-pass-filtered mixer output signal is

$$B(t) = \cos\left[\theta + \frac{4\pi x(t)}{\lambda} + \frac{4\pi y(t)}{\lambda} + \Delta\phi(t)\right] \tag{5.3}$$

where $\Delta\phi(t)$ is the residual phase noise and θ is the constant phase shift related to the nominal distance to the subject with a factor θ_0, which compensates for the phase change at the surface of a target and phase delay between the mixer and antenna. Each is expressed as

$$\Delta\phi(t) = \phi(t) - \phi\left(t - \frac{2d_0}{c}\right) \tag{5.4}$$

$$\theta = \frac{4\pi d_0}{\lambda} + \theta_0 \tag{5.5}$$

There are two extreme cases for the output signal with respect to θ. The first case occurs when θ is an odd multiple of $\pi/2$. In this, the optimum case, we can apply the small-angle approximation, which is valid when both $x(t)$ and $y(t)$ are much smaller than $\lambda/4\pi$. Thus, Equation 5.3 becomes

$$B(t) \approx \frac{4\pi x(t)}{\lambda} + \frac{4\pi y(t)}{\lambda} + \Delta\phi(t) \tag{5.6}$$

In order to simplify, assuming the displacements $x(t)$ and $y(t)$ associated with respiration and heart activity are sinusoidal movements with corresponding frequencies and amplitudes as a first-order approximation. Now, Equation 5.6 is modified as

$$B(t) \approx A \sin 2\pi f_1 t + B \sin 2\pi f_2 t + \Delta\phi(t) \tag{5.7}$$

where $f_1 \ll f_2$ and $A \gg B$, because, in general, breathing movement has lower frequency and bigger displacement than heart beat. Thus, the output signal is linearly proportional to the chest motion, and with appropriate filtering, it should be possible to obtain the desired data accurately.

The second case occurs when θ is an integer multiple of π, and the output data are given as

$$B(t) \approx 1 - [A \sin 2\pi f_1 t + B \sin 2\pi f_2 t + \Delta\phi(t)]^2 \tag{5.8}$$

When $\Delta\phi(t)$ is much smaller than the other components in the bracket, Equation 5.8 can be written as

$$\begin{aligned} B(t) &\approx 1 - [A\sin 2\pi f_1 t + B\sin 2\pi f_2 t]^2 \\ &= 1 - [A^2\sin^2 2\pi f_1 t + B^2\sin^2 2\pi f_2 t + 2AB\sin 2\pi f_1 t\sin 2\pi f_2 t] \end{aligned} \quad (5.9)$$

Thus

$$\begin{aligned} B(t) \approx 1 - \frac{1}{2}[(A^2 + B^2) - A^2\cos 2\pi(2f_1)t - B^2\cos 2\pi(2f_2)t \\ -2AB(\cos 2\pi(f_2 + f_1)t - \cos 2\pi(f_2 - f_1)t)] \end{aligned} \quad (5.10)$$

where first term $(A^2 + B^2)$ is DC information, second and third terms $A^2\cos 2\pi(2f_1)t - B^2\cos 2\pi(2f_2)t$ are harmonics of breathing and heart beat signals, and fourth and fifth terms $\cos 2\pi(f_2 + f_1)t - \cos 2\pi(f_2 - f_1)t$ are the sum and deference frequency components of breathing and heart beat signals. In this case, the output signal is no longer linearly proportional to the displacement of the chest, and this can result in two types of significant distortions in the measured data. This is referred to as the null case. First, when both A and B are less than one, which means that chest motion due to the respiration and heart beat is relatively small compared with the wavelength, the output data are proportional to the square of the signals and becomes much less sensitive to both respiration and heart motion. Second, in addition to the increased error resulting from poor sensitivity, frequency information for the subject's movement is distorted regardless of the magnitudes of A and B. From Equation 5.10, after low-pass filtering, only the second term in the bracket, $A^2\cos 2\pi(2f_1)t$, would remain associated with the DC offset, and its frequency is double that of the original respiring motion, f_1. In the normal case of a heart motion signal whose magnitude is much smaller than that of the respiration, not only frequency self-mixing occurs for the null case but there is also mutual coupling. In other words, the last term in the bracket of Equation 5.10, starting with $2AB$, which is related with mutual coupling, is much larger than the third term, that refers to self-coupling in the bracket since $AB \gg B^2$, and as a result different types of distortion can occur depending on the frequency of respiration. In the extreme, these are the cases of continuous breathing, where f_1 is nonzero, and the case of breath holding, where f_1 is zero. As described in Equation 5.10, when f_1 is nonzero, the last term is much larger since $AB \gg B^2$, thus the difference of the two frequencies is obtained instead of double of the actual frequency. On the other hand, when f_1 is zero, the output frequency can be either doubled or the same as the actual frequency. For the first case, since f_1 is zero, only the third term, whose frequency is double that of the actual motion frequency, in the bracket remains after high-pass filtering. The second case occurs when the constant phase shift θ is slightly different from multiples of π, and the output data becomes

$$B(t) \approx 1 - [\theta_{\text{offset}} + B\sin 2\pi f_2 t]^2 \quad (5.11)$$

where θ_{offset} is the phase difference between θ and the multiple of π. From Equation 5.11, when the difference, θ_{offset}, is much larger than the magnitude of the

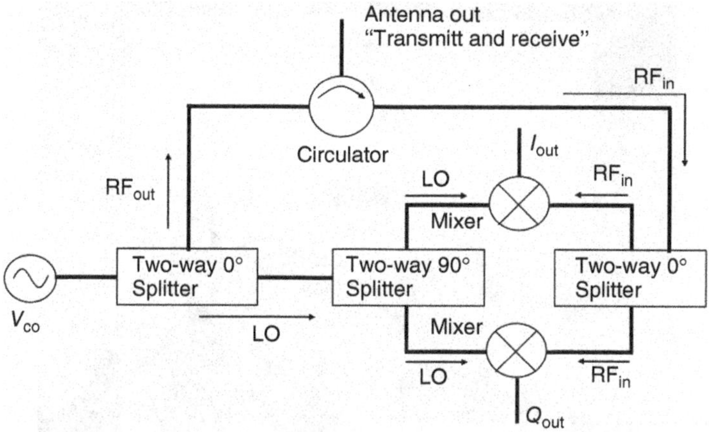

Figure 5.2 Block diagram of Doppler radar transceiver. © 2006 IEEE, Reprinted, with permission, from Park et al. [2006].

heart displacement B, only the component that corresponds to the actual frequency remains.

A block diagram of the Doppler radar transceiver is shown in Fig. 5.2. The receiver chain has two independent channels, one with a 90° phase shift with respect to the other, which can be monitored independently to assess optimum and null case signals simultaneously for any given target position.

The RF_{out} and LO signals are derived from the voltage-controlled oscillator (VCO) using a two-way 0° power splitter. The LO signal is further divided by another two-way 90° power splitter for the two receiver channels. The RF_{out} signal is routed to the antenna via a circulator to allow isolation of the incoming and outgoing signals from the antenna.

Based on the block diagram shown in Fig. 5.2, a transceiver system was fabricated as a compact microstrip circuit with surface mount components on a 10.2 cm by 11.2 cm printed circuit (PC) board, with a coaxially connected patch antenna and separate power supply. A commercially available Antenna Specialists ASPPT2988 2.4 GHz ISM-band patch antenna was used, connected by coaxial cable to an SMA (SubMiniature version A: coaxial RF connector) connector on the PC board. The reflected signal also comes from the antenna along the same path and is separated by the circulator to create RF_{in}. The RF_{in} signal is divided using a two-way 0° power splitter to feed the two receiver chains. Each chain has its own mixer and produces the separate output channel signals as I_{out} and Q_{out} at corresponding SMA connectors. The two quadrature outputs allow simultaneous evaluation of optimum and null case signals and ensure that the system is never completely restricted by a null case.

A photograph of the Doppler radar board is shown in Fig. 5.3. The RF and LO signals are routed via 50 Ω microstrip lines and the design frequency is 2.4 GHz. An FR4 substrate is used for the PC board with the dielectric constant of 4.5, a substrate thickness of 1.57 mm, a conductor thickness of 35 μm, a metal conductivity of

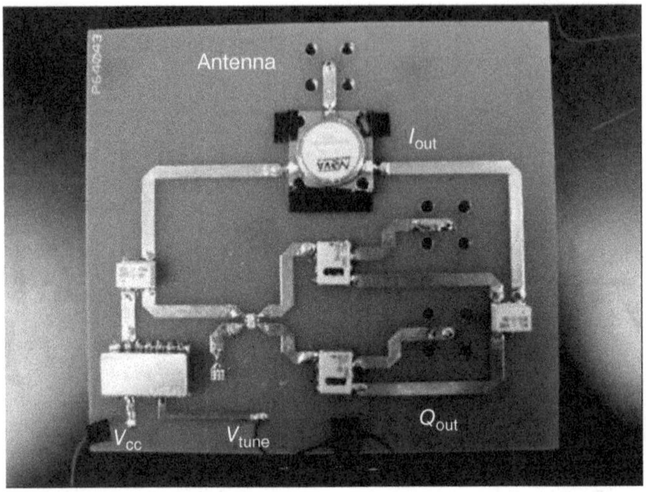

Figure 5.3 Photograph of Doppler radar transceiver board. © 2006 IEEE, Reprinted, with permission, from Park et al. [2006].

5.5e7 S/m, and loss tangent of 0.018. A Mini-Circuits JTOS-2700V VCO was used as the signal source, which delivers 0.8 dB m at 2.4 GHz signal to the antenna out connector, and consumes 170 mW. A Mini-Circuits RPS-2-30 was used for the two-way 0° power splitter and Mini-Circuits QCN-27 was used for the two-way 90° power splitter. Mini-Circuits SKY-42 mixers were used for downconversion.

The measurement setup is shown in Fig. 5.4. An HP E3630A was used to supply voltage. It provided 5 V to V_{cc} and 9.16 V to V_{tune} of the VCO in order to obtain a 2.4 GHz signal. The baseband output signals were amplified and filtered with SR560

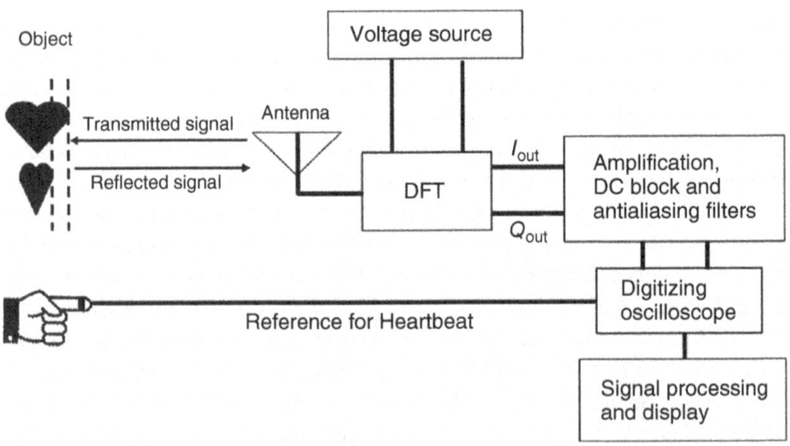

Figure 5.4 Measurement setup. © 2006 IEEE, Reprinted, with permission, from Park et al. [2006].

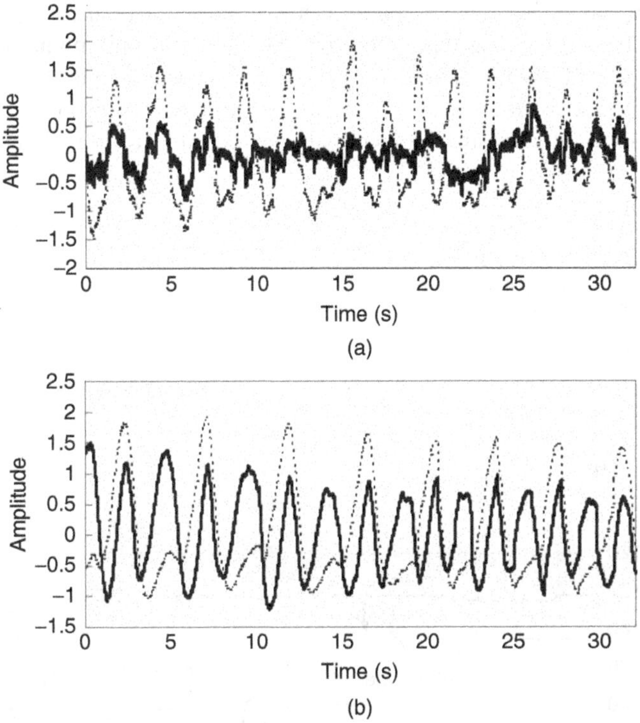

Figure 5.5 Measured respiration signals at *optimum* (dashed lines) and *null* (solid lines) positions. When at the null point and displacement of the target is much smaller than λ (a), sensitivity decreases significantly making accurate rate measurements difficult. Even with exaggerated deep-breathing displacement (b), error still occurs with the frequency of the output signal double that of actual motion as measured at the optimum position. © 2006 IEEE, Reprinted, with permission, from Park et al. [2006].

LNAs and then digitized with a Tektronix 3014 digital oscilloscope. A wired finger pressure pulse sensor was used only as a reference to compare with the heart rate data obtained with the Doppler radar.

Figure 5.5 shows the measurement results for the respiration signal at both optimum and null case positions. To demonstrate these two different distortions, measurements were performed under two different conditions. One measurement was performed under normal breathing conditions with displacement much smaller than the wavelength (Fig. 5.5(a)). The other measurement was performed under exaggerating deep-breathing conditions, with displacement comparable to the wavelength (Fig. 5.5(b)). As shown in Fig. 5.5(a), when displacement of the target is much smaller than λ, sensitivity or signal-to-noise ratio (SNR) of the output becomes much smaller due to the square relationship and thus increases the difficulty in resolving rate data. However, even in the case of relatively large exaggerated respiration displacement frequency coupling occurs and results in inaccuracy.

Figure 5.6 shows heart and respiration measurement data at the optimum point and that of three different types of error cases at the null point described by Equations 5.8–5.11. While it is clear from these measurements that approaching the null position increases the likelihood of inaccuracy, it is also clear that the use of a two-channel quadrature receiver makes it possible to completely avoid the

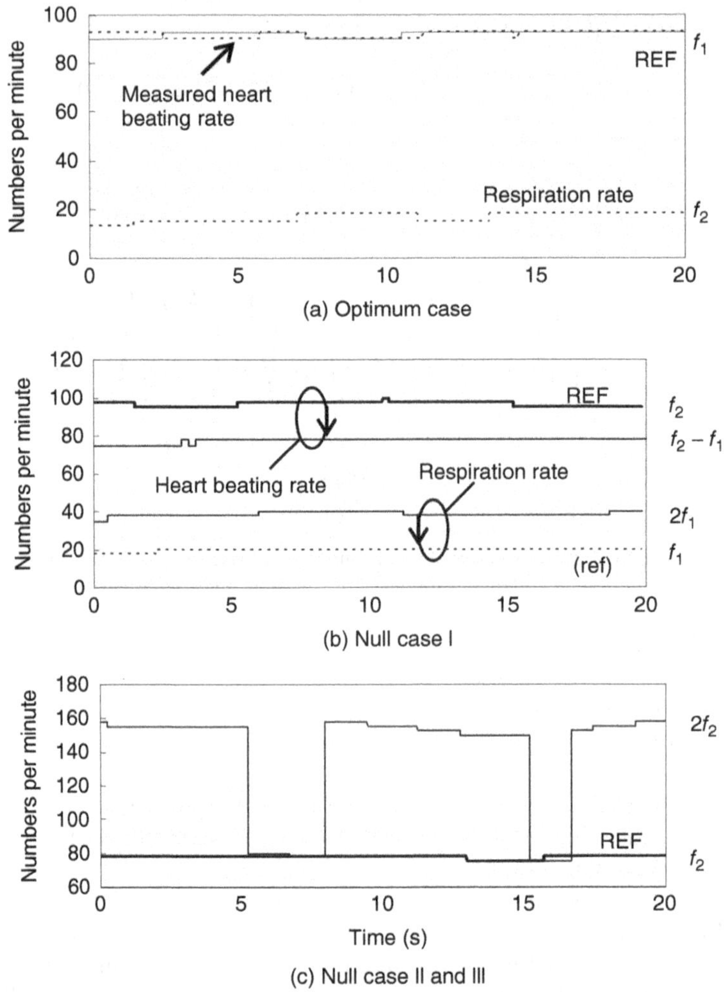

Figure 5.6 Measurement history data for both respiration and heart rate with quadrature receivers, at either optimum (a) or null points (b, c). At the optimum point (a), the Doppler-measured heart rate corresponds closely to the reference for all f_2. At the null point during continuous breathing (b), the Doppler measured heart rate and reference differ by the respiration reference frequency (f_1), while with breath-holding (c) it jumps between either double (case II) or equal to the actual frequency (case III). The "(ref)" signal in (b) is that measured from the quadrature (not null) channel. © 2006 IEEE, Reprinted, with permission, from Park et al. [2006].

worst-case scenario. With two receivers that are half-lambda out of phase from each other, whenever one of the channels receives the reflected signal at the null point, the other will be at the optimum point. Moreover, thanks to its double frequency, null point data can be distinguishable.

5.1.2 LO Leakage Cancellation

Fundamental limitation of CW radar is continuous signal transmission, which presents interference for a weaker received signal. Direct conversion transceivers are commonly used in wireless communications to avoid image rejection filters, and enable higher level of integration with lower power dissipation [Abidi, 1995; Razavi, 1997]. It is also used for microwave Doppler-radar for noncontact cardiopulmonary monitoring [Lubecke et al., 2000; Droitcour et al., 2004; Park et al., 2006; Redman-White and Leenaerts, 2001]. In direct conversion systems, performance limitations typically stem from quadrature channel imbalance, LO leakage, Flicker noise, and DC offset at the receiver output [Abidi, 1995; Razavi, 1997]. Passive mixers exhibit significant lower Flicker noise compared with active mixers due to the fact that there is no DC current and are often used in direct down-conversion receivers for that reason. However, due to the finite mixer port-to-port isolation, and mismatches between receiver components, significant DC offset at the mixer output may result from the self-mixing of LO leakage and LO signal. This DC offset results in Flicker noise [Redman-White and Leenaerts, 2001; Margraf and Boeck, 2004], which limits system sensitivity. In addition, LO leakage at the antenna end in receiver configurations may cause interference. In transceiver configurations, there is also transmit-to-receive signal leakage, which similarly reduces receiver dynamic range [Kim et al., 2006], in addition to the LO leakage. While the DC offset can be cancelled at the output of the transceiver [Park et al., 2007a; Furuta et al., 2007], the Flicker noise cannot be reduced using such techniques. In this section, LO leakage and DC offset reduction techniques in order to improve the sensitivity of direct conversion systems are described [Yamada et al., 2008].

Figure 5.7 shows the block diagrams of the direct conversion systems. Due to the finite mixer port-to-port isolation, the LO signal can leak through the RF port. To get low conversion loss from a passive mixer, typically a high LO power is needed, which may result in significant LO leakage. LO leakage can also be radiated by the antenna as "external LO leakage" (Fig. 5.7(a)), and may affect the transmitted signal as well as the operation of other receivers in the system. LO leakage is also reflected by other components such as the low noise amplifier (LNA) and antenna, and can thus get back to the mixer causing LO self-mixing. This generates second harmonic, $2f_{\mathrm{LO}}$, components and DC offset. The harmonics can be rejected by a low-pass filter (LPF) or band-pass filter (BPF). The DC offset will be directed to the baseband signal processing IC, and may result in measurement error. While the DC offset can be cancelled at the output of the mixer [Park et al., 2007a; Furuta et al., 2007] with the increase in baseband system complexity, this does not result in Flicker noise reduction. Therefore, it is more desirable to prevent the DC offset thus simplifying receiver design and

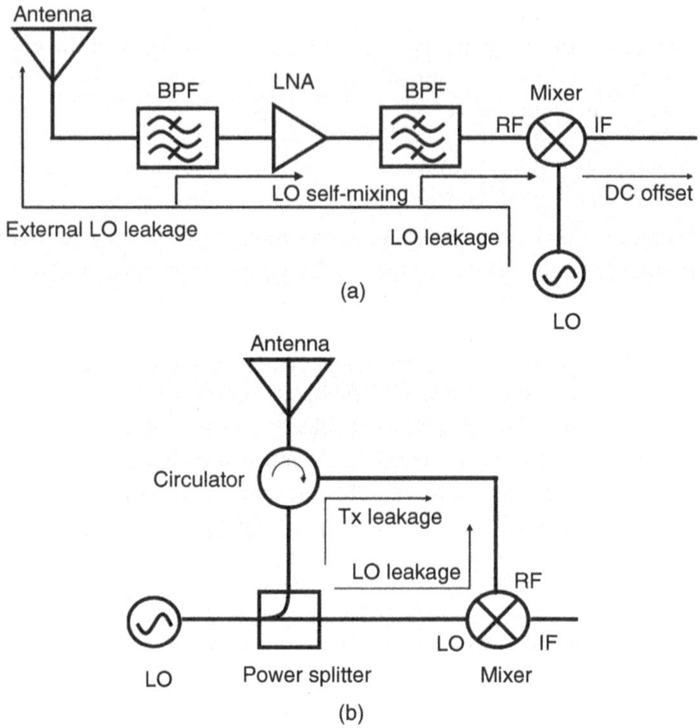

Figure 5.7 Block diagrams of direct conversion systems. The LO leakage from the RF port of the mixer in the receiver configuration, (a), is separated into two components. One is external LO leakage, which affects other receivers, and the other is an LO self-mixing signal which induces DC offset. In a transceiver configuration (b), there is Tx leakage to the receiver chain and a self-mixing component, in addition to receiver leakage problems. © 2008 IEEE, Reprinted, with permission, from Yamada et al. [2008].

reducing the noise floor of the system. In a transceiver configuration (Fig. 5.7(b)), in addition to the LO leakage, there is also Tx leakage due to finite Tx-to-Rx isolation.

In Fig. 5.8, the block diagram of the mixer DC offset canceller and the measurement setup is shown. In order to eliminate the DC offset, a bypass circuit is connected to the RF port, delivering the same signal amplitude as the leakage signal, with a 180° phase difference. The signal is directed to the RF port where it cancels the leakage signal. When two different signals that have same frequency were injected to a mixer, the output signal can be expressed as

$$\text{Out} = \{A\cos\omega t\}\{B\cos(\omega t + \varphi_1)\} = \frac{AB}{2}\{\cos(2\omega t + \varphi_1) + \cos(-\varphi_1)\} \tag{5.12}$$

Clearly, this equation shows that the mixer induces DC offset and second harmonics. In order to cancel this self-mixing signal, in this case, $B\cos(\omega t + \varphi_1)$, a signal which

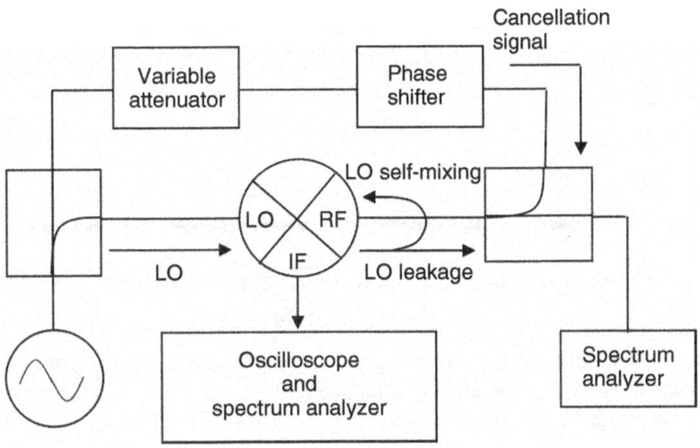

Figure 5.8 Block diagram of DC offset canceller. © 2008 IEEE, Reprinted, with permission, from Yamada et al. [2008].

has 180° phase difference from the self-mixing signal can be used as the cancellation signal. Due to cosine function (i.e., $\cos(\varphi + \pi) = \cos(-\varphi)$), two different phase signals can be used as the DC offset cancellation signals (i.e., either $B\cos(\omega t + \varphi_1 + \pi)$ or $B\cos(\omega t - \varphi_1)$). Furthermore, based on transmission line theory, if there are incident wave in a lossless transmission line at point X, the signals can be expressed as

$$V = e^{\gamma x} = \cos(\varphi) + j\sin(\varphi) \tag{5.13}$$

In order to cancel both cosine and sine functions, only a cancellation signal that has 180° phase difference can be used as the LO leakage cancellation signal (i.e., $\cos(\varphi + \pi) + j\sin(\varphi + \pi)$). Therefore, two phase values will result in minimum DC offset, while only one yields the minimum external LO leakage signal.

Mixer DC offset and Flicker noise was measured with and without the canceller. The measured DC offset with and without canceller is shown in Fig. 5.9. A Mini-Circuits ZFM-4212 passive mixer was used for all measurements. Agilent E4433B signal generator was used for the signal source. The signal was divided first by a two-way 0° power splitter (Mini-Circuits ZFSC-2-2500) to separate +7 dB m of the LO input signal and bypass signal. An attenuator (Broad Wave 751-002-030) was used to match the amplitude, and a phase shifter (Pulser ST-21-444A) was used to create the 180° phase difference. The RF port was terminated with 50 Ω, and the IF port was connected to an oscilloscope (Tektronix TD8 3014B) to measure DC offset and Flicker noise. The manufacturer's specified level of 7 dB m was used to drive the LO port to get an optimum conversion loss of 5.9 dB. From Fig. 5.9, the mixer induces 20 mV of the DC offset even with the RF port terminated with 50 Ω; however, using the canceller, the DC offset could be set close to zero. The Flicker noise was measured and is shown in Fig. 5.10 for mixer with and without LO leakage canceller. From this result, the canceller configuration reduces the Flicker

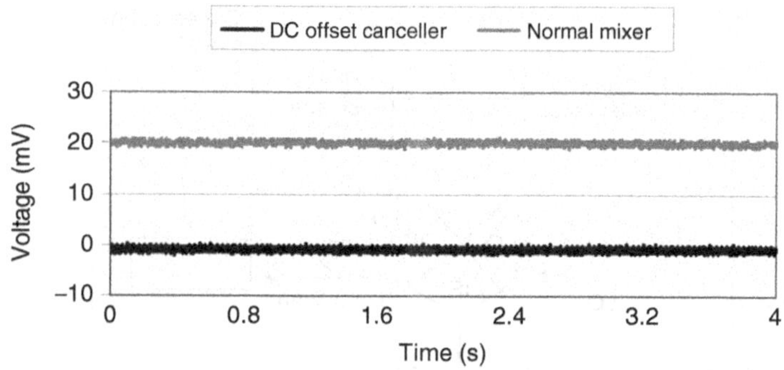

Figure 5.9 Measured DC offset with and without compensation. The DC offset could be set to zero using the LO leakage cancellation technique. © 2008 IEEE, Reprinted, with permission, from Yamada et al. [2008].

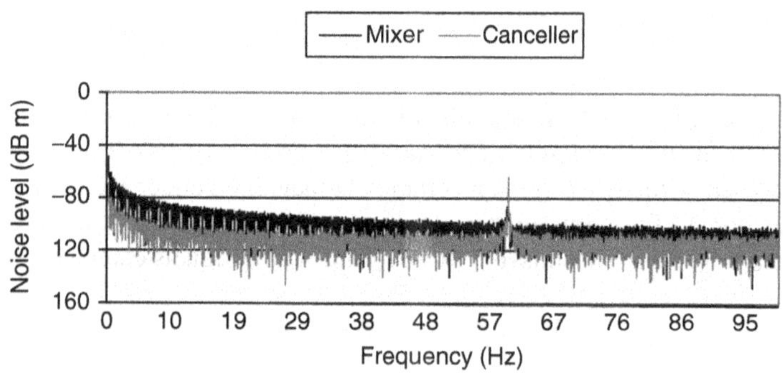

Figure 5.10 Measured receiver Flicker noise reduction. © 2008 IEEE, Reprinted, with permission, from Yamada et al. [2008].

noise by 19.3 dB at 1 Hz. LO leakage power was measured by connecting RF port to the spectrum analyzer, and it was determined that it was reduced by 17 dB, from −19.3 to −36.7 dB m by using the canceller.

This canceller can be realized using a single unbalanced power splitter for both the 0° power split and attenuation (set according to the LO leakage signal power) with the phase relationship set by transmission line length. Figure 5.11 shows the measured results for variation of DC offset at the IF port and LO leakage power at the RF port. As predicted in Section 5.2, there are two optimum phase points for DC offset and one for LO leakage. For LO leakage, the minimum leakage power of −52 dB m could be achieved at the 210° phase shift point. In this case, 59 dB of LO-RF isolation was achieved, which is a 33 dB improvement over the typical value of 26 dB for this mixer. On the other hand, near-zero DC offset was achieved at 120° and 224° phase shift points. This result shows that there is a trade-off relationship between the optimum case for LO leakage and DC offset. In Fig. 5.12, the result of DC offset and Flicker noise-level measurements at 3 Hz is shown. In this measurement,

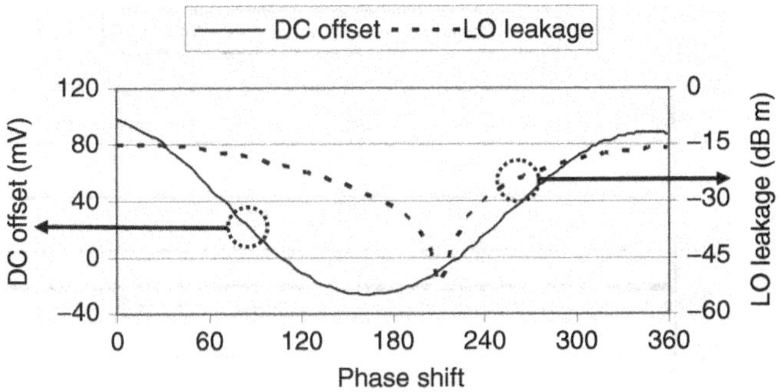

Figure 5.11 Measured DC offset and LO leakage power variation. © 2008 IEEE, Reprinted, with permission, from Yamada et al. [2008].

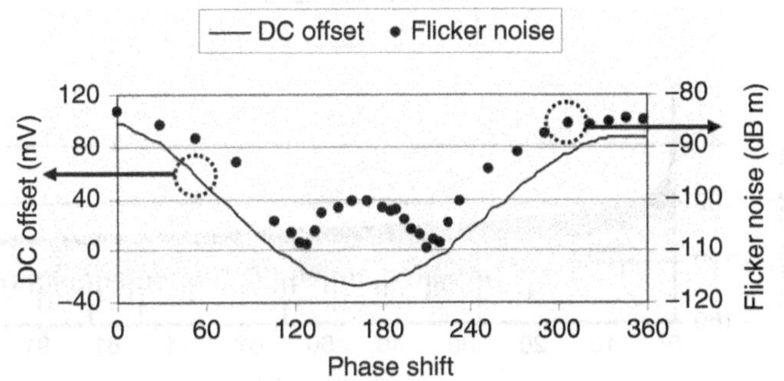

Figure 5.12 Measured DC offset and Flicker noise levels at 3 Hz. © 2008 IEEE, Reprinted, with permission, from Yamada et al. [2008].

the averaging function from a spectrum analyzer (Agilent E4448A) was used. The optimum point for DC offset is the same as shown earlier. A minimum Flicker noise level of −109.7 dB m was achieved at the 210° phase shift point.

In addition to the leakage issues found in direct conversion receivers, there is also the issue of Tx leakage due to finite Tx to Rx isolation. This Tx leakage signal is delivered to the mixer RF port, also resulting in LO self-mixing and DC offset. To avoid this leakage, high Tx–Rx isolation is required [Kim et al., 2006]. However, there is also LO signal leakage from the RF port of the mixer, which can be used advantageously for Tx leakage compensation by adjusting the phase of either the LO chain or the RF chain signal.

In Figs 5.13 and 5.14, measured DC offset and Flicker noise with minimum and maximum DC offset conditions are shown. In this measurement, an Agilent E4433B signal generator was used for the signal source. The signals were divided first by a two-way 0° power splitter (Mini-Circuits ZFSC-2-2500) to separate the LO and RF signals. A passive coaxial mixer (Mini-Circuits ZFM-4212) was used and +7 dB m of signal power (the specified ideal mixer drive level) was directed to

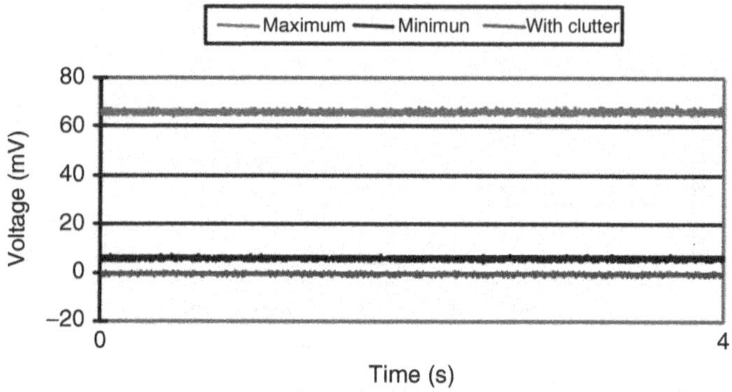

Figure 5.13 Measured DC offset.

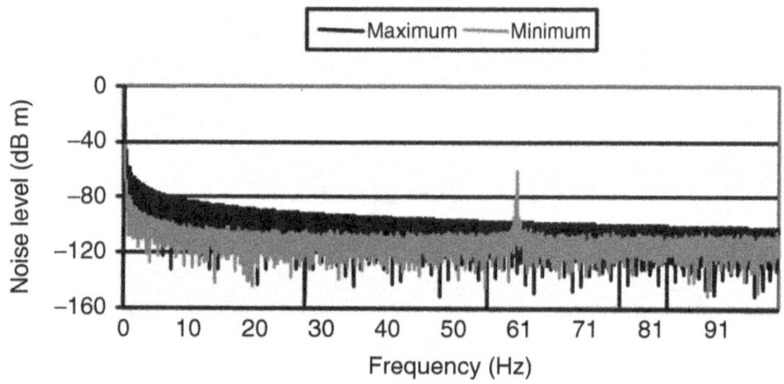

Figure 5.14 Measured Flicker noise. © 2008 IEEE, Reprinted, with permission, from Yamada et al. [2008].

the LO port. An electrical phase shifter (Pulser ST-21-444A) was installed in the Rx path to change the phase of the Tx leakage, and the DC offset was measured with an oscilloscope (Tektronix TD8 3014B) connected to the IF port of the mixer. The antenna port of the circulator (Narda 4923) was terminated with 50 Ω, so that only the Tx leakage signal was delivered to the Rx path. The maximum and minimum DC offset levels observed were 67 and 6 mV (Fig. 5.13), respectively, Flicker noise was also measured for maximum and minimum cases. In this measurement, the FFT function of an oscilloscope (Agilent Infiniium 54832D MSO) was used. In Fig. 5.14, at 1 Hz, 21.7 dB noise level difference between the maximum and minimum cases was achieved, where the maximum and minimum phase points were exactly the same as those for DC offset. With additional circuit modification to allow amplitude compensation between Tx and LO leakage, DC offset could be completely eliminated. These results indicate that DC offset and Flicker noise can be minimized in a direct conversion transceiver, through appropriate measures taken to ensure the optimum phase and amplitude relationship between Tx and LO leakage signal paths.

5.1.3 *IQ* Imbalance Assessment

In the previous section, the single-channel system limitations are described. Since a quadrature receiver system and the two orthonormal output signals enable to measure the relative phase information accurately, it is used in various applications, including digital communications and Doppler radar [Droitcour et al., 2004; Umstattd, 1993]. However, imperfections of the quadrature in circuit components used in real systems introduce amplitude and phase imbalance that adversely affect the recovery of output data [Huang, 2000; Moraes and Evans, 1996]. These imbalance factors can be estimated by comparing each channel's output power and compensated using a linear transform known as the Gram–Schmidt procedure [Huang and Caron, 2002; Nakagawa et al., 2004]. Through advanced digital signal processing (DSP), the imbalance factors in digital communications receivers can even be self-corrected after some iteration [Nakagawa et al., 2004; Chen et al., 2004; Zhu and Huang, 2004, 2005; Noon et al., 1999; Finol and Buchholz, 2004]. However, in the case of direct conversion quadrature Doppler radar systems, where the continuous analog variation of carrier phase must be recovered, such DSP compensation cannot be applied and other means must be sought for dealing with the imbalance. Direct measurement of imbalance factors in such phase-modulated systems can be performed through a comparison of output signals resulting from the mixing of two input signals of different frequency, generated by two synchronized signal generators. However, in the case of a single-antenna quadrature Doppler radar (Fig. 5.2), a major modification to the hardware would be required to perform this measurement. This could be the removal of a circulator or coupler, which is used to isolate transmitting and receiving signals or provide the same signal to the RF and LO ports, respectively, thus introducing new unknown imbalance factors associated with these components.

In this section, a measurement method in order to measure the imbalance factor in quadrature system is described, which instead of using two signal generators uses an external voltage controllable phase shifter to provide a phase delay equivalent to the Doppler shift produced by an object moving at a constant velocity [Park et al., 2007b]. This Doppler shift results in sinusoidal *I* and *Q* outputs, which can be easily compared with determine phase and amplitude imbalance factors. Using this technique, imbalance measurement can be performed without modifying the radar transceiver board, resulting in highly accurate phase and amplitude correction factors.

Figure 5.2 shows the block diagram of quadrature Doppler radar used for sensing cardiopulmonary motion [Park et al., 2006]. A single signal source provides both the RF output and LO signals. The LO signal is further divided using a 90° power splitter to provide two orthonormal baseband outputs. Assuming that heart and lung motion is given by $x(t)$ and $y(t)$, the quadrature baseband outputs can be expressed as

$$B_I(t) = \sin\left[\theta + \frac{4\pi x(t)}{\lambda} + \frac{4\pi y(t)}{\lambda} + \Delta\phi(t)\right] \tag{5.14}$$

and

$$B_Q(t) = \cos\left[\theta + \frac{4\pi x(t)}{\lambda} + \frac{4\pi y(t)}{\lambda} + \Delta\phi(t)\right] \tag{5.15}$$

where $\Delta\phi$ is the residual phase noise and θ is the constant phase shift related to the nominal distance to the subject including the phase change at the surface of a target and the phase delay between the mixer and antenna. In the system, the difference between I and Q mixers and signal paths, as well as inaccuracy of the 90° power splitter contributes to phase and amplitude imbalance. Those factors create an undesired linear transform on the I and Q output signal components, and adversely affect the orthonormal properties assumed for a quadrature system. Thus, the baseband signal for each channel can be expressed as

$$
\begin{aligned}
B_I &= A\sin(\theta + p(t)) \\
B_Q &= A_e A\cos(\theta + p(t) + \phi_e)
\end{aligned}
\qquad (5.16)
$$

where A_e and φ_e are the amplitude and phase imbalance factors, θ is constant phase delay for the traveling wave, $p(t) = (4\pi/\lambda)(x(t) + y(t))$ is the Doppler modulated signal, and assumed that there is no residual phase noise $\Delta\phi$ in the system for simplification.

It is possible to correct for a known phase and amplitude imbalance by a simple transformation known as the Gram–Schmidt procedure, shown in Equation 5.17, which produces two orthonormal vectors:

$$
\begin{bmatrix} B_{I\_\text{ort}} \\ B_{Q\_\text{ort}} \end{bmatrix} = \begin{bmatrix} 1 & 0 \\ -\tan\phi_e & \frac{1}{A_e\cos\phi_e} \end{bmatrix} \begin{bmatrix} B_I \\ B_Q \end{bmatrix} \qquad (5.17)
$$

Imbalance factor measurements for a quadrature receiver homodyne system can be made by injecting two sinusoidal waves with slightly different frequencies to the LO and antenna ports, respectively, using two external sources. However, in the case of the system shown in Fig. 5.2, a major hardware modification is required to perform this measurement, including a bypass of the LO, and removal of the antenna and the circulator. Instead of the modification on the hardware, external voltage controllable phase shifters connected between the antenna and the radar board to provide similar conditions to those achievable through the use of two external sources, but with certain distinct merits. These measurements can be accomplished without any modification to the original radar transceiver thus creating much closer conditions to those in a practical homodyne radar system where the same source is used to produce both the RF and LO signals.

The imbalance measurement system is illustrated in Fig. 5.15(a). Two external circulators and phase shifters are connected between the radar board and the antenna. A metal plate is placed at a fixed distance in front of the antenna beam, while phase shifters simulate the phase delay that would result from an object moving at a constant velocity. According to Doppler radar theory, when a transmitting signal is reflected from an object with constant velocity, v_r, the frequency of the reflected signal, $R_{\text{receive}}(t)$, is shifted by a Doppler frequency, f_d, where the polarity of the

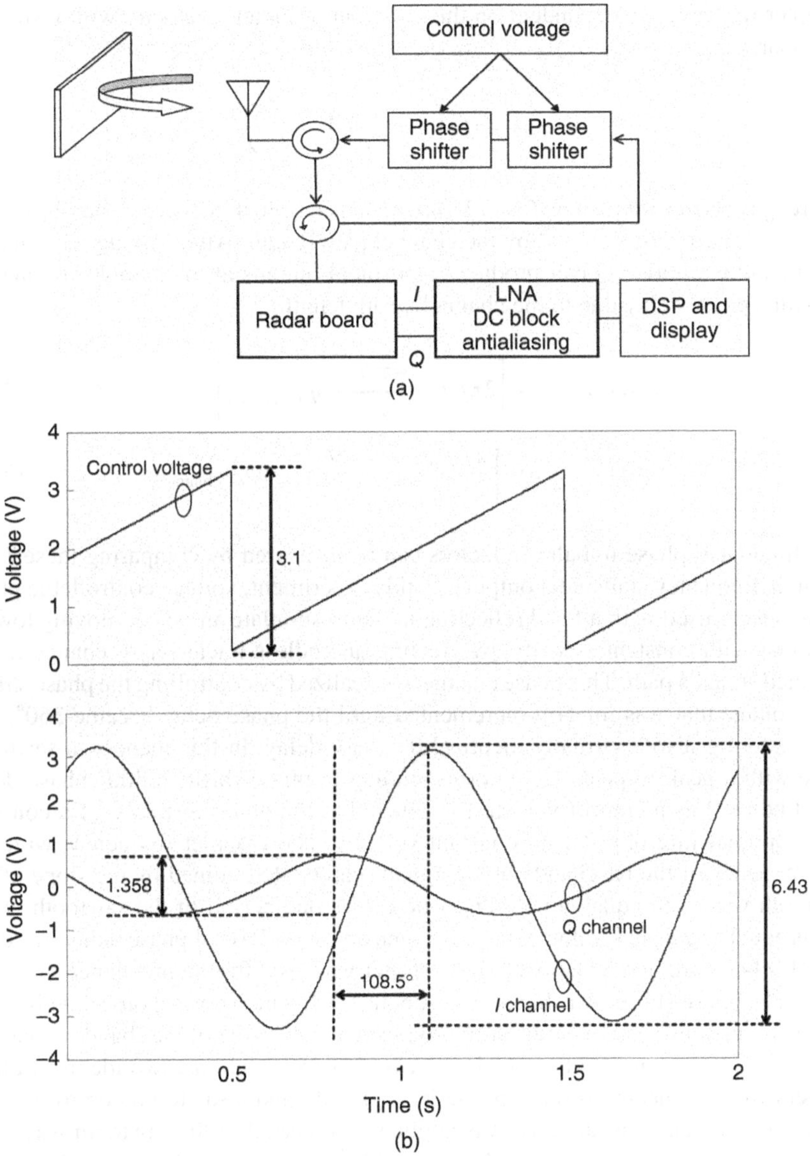

Figure 5.15 Measurement setup (a) and measured control voltage and imbalance factors (b). Using phase shifters, *I* and *Q* imbalance factors for a homodyne radar system can be measured without circuit board modification. An object moving with constant velocity is simulated by using a sawtooth wave to linearly sweep a set of phase shifters through 360° (3.1 V). The resultant *I* and *Q* baseband output signals are sinusoidal, with a single frequency that corresponds to the velocity simulated by the slope of the sawtooth wave. Amplitude and phase imbalance factor were measured here as 4.7° and 18.5°, respectively. © 2007 IEEE, Reprinted, with permission, from Park et al. [2007a].

Doppler frequency is dependent on the direction of target's velocity with respect to the radar:

$$R_{\text{receive}}(t) = A_r \cos\left[2\pi f_0 \left(1 \pm \frac{2v_r}{c}\right) t - \frac{4\pi d_0}{\lambda} - \phi_{\text{channel}}\right] \tag{5.18}$$

where f_0 is carrier frequency, $f_d = (2v_r f_0)/c$, d_0 is nominal distance of an object, and ϕ_{channel} is phase delay caused by the channel path length. After mixing with the LO signal, a quadrature receiver produces sinusoidal outputs at the Doppler frequency, f_d, with a phase delay due to the channel's path length:

$$B_I(t) = A_I \cos\left[2\pi f_d t - \frac{4\pi d_0}{\lambda} - \phi_{I\_\text{channel}}\right] \tag{5.19}$$

$$B_Q(t) = A_Q \cos\left[\frac{\pi}{2} + 2\pi f_d t - \frac{4\pi d_0}{\lambda} - \phi_{Q\_\text{channel}}\right] \tag{5.20}$$

Amplitude and phase imbalance factors can be measured by comparing these *I* and *Q* single frequency sinusoidal outputs. In this experiment, voltage controllable phase shifters were used with a fixed reflecting target to simulate an object moving toward the radar with constant velocity, by creating an endless linear phase change in the reflected signal's path. This phase change was realized by controlling the phase shifter with voltage that was linearly incremented until the phase delay became 360°, and then restoring it to a virtually identical 0° phase delay. In this manner, a sawtooth wave with a peak-to-peak value corresponding to phase shifter's 360° phase delay could be used as a control voltage for generating the phase response of a continuously approaching object with constant velocity. The Doppler frequency, which is the frequency of the baseband output signals, can be determined by the slope of the sawtooth wave and equals to $V_{2\pi}/t_d$, where t_d is one period of the sawtooth wave, and is equal to the peak value of the wave that achieves 360° of phase delay. A Pulsar ST-21-444A commercial coaxial phase shifter was used for the imbalance measurement. The phase shifter was linear up to about 180°, which corresponded with 3.1 V. To ensure that the system could fully produce the half-cycle of baseband output signal under linear phase control needed to avoid approximations, two identical phase shifters were connected serially in the RF-out path, and a sawtooth control voltage with a 3.1 V peak-to-peak value was applied. The period of the sawtooth wave was set to 1 s in order to get sinusoidal waves with a frequency of 1 Hz at each channel output, which closely approximates a cardiac signal.

The method was applied to a custom radar circuit board [Park et al., 2006]. Figure 5.15(b) shows the phase shifter control voltage with an amplitude of 3.1 V and resulting *I* and *Q* sinusoidal outputs at a Doppler frequency of 1 Hz. The period of the *I* and *Q* sinusoidal waveforms corresponds to the target velocity simulated by the sawtooth control voltage. By comparing amplitude and phase delay of *I* and *Q* waveforms, the measured amplitude and phase imbalance factors were determined to be 4.7° and 18.5°, respectively.

5.2 BASEBAND MODULE

To maximize the acquisition of the signal, it is very important to maximize the effective dynamic range of analog-to-digital converter (ADC). The difference in the magnitude between the respiration and heart beat signals is due to respiration having a larger cross section and a larger displacement than heart activity. Even with proper demodulation, it is difficult to separate the two signals. In addition, the large DC component is always close to the cardiopulmonary signal spectrum. Thus, although the heart and respiratory signals should be amplified as much as possible, effective removal of DC components is required in order to avoid saturating the system.

5.2.1 AC and DC Coupling

Although the AC coupling method of DC cancellation is common, it is generally unsuitable for low frequency of physiological signals in which the signal's DC offset is critical for demodulation. The AC coupling filter's time constant, τ, is the amount of time in which the step function response moves to $1/e$ or 36.8% of the current state:

$$\tau = RC = \frac{1}{2\pi f_c} \tag{5.21}$$

Equation 5.21 shows time constant, τ, with a dimension in seconds. Assuming the first-order filter, R and C are the values of a high-pass filter resistor and capacitor respectively and f_c is the filter cutoff frequency. The natural and step response of the system will be

$$x(t) = x(0)e^{-t/\tau} \tag{5.22}$$

$$x(t) = x(\infty)\left(1 - e^{-t/\tau}\right) \tag{5.23}$$

respectively. A step response with a τ of 5.31 s (due to a cutoff frequency of 0.03 Hz) will require $2.3 \times \tau$ or 12.2 s to reach within 10% of the maximum value. To reach within 1% of the maximum value, $4.6 \times \tau$ or 24.4 s is required. Even small movements of the subject or clutter will result in large changes in the DC offset due to large required baseband gains, typically on the order of 1000 V/V. These changes can be seen as a sudden step function in the I and Q signals resulting in the large settling times before a usable signal is available. Additional signal distortion effects due to AC coupling are described in Chapter 4.

The purpose of the DC canceller is to cleanly remove the DC components of the I and Q outputs of a quadrature system with minimum time delay, distortion, and noise. Then the remaining time-varying signal can be amplified and sampled with maximum resolution. Each binary increase in gain is equivalent to an added bit in acquisition. For example, a 0.2 mV_{pp} signal to a 16-bit, ± 2 V ADC is resolvable to about 3 least significant bit (LSB), equivalent to a 2-bit ADC. To reach a full-scale of ± 2 V, a 20,000 V/V amplifier can be used, which is a gain of more than 2^{14}. This is, of course, an ideal situation without electrical noise and variations in the radar signal. However, this highlights the concern with dynamic range and the importance of amplification after the removal of large DC offsets.

The DC cancellation system will also need to compensate for a changing DC offset. For the vital-sign monitor, Vergara et al. [2008] considered two types of changes: a gradual shift in nominal displacement of the oscillating reflector (representing the chest motion) or a chaotic environment caused by the introduction of moving objects into the radar environment. A gradual baseline shift in DC offset could be a slow moving object in the radar field-of-view or voltage drift due to low-frequency noise effects (e.g., due to $1/f$ noise or ambient temperature variations). In the chaotic environment, stationary objects producing clutter noise may suddenly move and show as a large rotating vector in the complex IQ space or as a sudden change in DC offset in both I and Q channels.

The system must be designed to accept the two types of large signal changes and wait for the oscillating motion of the chest to re-emerge to continue the measurement. At that moment of radar quiescence, reacquisition of the DC offset should begin instantly for compensation and the process for the vital sign extraction can begin again.

5.2.2 DC Canceller

If an instantaneous value was sampled and then used as a reference in the negative port of a difference amplifier, the DC offset would effectively be removed and the idea of a digital DC canceller would be of more practical for use than AC coupling alone. Figure 5.16(a) shows a sample and hold (S/H) element and amplifier used before data acquisition. The system shown is for a single channel, and a pair will be required to acquire both in-phase and quadrature signals. A problem with traditional S/H is susceptibility to voltage droop due to transistor parasitic resistances that will drain the charge away from a holding capacitor. Therefore, the next step is to look for an S/H with zero-droop: an infinite S/H. The simplest way to implement this is to use an ADC–DAC (digital-to-analog converter) pair (Fig. 5.16(b)).

With changes in the environment, whether gradual or chaotic, occasionally deviations from the estimated DC offset will require a new reference point. Algorithms and signal processing will determine the appropriate time to resample. By tracking the signal at the end of the chain, a feedback control loop can be established. A set of comparators can send signals to warn of signal clipping. These signals will trigger the ADC to read and the DAC to write a new DC value, as shown in Fig. 5.16(b). Although an instantaneous value may be used, low-pass filtering before acquisition or digital averaging before analog output from the DAC will smooth out the signal. It should be noted that the sampled values are an estimate of the DC offset and will probably include some of the DC information.

Although this system is adequate for DC cancellation, and will allow further amplification, it wastes the possibilities of digital control. Of course, taking advantage of these possibilities requires an increase in the complexity of the system.

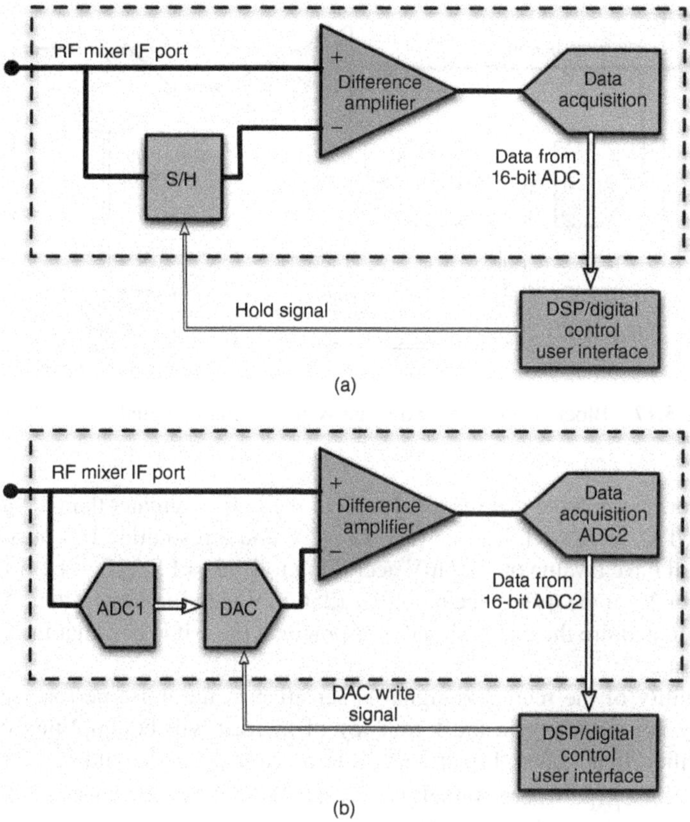

Figure 5.16 (a) DC cancellation using a sample-and-hold and (b) using an ADC–DAC pair as an infinite S/H in the two-stage system.

A two-stage architecture is most similar to the infinite S/H. It is called a two stage because it requires two ADCs, ADC1 and ADC2, for signal control and acquisition (see Fig. 5.17). Input to the first signal stage includes the large DC offset as well as the small signal that provides the important bioinformation. At the start of the acquisition cycle, ADC1 instantly acquires a value from the signal. This value is the initial estimated DC offset. This initial value is given to the DAC and the DAC outputs the initial estimated DC offset as a reference voltage level. In essence, this ADC–DAC pair acts as the zero-droop sample-and-hold unit.

Next, the difference amplifier subtracts the estimated DC offset output from the input signal and then the result is amplified and sampled by ADC2. Once the amplified and DC cancelled signal is acquired, the original signal can be faithfully recreated:

$$\text{Recreated signal} = \text{Signal}_{\text{ADC1}} + \frac{\text{Signal}_{\text{ADC2}}}{G_{\text{amplifier}}} \tag{5.24}$$

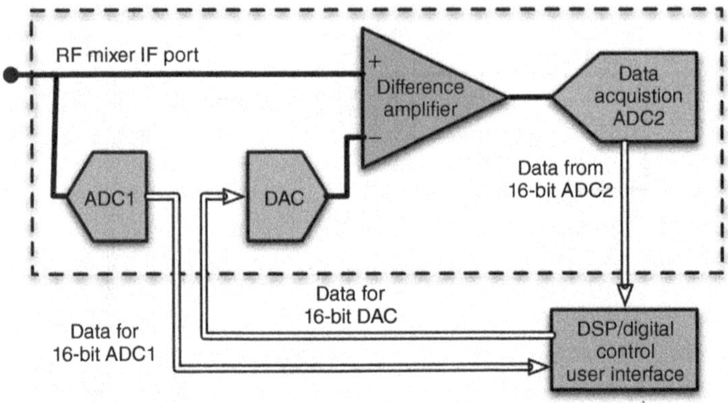

Figure 5.17 Block diagram of two-stage system utilizing digital feedback system.

It is important to note that the resolution of the DAC is greater than the gain of the amplifier. If DAC has a full scale (FS) of ±2 V and a resolution (N) of 8-bits, then its LSB will have a value of 15.7 mV according to $\text{LSB} = \text{FS}/(2^N - 1)$. If the gain is greater than 2^N of the difference amplifier (256 V/V gain for 8-bits), then ADC2 will be unable to acquire the entire signal. For practical uses, it is best that the gain is not greater than 2^{N-2}.

The quality of the recreated signal is sensitive to the noise performance of the DAC. Any noise present in the frequency of interest will be amplified due to the gain amplifier. In actuality, Equation 5.24 term: $\text{Signal}_{\text{ADC2}}$ should be $\text{Signal}'_{\text{ADC2}} + G_{\text{amplifier}} \cdot \text{Noise}_{\text{DAC}}$, where $\text{Signal}'_{\text{ADC2}}$ is the DC offset compensated signal. The noise introduced by the DAC must be less than the signal of interest, preferably by an order of magnitude. The DAC should also reject power supply variations and digital feed-through.

Taking advantage of digital acquisition (DAQ) and control allows for further optimizations to maximize the system dynamic range. For example, as the initial reference value from the DAC is only an estimation of the DC offset, analysis of the acquired signal from ADC2 can be used to refine the estimation. If a few cycles of respiration are recorded by ADC2, and there is no change in the DC offset baseline, a midpoint function will find a new reference value and a better DC offset estimation. The midpoint function is preferred over an averaging function as the concern is over the extremes of the voltage swing. Further optimization can be achieved through implementing system calibration to compensate for additional DC offsets introduced by the DAC and amplification stage.

If automatic gain control (AGC) is implemented, in order to retain signal integrity for reconstruction, the gain amount should be recordable. This makes it necessary for a digitally controlled variable gain amplifier (VGA). For ease of use, a VGA with binary steps is best. This provides linearity on the logarithmic scale. Controlling the AGC is determined by another set of comparators reading the signal at ADC2. These

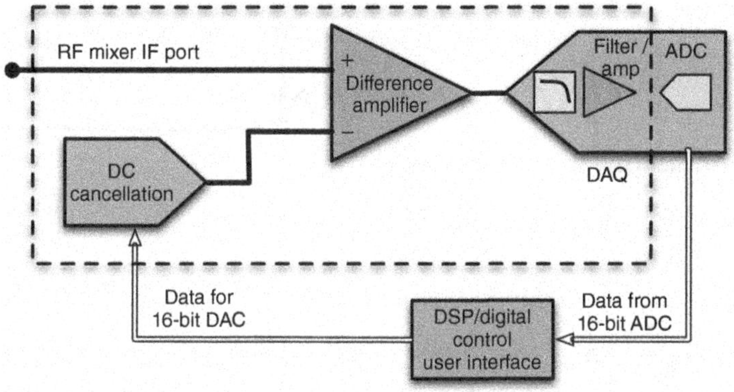

Figure 5.18 Block diagram of DC offset compensation system.

comparators check for gain-increase window condition, the value determined by the next gain step. As long as the signal remains in the window, the AGC will attempt to increase the gain until the maximum gain step is reached. On any condition of clipping, the gain will reset to zero with a reset of the timing buffer.

The data acquisition system with DC canceller is similar to the two-stage DC cancellation except that there is no ADC1 [Vergara et al., 2008]. All of the data acquisition occurs with the ADC behind the difference amplifier, as shown in Fig. 5.18. Again, two identical channels will be required to acquire both I and Q. Because there is no full-scale ADC, initial estimation of the DC offset is attempted through an incremental search function. Clipping of the signal to the ADC will tell the DAC to incrementally increase or decrease in value until a usable signal is present to the ADC. This process may be improved with use of a VGA.

Figure 5.19 shows the signal of a sitting subject moving his hand from his lap to his head. Figure 5.19(a) shows a zoomed in section of the full-scale signal at 20 V/V gain, panel (b) shows the AC coupling acquisition at 1000 V/V gain and panel (c) shows the I and Q recording using the DC cancellation system. As seen in Fig. 5.19(a), the DC offset changes in a manner similar to a step function. Figure 5.19(b) shows that the AC coupling requires a period of time to respond, as the high-pass filter relaxes. The digital DC canceller also has a time delay in Fig. 5.19(c) as the DC offset search function attempts to provide an estimation of the DC offset.

If DC offset values and intrinsic noise of the electronics are reasonably low, a high-resolution ADC may be used to acquire the whole signal without AC coupling or other baseband DC offset compensation techniques. Gain amplifiers may be applied but one will need to be conscientious of exceeding the input range of the ADC due to the large DC offset. After acquisition, DSP can remove the DC offset and deal with changes in the signal profile while still providing a usable signal. The greater the resolution of the ADC and the ability of the system to reject noise, respiration and even heart signals should be recoverable.

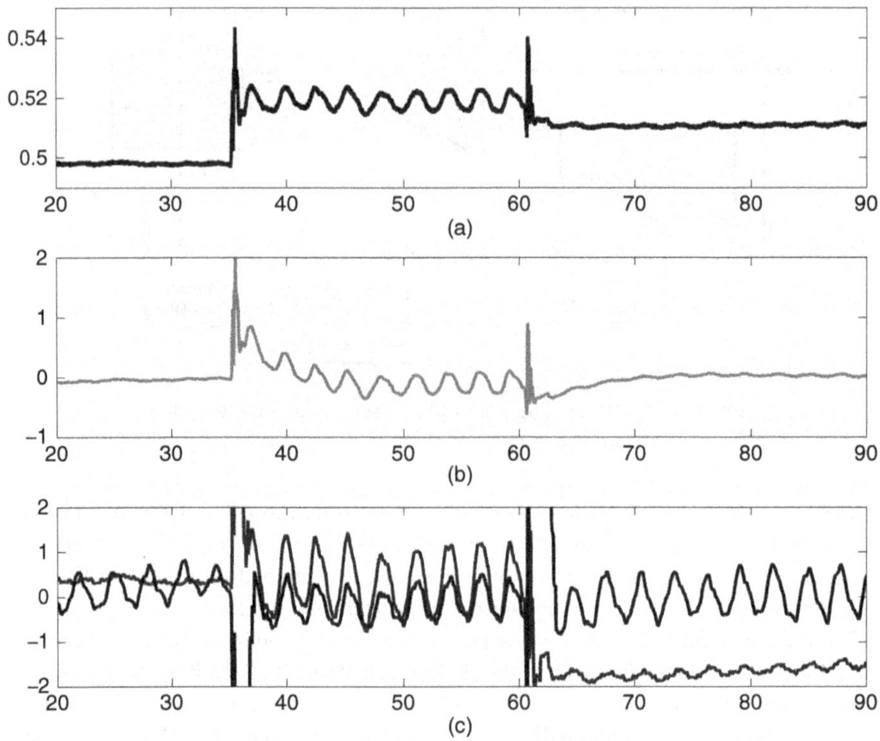

Figure 5.19 Time plots of a subject 1 m away with arm movement. (a) *I* plot of the magnified raw signal, (b) *I* plot of the AC coupling response, (c) *I* and *Q* plot of the DC cancelled output.

5.3 SIGNAL DEMODULATION

Linear and nonlinear demodulations are the two methods used to recover the phase of Doppler radar signals modulated by physiological motion. The amount of phase modulation (arc length in the polar plot), and signal quality in terms of SNR (arc width) and presence of distortion (arc irregularities) will determine the optimum demodulation method. Assuming that arc length is relatively small, as is the case for 2.4 GHz system monitoring cardiorespiratory motion, both demodulation methods may yield adequate results. However, if the arc lengths become longer, due to higher frequency or larger displacement (e.g., subject walking), small signal approximation will no longer be valid, and nonlinear demodulation must be used for phase recovery. However, nonlinear demodulation is more sensitive to noise and distortion.

5.3.1 DC Offset and DC Information

One challenge in providing robust sensing is detection sensitivity to target position due to the periodic phase relationship between the received signal and LO, resulting in "optimum" and "null" extreme target positions [Park et al., 2006]. A quadrature Doppler radar receiver with channel selection has been proposed to alleviate this problem [Droitcour et al., 2004]. This method selects the better of the quadrature (I and Q) channel outputs and is thus limited to the accuracy of a single channel. A frequency tuning technique with double-sideband transmission has also been proposed for Ka-band radar [Xiao et al., 2006]; however, this technique requires more complex hardware with a tunable intermediate frequency. In this section, arctangent demodulation with DC offset compensation to combine quadrature outputs is described [Park et al., 2007a]. Arctangent demodulation overcomes position sensitivity issues while removing the small-angle limitation on the range for phase deviation detection, which can be significant in single-channel systems operating at high frequencies. The use of DC offset compensation ensures that unwanted DC components produced by receiver imperfections and clutter reflections are removed, while DC information required for accurate arctangent demodulation is preserved.

Several DC offset compensation techniques have been proposed for communications receivers [Svitec and Raman, 2005; Mashhour et al., 2001; Matinpour and Laskar, 1999], where all of the DC signal is assumed to be undesired. The simplest solution for DC offsets is to remove them by using a high-pass filter. However, several modulation methods, such as the phase modulation method contain critical "DC information" which must be distinguished from unwanted "DC offsets" caused by imperfections in circuit components and reflections from stationary objects. The DC information component, associated with target position in Doppler radar, is typically several orders of magnitude larger than the amplitude of the periodic baseband signal related to heart activity, making it impractical to simply digitize the full signal with reasonable resolution. Thus, it is required that techniques isolate DC offset, DC information, and the AC motion signal to overcome dynamic range limitations for pre-amplifiers and ADCs, without discarding important components of the desired data. The results of arctangent demodulation experiments with a target at several different positions are described here, demonstrating proper preservation of cardiopulmonary-related motion information, and verifying accuracy insensitivity to target position.

As shown in Equations 5.14 and 5.15, the I and Q outputs are the cosine and sine of a constant phase delay caused by the nominal distance to a target, with a time-varying phase shift that is linearly proportional to the chest displacement. By applying the arctangent operation to the I and Q output data ratio, accurate phase demodulation can always be obtained regardless of the target's position as

$$\phi(t) = \arctan\left(\frac{B_Q(t)}{B_I(t)}\right) = \arctan\left(\frac{\sin(\theta + p(t))}{\cos(\theta + p(t))}\right) = \theta + p(t) \tag{5.25}$$

where $p(t) = 4\pi(x(t) + y(t))/\lambda$ is the superposition of the phase information due to respiration or heart signals.

However, quadrature channel imbalance and DC offset act as a linear transform on the I and Q components, thus modifying Equation 5.25 to

$$\phi'(t) = \arctan\left(\frac{B_Q(t)}{B_I(t)}\right) = \arctan\left(\frac{V_Q + A_e \sin(\theta + \phi_e + p(t))}{V_I + \cos(\theta + p(t))}\right) \tag{5.26}$$

where V_I and V_Q refer to the DC offsets of each channel, and A_e and ϕ_e are the amplitude error and phase error, respectively.

Correction for a known phase and amplitude imbalance is straight forward using the Gram–Schmidt procedure [Moraes and Evans, 1996]. The DC offset issue is more complex, however, due to the fact that the total DC signal contains DC information required for accurate demodulation. The DC offset is caused by two main sources: reflections from stationary objects (clutter), and hardware imperfections. Hardware imperfections include circulator isolation, antenna mismatch, and mixer LO-to-RF port isolation, resulting in self-mixing that produces a DC output. On the other hand , as indicated by Equation 5.26, DC information associated with the target's position is also part of each baseband signal. The magnitude of this DC level is dependent on the target's position, such that the DC level is higher for target positions closer to the "null" case. Consequently, before arctangent demodulation is performed the DC information must be extracted from the total DC output, and preserved.

A coaxial quadrature radar system, as shown in Fig. 5.2, was used to examine arctangent demodulation issues by the measurement method described in Park et al. [2007b]. The DC offset caused by hardware imperfections was measured by terminating the antenna port with a 50 Ω load. The main contribution to this DC offset is caused by self-mixing with circulator leakage power, dependent on the phase difference between the LO and antenna feed line. By connecting a phase shifter between the LO feed line and varying the phase delay, the DC offset range for each channel can be measured at the corresponding mixer's IF port. This was determined to be 19.4 mV for the I channel and 19.8 mV for the Q channel with an LO power of 0 dB m. The DC offset due to reflections was estimated by putting a large metal reflector at a distance of 1 and 2 m from the radar, with a half-wavelength position variation used to find the maximum and minimum DC values. The DC offset range for the I and Q channels from a reflector at 1 or 2 m distance are 3 and 3.4 mV, and 0.6 and 0.8 mV, respectively. As expected, experimental results show that the DC offset is dominated by the contribution from imperfections in the circuit components rather than from clutter located 2 m away from radar. The measurement setup for DC compensation is shown in Fig. 5.20. The coaxial radar described in the previous section was used to collect data from a seated subject facing the antenna at a distance of about 1 m. A wired finger pressure pulse sensor was used to provide the reference for heart rate. Once the DC offset components were determined as described earlier, they could be subtracted from the output signal.

The remaining challenge was to preserve the relatively large DC information level while sufficiently amplifying the weak time-varying heart-related signal. In this coaxial radar system, the maximum DC information which occurs at the null case, reaches about 3.8 mV, while peak-to-peak voltage for heart motion typically results in less than 25 μV. In other words, the DC information is 2–3 orders of magnitude larger than the signal amplitude. This makes it difficult to amplify the signal associated with heart displacement sufficiently for high-resolution digitization without saturating the amplifiers or the ADC. Details for the method used for achieving high amplification without saturation is shown in Fig. 5.20(b). With no object within 1 m in front of the radar, the internally or externally induced DC offset of each channel could

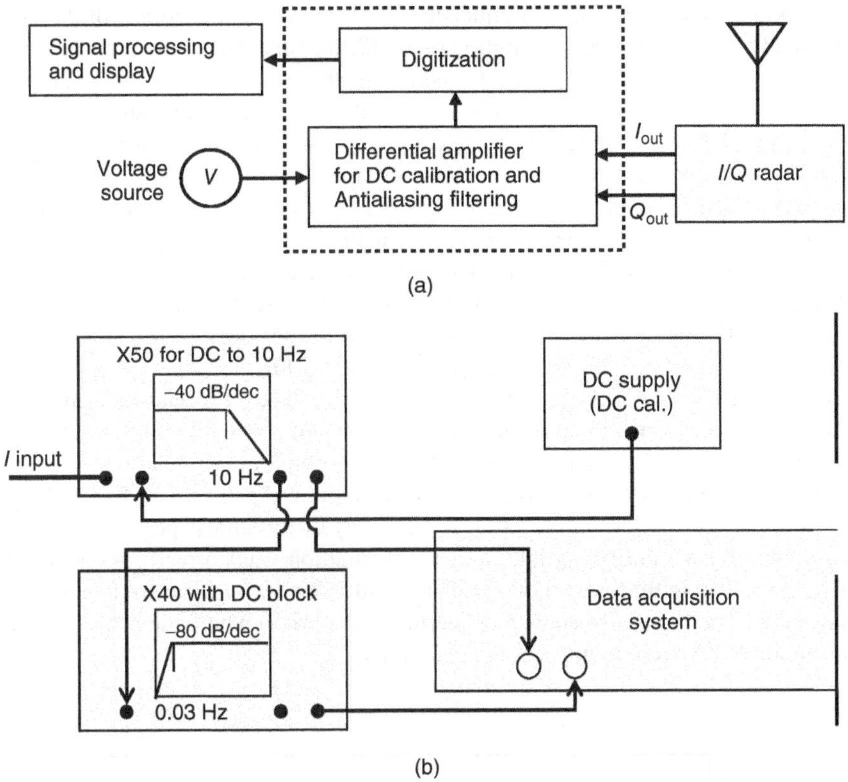

Figure 5.20 Measurement setup for DC compensation. Overall radar setup is shown (a), with data acquisition (dashed region in (a)) details provided for the *I* channel (the *Q* channel is exactly the same) (b). The clutter- and circuit-based DC offset measured with no target present is reproduced (DC supply) and subtracted from the response for a human subject, so that the heart motion signal (which includes a DC component) can be digitized with maximum resolution. © 2007 IEEE, Reprinted, with permission, from Park et al. [2007a].

be measured. These DC offsets were then calibrated by using differential amplifiers, each with one input port connected to a DC power supply. The DC supplies were then used to generate the same voltage as the DC offset of each channel, thus producing a zero DC level at the output. While preserving this condition, a human subject was then located at a distance of about 1 m from the radar. In this experiment, the full DC level, including the heart motion signal, was detected at each channel. To achieve sufficient amplification of the signals, three amplifiers were used at the baseband stage of the *I* and *Q* channels. The first one was a differential amplifier with a gain of 50 that amplified both the DC and the heart motion signal and calibrated the DC offset. Subsequently, the output of the first amplifier was divided into two outputs, one of which was saved in the data acquisition system and the other was saved after the DC was removed and the AC content was amplified. Two amplifiers were used for the

DC blocking filter with a cutoff frequency of 0.03 Hz and gain settings of 20 and 2, respectively, in order to obtain a high-Q (−80 dB/dec) and thus a sharp cutoff.

Arctangent demodulation was performed using these signals with and without DC content using MATLAB software. The signal with DC content was multiplied by 40 in the MATLAB code before summation with the AC signal that was pre-amplified before the ADC. At the same time, the AC-only signal was filtered with a Butterworth filter that passed frequencies between 0.9 and 2 Hz to eliminate the still-detectable low-frequency component due to respiration and thus avoid including this effect twice when summing with the DC-included signal. Consequently, a high-resolution heart motion signal combined with a virtual DC component was created. Without this procedure, the DC component would saturate the amplifiers before the smaller heart motion signal could be sufficiently amplified for recording.

To verify that the DC information was properly preserved, I/Q data after imbalance and DC offset compensation was plotted on a polar plot. Two orthonormal sinusoidal functions of the same phase information will compose part of circular trace centered at the origin, corresponding to the phase information. As shown in Fig. 5.21, the I/Q baseband signals DC information form a part of an almost perfect circle centered at the origin, confirming that the DC information was correctly accounted for (it would be a circle for two orthonormal sinusoids). The same measurement with the DC portion removed is also shown, appearing at the origin where the phase information cannot be recovered with the same certainty.

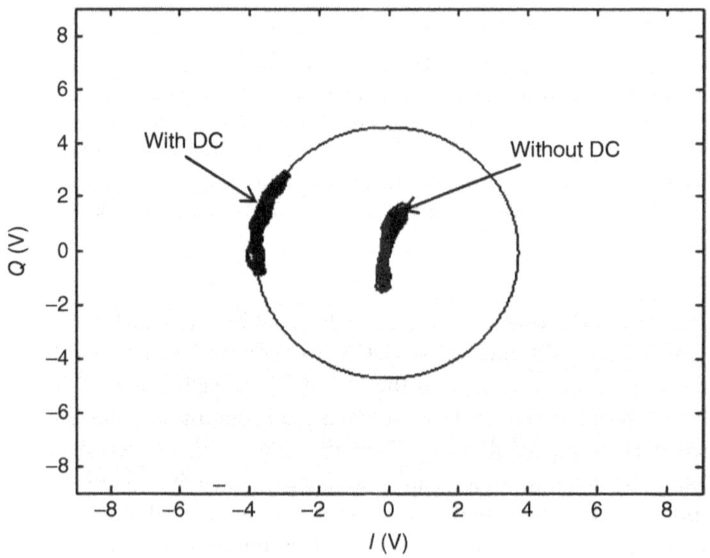

Figure 5.21 Polar plot of I/Q data. The I/Q data with DC preserved forms a portion of a circle centered at the origin, verifying preservation of all phase information, while the I/Q signals without DC information form a line near the center for which phase information cannot be accurately recovered. © 2007 IEEE, Reprinted, with permission, from Park et al. [2007a].

Figures 5.22–5.24 show the *I*, *Q*, and arctangent-demodulated signals obtained using the measurement setup shown in Fig. 5.20, for the subject in an intermediate position for both channels (Fig. 5.22), close to a null position for the *Q* channel (Fig. 5.23), and close to a null position for the *I* channel (Fig. 5.24). The null and optimum positions cannot be set exactly for heart rate measurements, as the nominal distance (and associated phase) varies as a result of respiration and effects rate data accordingly. To examine the effectiveness of arctangent demodulation, standard deviation was used to provide a quantitative comparison of accuracy. As shown in Figs. 5.23 and 5.24, a dropout region occurs at the null point due to degradation in signal power, and this region is excluded when calculating standard deviation.

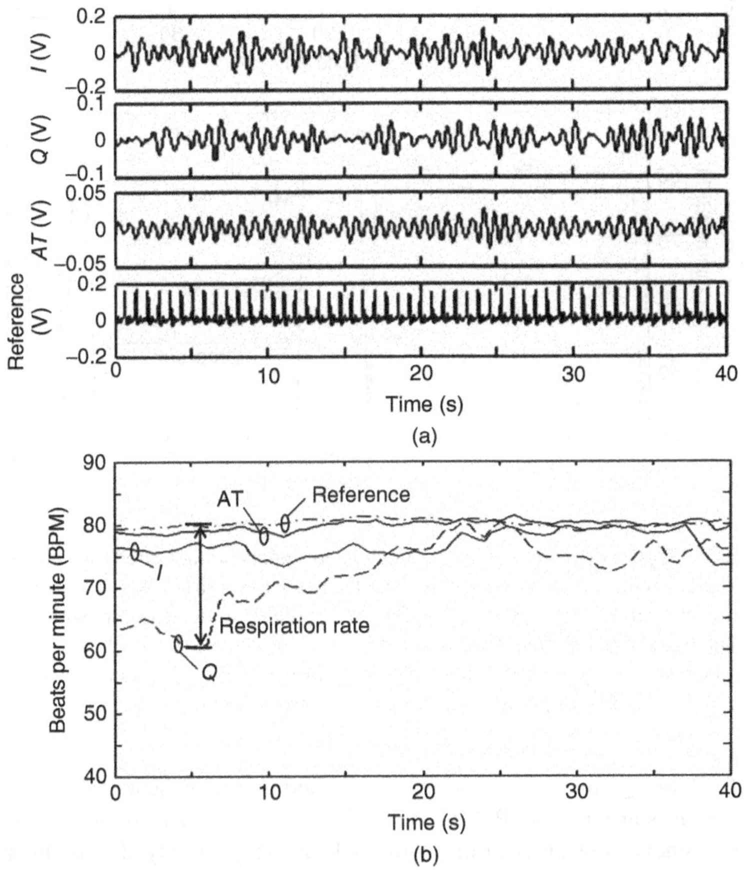

Figure 5.22 Heart rate measurements for both channels in an intermediate position. Band-pass-filtered (0.9–2 Hz) Doppler radar *I* and *Q* signals are shown along with the combined arctangent demodulated output (AT), and a wired finger pulse reference (a). Heart rate history (using autocorrelation) is also shown (b), where the *Q* channel data are at times off by the respiration rate value, as predicted. Standard deviation is less than 1 beat over the full 40-s interval for the AT data, while it is 3.9 and 9.8 beats for the *I* and *Q* channels, respectively. © 2007 IEEE, Reprinted, with permission, from Park et al. [2007a].

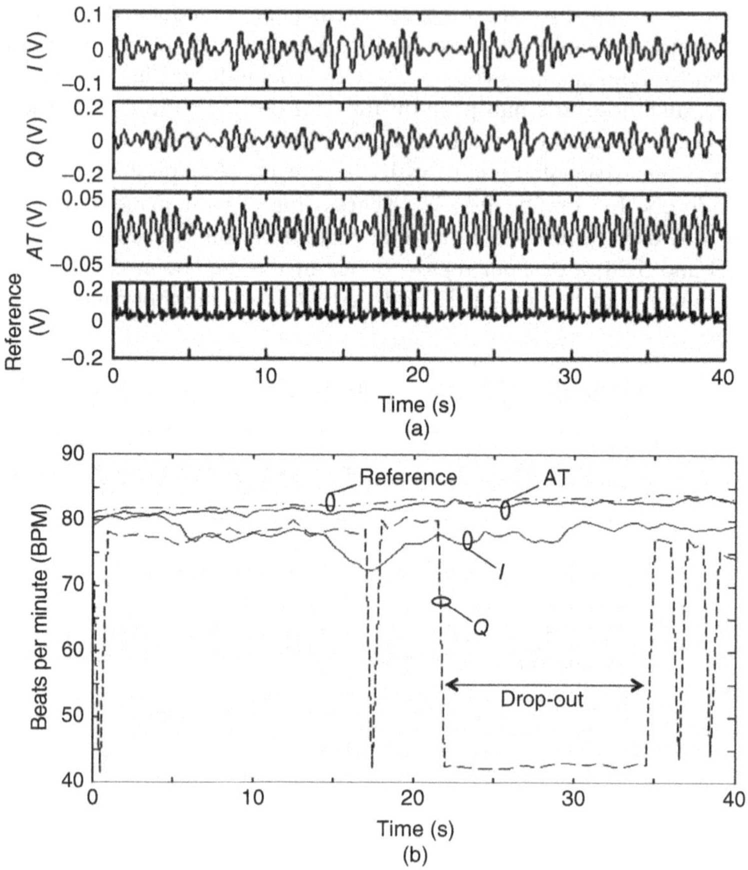

Figure 5.23 *I*, *Q*, and arctangent (AT) demodulated signals (a) measured for a position where the *Q* channel is close to a null condition. The *Q* channel rate (b) shows drop-out regions (in 35% of the interval) when the SNR is insufficient for digitization, as occurs with the squaring effect when in the null position. Excluding drop-outs, the *I* and *Q* channels have errors of 4.8 or 5.2 beats, respectively, over the same 40-s interval where the AT data has an error of only 0.9 beats. © 2007 IEEE, Reprinted, with permission, from Park et al. [2007a].

In Fig. 5.22, the *Q* channel heart signal is affected by the presence of the respiration signal, which is around 20 BPM, at the beginning of the measurement interval. The *I* and *Q* channels show an error of 3.9 or 9.8 beats, respectively, during the 40 s time interval while the arctangent combined output has an error of only 0.95 beats. In Fig. 5.6, 35% of the *Q* channel data could not be acquired or, dropped out, and the rest has an error of 4.8 beats. The more stable *I* channel data still has an error of 5.2 beats, while the arctangent combined output has an error of only 0.9 beats. In Fig. 5.24, both *I* and *Q* channels drop out for 23% and 5% of the total time interval, respectively. The *I* channel data have an error of 7.5 beats and the *Q* channel data have an error of 1.7 beats, while the arctangent combined output has an error of only

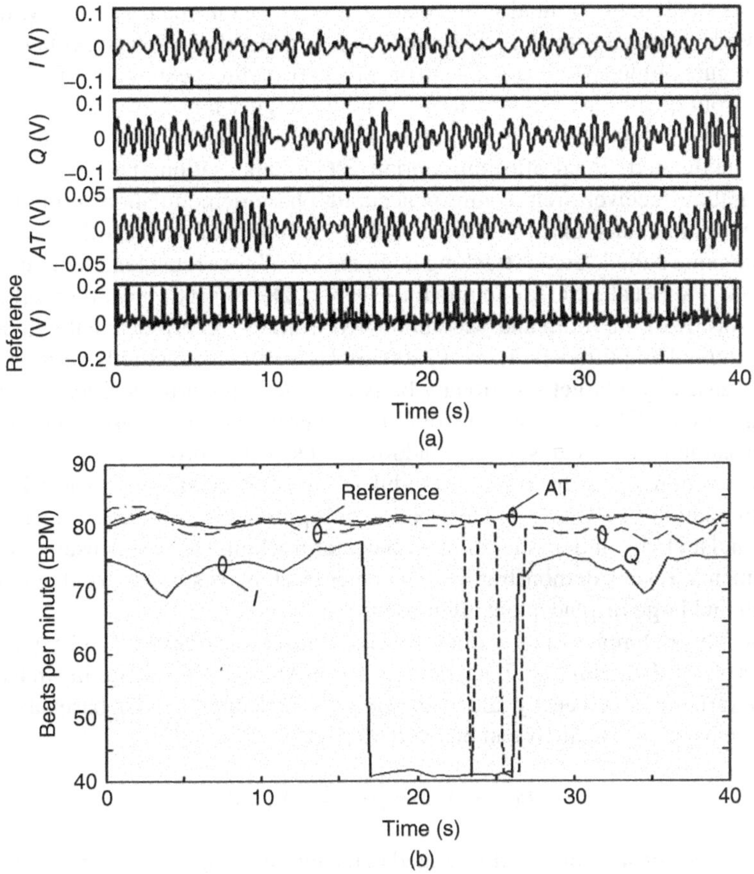

Figure 5.24 *I*, *Q*, and arctangent demodulated signals (a) measured for the *I* channel close to a null position. Data dropout regions occur for both *I* (23% of the interval) and *Q* (5%) channels. Standard deviation is 7.5 or 1.7 beats for the *I* and *Q* channels, respectively, and only 0.6 for the arctangent output. © 2007 IEEE, Reprinted, with permission, from Park et al. [2007a].

0.6 beats. From the measurement results described earlier, it is evident that arctangent demodulation results are significantly more accurate than any single-channel output, with an error that is consistently less than 1 beat in standard deviation over the 40-s monitoring interval, and when using this data there is no drop-out region. Thus, arctangent demodulation produces robust and accurate data for rate tracking regardless of a target's position, without need for channel selection.

5.3.2 Center Tracking

While simple channel selection in quadrature receiver architecture provides some accuracy gains for small displacement compared with the wavelength, full quadrature demodulation (arctangent demodulation) is required for larger displacement or higher frequency systems. Arctangent demodulation overcomes a target position limitation as well as small angle limitation since it extracts phase information directly, which is linearly proportional to the target's actual motion, rather than choosing either in-phase or quadrature output. Moreover, due to the property of direct extraction of phase information from quadrature outputs, target's motion bigger than half wavelength of the carrier signal, which results in two-phase change, still can be recovered by unwrapping demodulated signal after appropriate DC compensation. The main challenge of quadrature receiver architecture is accurate phase demodulation in the presence of channel imbalance and DC offset. In this section, another DC offset compensation technique called "center tracking compensation" for arctangent demodulation is described [Park et al., 2007a]. In previous section, DC offsets were measured without a target inside of the room and was subtracted from the received signal with a target via differential amplifiers [Park et al., 2007a]. However, this method overlooked clutter DC caused by a target's stationary body part including arms and legs, which can cause additional DC offset to that from the background of the room. Center tracking compensation method can solve this additional DC offset problem since it estimates DC information signal from phase-modulated quadrature signals' properties rather than elimination of DC offset. This section will present relevant quadrature receiver theory and arctangent demodulation with center tracking DC compensation method. Experimental results demonstrating that center tracking is suitable for demodulation of small and large displacement are presented.

Typically, a Doppler radar motion sensing transceiver transmits a CW signal and receives phase-demodulated signal reflected from a target. Now assuming that target's motion variation is given by $\Delta x(t)$, the quadrature Doppler radar baseband output assuming balances channels can be expressed as

$$B(t) = A_r \exp\left[\theta + \frac{4\pi}{\lambda}\Delta x(t)\right] \tag{5.27}$$

where θ is the constant phase shift related to the phase change at the surface of a target and the phase delay between the mixer and antenna. As shown in Fig. 5.26(a), when there is only one target, and reflected signal is phase-modulated by variation from it, complex plot of quadrature outputs forms fraction of the circle that has a radius of signal amplitude, A_r, with center offset by DC offset of each channel. This property allows elimination of DC offset and preservation of DC information properly, which is the magnitude of radius projected on each axis, if the center of arc formed by motion of a target is tracked back to the origin of the complex plot. Now, arctangent demodulation of quadrature outputs, whose complex plot is centered at the origin, produces phase information which corresponds to actual motion of a target, thus real-time target motion monitoring can be achieved. These properties can extend their validation to the larger phase-modulated signal that happens when a target's motion variation becomes bigger than wavelength of the carrier frequency. Complex plot of the I and Q outputs is related mainly with both received signal power and phase deviation due to a target's motion. From Equations (5.27) and (5.25), received signal power becomes A_r^2, square root of which is the radius of the arc formed by phase deviation from a target's motion. Phase variation, which is proportional to the arc length, is proportional to the ratio of target's motion over wavelength of the carrier signal. In other words, arc length becomes longer either due to the increase of target's actual motion or due to the increase of the carrier frequency. Consequently, when a target is moving with large deviation resulting in changing received signal power, the radius of the arc will vary while its center is located at the same point, thus forming spiral-like shape rather than a circle. However, when operating frequency is increasing so that small physical motion of a target is converted in large-phase variation, longer arc length on a circle can be obtained.

A coaxial quadrature radar system and measurement setup for DC compensation are shown in Fig. 5.25(a) and (b), respectively.

Data are collected from a seated subject facing the antenna at a distance of about 1 m for the stationary target, while for tracking moving target data are collected from a subject walking back and forth with 200-cm deviation from 100 cm away from the antenna. As described in Fig. 5.25, to preserve the relatively large DC information level while sufficiently amplifying the weak time-varying heart-related signal without saturating neither preamplifiers nor ADC, two serially connected preamplifiers are employed. First amplifier has gain of 50 times from DC to 10 Hz in order to preserve DC information, while second amplifier further amplifies by 40 times from 0.03 to 10 Hz to provide more SNR to small cardiac signal. Each output is digitized with an ADC card and saved in data acquisition system. Subsequently, those two outputs are combined together in MATLAB after multiplication of DC-included signal by 40 times to compensate amplification difference between both outputs. At the same time, the AC-only signal was filtered with an FIR, which has linear phase delay, flat-top filter that passed frequencies between 0.8 and 10 Hz to eliminate the still-detectable low-frequency component due to respiration and thus avoid including this effect twice when summing with the DC-included signal. Consequently, high heart-related signal power with DC information can be obtained. These reconstructed DC-included signals still require more signal processing to exclude DC offset caused by either clutter or leakage LO power in the system. As explained at the end of Section 5.2, chest motion from a target forms arc in the complex plot that is centered away from the origin by the amount of DC offset. Center estimation is done before arctangent demodulation. First 3 s of data is used for estimating center of arc, which can be one cycle of respiration and can form enough arc length. Finally, quadrature signals that form arc centered at origin in complex plot are combined by using arctangent demodulation. Demodulated output is then digitally filtered by a flat-top filter with frequency range of 0.8–10 Hz to obtain heart signal. Heart rates are extracted in real time with custom software based on an autocorrelation algorithm described in Lohman et al. [2001], and heart rate was compared with that obtained from a wired finger pressure pulse sensor (UFI 1010) used as a reference. In addition, walking subject's movement tracking measurement has also been done with same arctangent demodulation method explained earlier. However, in this case, since phase variation caused by a target's motion is much bigger than 2π or half wavelength, which is 6.25 cm at 2.4 GHz, arctangent-demodulated output need to be unwrapped, and complex plot is no more a small fraction of the circle but spiral-like shape, which has the same center point. This is to be expected, because DC offset caused by clutter or leaking within the device is fixed while receiving signal power, which corresponds to the radius of the complex signal circle varies associated with a target's distance from the antenna. Figure 5.26 shows the I, Q, and arctangent-demodulated signals

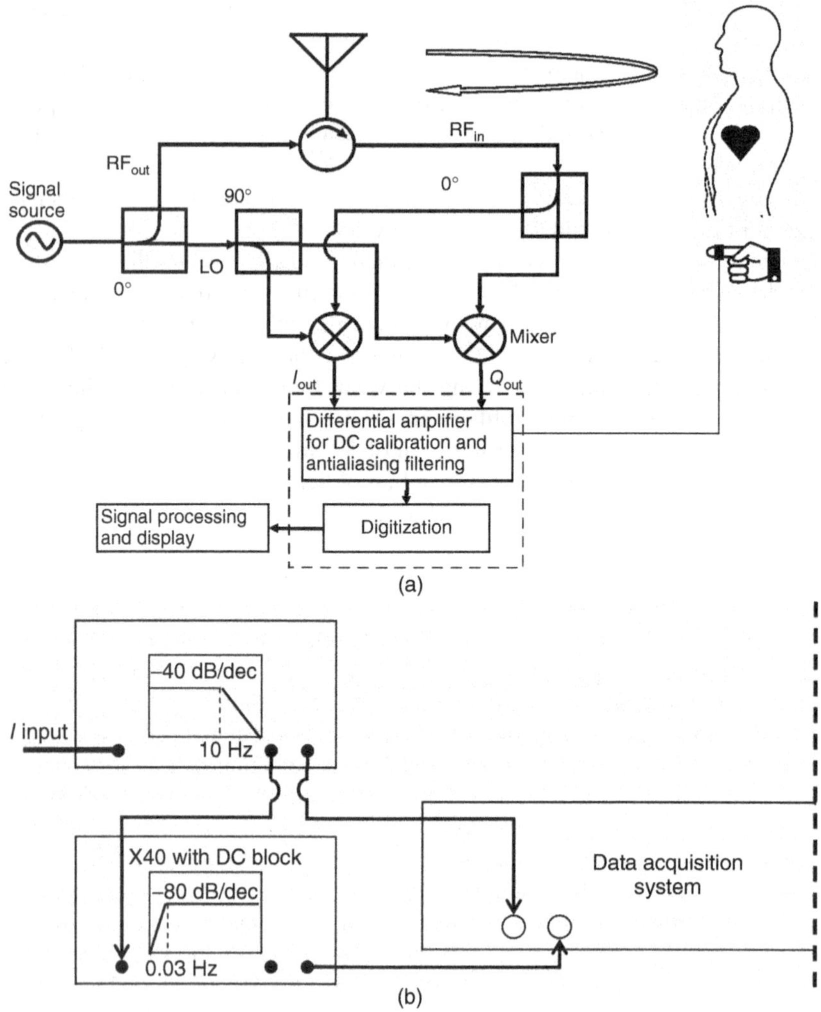

Figure 5.25 Block diagram of a quadrature direct conversion Doppler radar system in a measurement setup for heart rate extraction (a), with data acquisition (dashed region in (a)) details provided for the I channel (the Q channel is exactly the same) (b). Two stages of preamplifiers are used to obtain high power of heart signal without losing DC information. First preamplifier is for obtaining DC signal as well as antialiasing filtering, then second amplifier is AC coupled, thus only the chest motion signal is amplified and digitized with maximum ADC resolution. © 2007 IEEE, Reprinted, with permission, from Park et al. [2007a].

for a stationary subject seated at 1 m away from the antenna, obtained using the measurement setup shown in Fig. 5.1 with center-tracking method. In this case, subject is null position for Q channel, resulting in heart rate modulated by respiration [Park et al., 2007a], while I channel is in optimum position resulting in higher detection accuracy. While I and Q channel outputs show standard deviation of 1.7 or 5.1 beats,

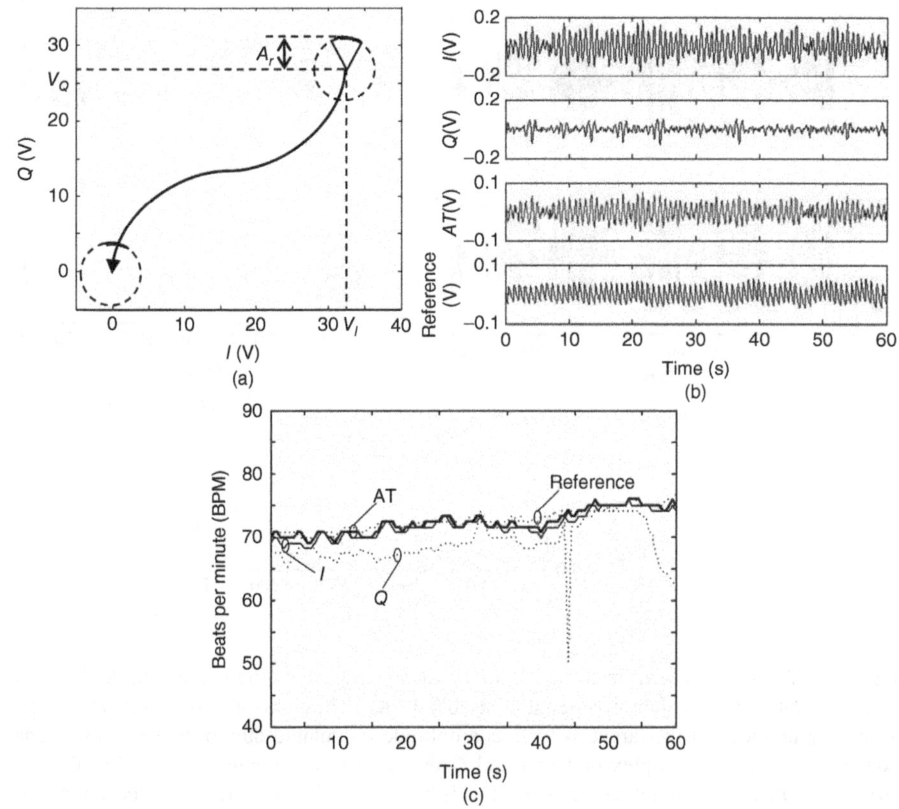

Figure 5.26 *I*, *Q*, and arctangent-demodulated signals measured for a position where the *Q* channel is close to a null condition. Arc formed by respiration motion of the chest tracked back to the origin in order to eliminate DC offset (a). Digitally band-pass-filtered data extracts heart signal from a raw data (b). The *Q* channel rate data (c) shows dropout regions when the SNR is insufficient for digitization, excluding dropouts, the *I* and *Q* channel data have an error of 1.7 or 5.1 beats, respectively, over the same 60-s interval where the arctangent output has an error of only 1.3 beats. © 2007 IEEE, Reprinted, with permission, from Park et al. [2007a].

respectively, during the 60-s time interval, arctangent combined output has an error of only 1.3 beats. This experiment was repeated at several difference target positions, with arctangent demodulation output consistently providing higher accuracy then either channel by itself. Figure 5.27 shows a walking subject's movement tracking outputs obtained by using arctangent demodulation with center tracking. For this measurement, subject was walking back and forth within 200 cm distance aligned with the antenna beam. As expected, complex plot forms spiral-like shape, due to the received signal power variation (Fig. 5.27(b)). Arctangent output is phase information, which is linearly proportional to the actual distance variation that was converted to distance using $\lambda/4\pi$ multiplication as indicated by Equation 5.27. As shown is Fig. 5.27(c), calculated subject displacement was 200 cm as expected.

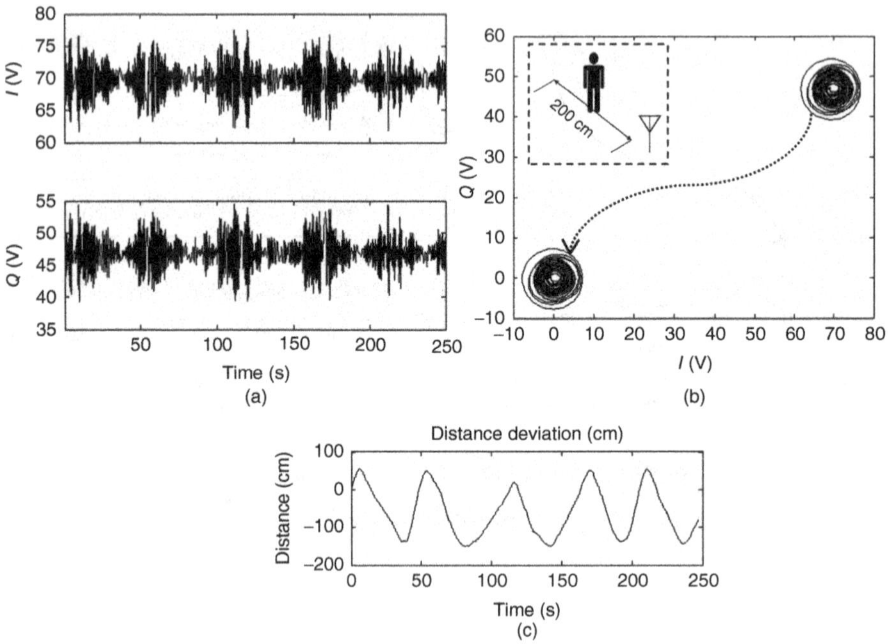

Figure 5.27 Measurement result of a target's movement. Since the movement deviation of 200 cm is much bigger than wavelength, baseband *I* and *Q* outputs are frequency modulated according to speed of the target as well as amplitude modulated due to the receiving signal power variation (a). Complex plot of *I* and *Q* outputs forms complete circle with different radius but same center point, and center offset is brought back to the origin in order to remove DC offset (b). Arctangent demodulation output can restore actual movement of a target by simply unwrapping output to compensate 2π singular effect (c). © 2007 IEEE, Reprinted, with permission, from Park et al. [2007a].

5.3.3 DC Cancellation Results

Another method, which is AC coupling, removes all static information thus also removing DC information, making it difficult for arctangent demodulation to properly extract phase information [Vergara et al., 2008]. A method to remove hardware and environmental DC offset while preserving DC information is the use of a digitally controlled voltage source and a real-time center-find function as shown in Fig. 5.28.

The digitally controlled voltage source, in this case a DAC, provides compensation for DC components when used in difference amplification with the input baseband signals. When the total DC offset for both *I* and *Q* is estimated and cancelled, small time-varying signal can be amplified without saturating the amplifiers or the DAQ system. When an appropriate period of acquisition is performed, any periodic movement will show as an arc on the *IQ* plot. A real-time center find algorithm can analyze this arc and provide the radius and a relative center point. By correcting for the error in the DAC output and the relative center of the arc, the arc center can be re-positioned to the origin on the *IQ* plot. V_I and V_Q have now been totally removed and the only

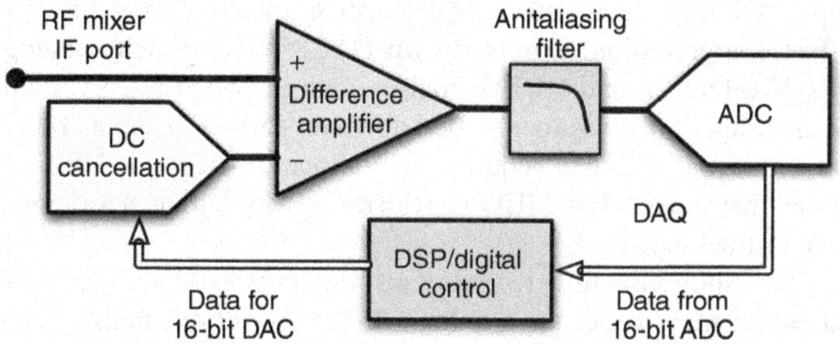

Figure 5.28 Block diagram of adaptive DC offset compensation. DC cancellation includes DAC and voltage divider to scale voltage to ±50 mV range. DSP includes center find function for DC information preservation. © 2008 IEEE, Reprinted, with permission, from Vergara et al. [2008].

information being acquired relates to the object in oscillatory motion. Phase information extraction is now a simple ratio of I and Q without any other signal processing.

For a simple case of a single subject sitting still a meter away from the antenna system, the only periodic motion source will be the chest. There may be small sources of signal on the arms and neck due to vascular movement. However, for the case of simplicity, we only consider the front surface of the chest as the source of both heart and respiration activity.

Figure 5.29 shows the setup in which the subject has a piezoelectric thumb sensor for pulse measurements and an airflow device for volumetric measurements provided

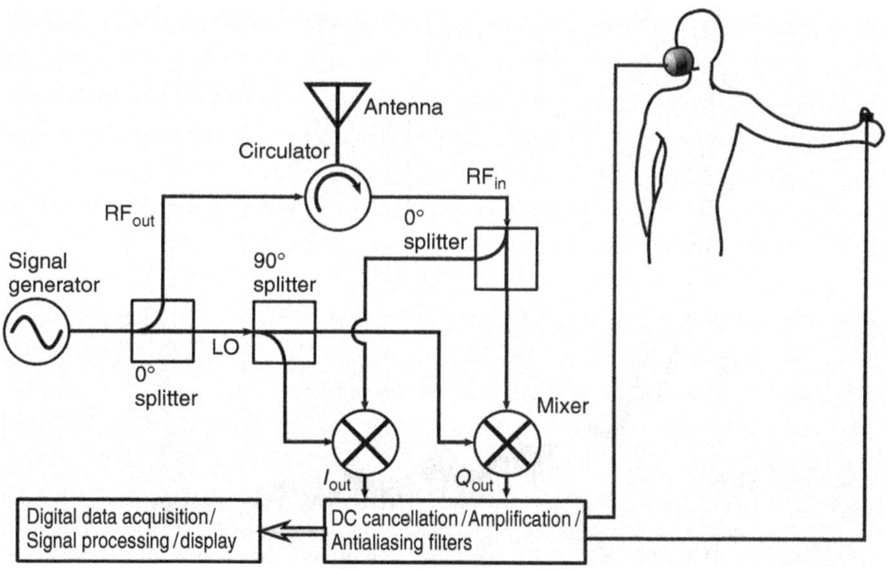

Figure 5.29 Diagram of test setup measuring respiration and pulse using quadrature direct conversion Doppler radar, airflow rate, and finger pressure transducers. © 2008 IEEE, Reprinted, with permission, from Vergara et al. [2008].

by the Biopac TSD117 Pneumotach Transducer in the MP150 acquisition system. For DC offset compensation, a Burr-Brown DAC8552 DAC with analog devices' ADR425 5.0 V reference and OP727 dual-op-amp provided a ±5 V digitally controlled voltage range. This was then scaled down to ±50 mV using a voltage divider. Radar baseband signal was then acquired by National Instruments USB-6259 16-bit DAQ. National Instruments LabVIEW provided data acquisition, a user interface, and real-time center tracking.

Figure 5.30(a) shows the time record of a sitting person on an *IQ* plot. This was a record of about 2.5 min. Notice the arc shape drifting over time. Figure 5.30(b) shows

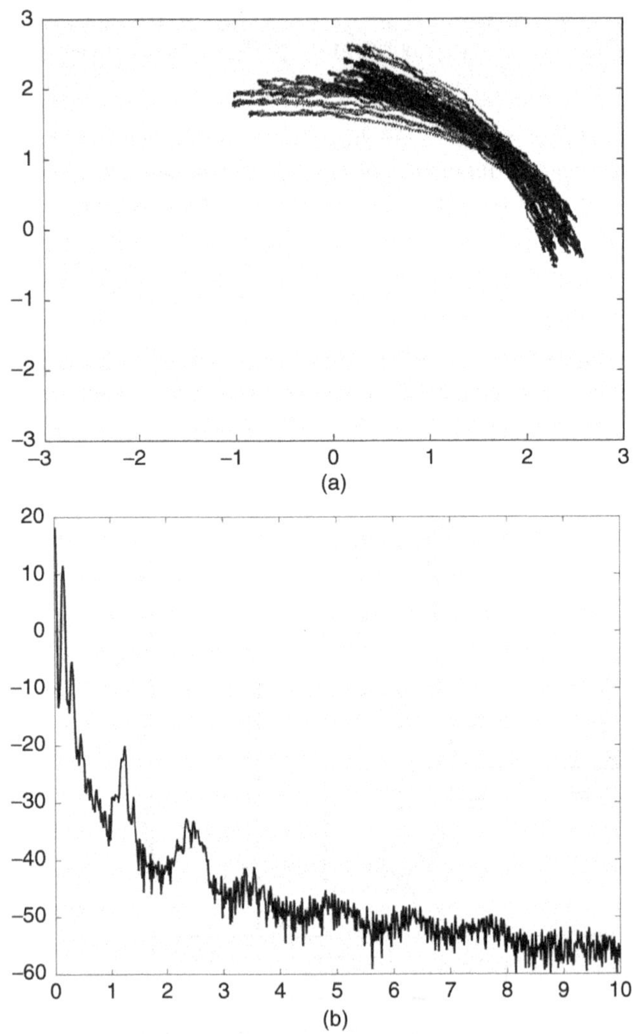

Figure 5.30 Time (a) and power spectral density (b) of Doppler radar off a human subject. © 2008 IEEE, Reprinted, with permission, from Vergara et al. [2008].

the power spectral density of the I channel with respiration rate at about 0.2 Hz and heart rate around 1.3 Hz. Also, notice the 15 dB difference between respiration and heart signals.

In Fig. 5.31, comparison between the AC coupling and the DC offset compensation is shown. The AC-coupled signal amplifier has a preamplifier high-pass filter (HPF) of 0.3 Hz and a postamplifier HPF of 0.1 Hz. The AC-coupled signal also requires additional processing before arctangent demodulation. This additional processing includes filter compensation and a re-introduction of DC information to the both I and Q. This is necessary to properly perform arctangent demodulation on the AC coupled signal. Heart rate was estimated utilizing an autocorrelation function. The black dotted line shows the measured heart rate in beats per minute (BPM) by using the finger pulse transducer and was used as a baseline reference. DC offset compensation in black follows the basic pattern of changes in the heart rate. The AC-coupled heart rate as a gray line also follows some of the trends in heart rate. However, there are some sections where the loss of the DC information due to AC coupling makes it difficult for arctangent demodulation. By comparing the heart rate over time with the reference, the DC offset compensation seems to track better than the AC coupling.

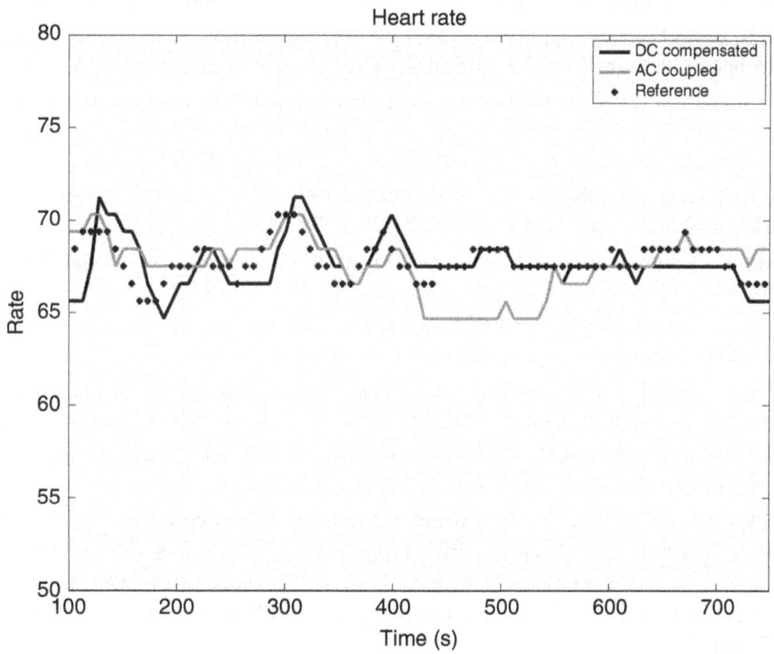

Figure 5.31 Extracted heart rate from AC-coupled and DC offset compensation signals with comparison to a wired finger pulse. Both utilized arctangent demodulation. © 2008 IEEE, Reprinted, with permission, from Vergara et al. [2008].

REFERENCES

Abidi AA. Direct-conversion radio transceiver for digital communications. IEEE J Solid-State Circuits 1995;30(12):1399–1410.

Chen H, Chen J, Huang P. Adaptive I/Q imbalance compensation for RF transceivers. IEEE Global Telecommunications Conference; 2004. Vol. 2, p 818–822.

Droitcour AD, Boric-Lubecke O, Lubecke VM, Lin J, Kovacs GT. Range correlation and I/Q performance benefits in single chip silicon Doppler radars for non-contact cardiopulmonary monitoring. IEEE Trans Microwave Theory Tech 2004;52(3):838–848.

Finol JL, Buchholz M, Design of an in phase and quadrature phase and amplitude imbalance compensation in quadrature receivers. IEEE International Caracas Conference; 2004. Vol. 1, p 254–258.

Furuta Y, Heima T, Sato H, Shimizu T. A low flicker-noise direct conversion mixer in 0.13 μm CMOS with dual-mode DC offset cancellation circuits. 2007 Topical Meeting on Silicon Monolithic Integrated Circuits in RF Systems; 2007 Jan 10–12. p 265–268.

Huang X. On transmitter gain/phase imbalance compensation at receiver. IEEE Commun Lett 2000;4:363–365.

Huang X, Caron M. Gain/phase imbalance and DC offset compensation in quadrature modulators. IEEE International Symposium on Circuits and Systems; 2002. Vol. 6, p 76–86.

Kim C-Y, Kim JG, Oum JH, Yang JR, Kim D-K, Choi JH, Kwon S-W, Jeon S-H, Park J-W, Hong S, Tx leakage cancellers for 24 GHz and 77 GHz vehicular radar applications. IEEE MTT-S International Microwave Symposium Digest, 2006; 2006 Jun. p 1402–1405.

Lohman B, Boric-Lubecke O, Lubecke VM, Ong PW, Sondhi MM. A digital signal processor for Doppler radar sensing of vital signs. IEEE Eng Med Biol Mag 2002;21(5):161–164.

Lubecke VM, Boric-Lubecke O, Awater G, Ong P-W, Gammel P, Yan R-H, Lin JC, Remote sensing of vital signs with telecommunications signals. World Congress on Medical Physics and Biomedical Engineering (WC 2000), Chicago, IL, USA; 2000 Jul.

Margraf M, Boeck G. Analysis and modeling in low-frequency noise in resistive FET mixers. IEEE Trans Microwave Theory Tech 2004;52(7):1709–1718.

Mashhour A, Domino W, Beamish N. On the direct conversion receiver – a tutorial. Microwave J 2001;44(6):114–128.

Matinpour B, Laskar J. A compact direct-conversion receiver for C-band wireless applications. IEEE RFIC Symposium Digest; 1999. p 25–28.

Moraes R, Evans DH. Compensation for phase and amplitude imbalance in quadrature Doppler signals. Ultrasound Med Biol 1996;22:129–137.

Nakagawa T, Matsui M, Araki K. Gain/phase imbalance compensation for multi-band quadrature receivers. IEEE VTC; 2004. Vol. 3, p 2034–2037.

Noon DA, Longstaff ID, Stickley GF. Wideband quadrature error correction (using SVD) for stepped-frequency radar receivers. IEEE Trans Aerosp Electron Syst 1999;35:1444–1449.

Park B-K, Boric-Lubecke O, Lubecke VM. Arctangent demodulation with DC offset compensation in quadrature Doppler radar receiver systems. IEEE Trans Microwave Theory Tech 2007a;55(5):1073–1079.

Park B-K, Yamada S, Lubecke VM. Measurement method for imbalance factors in direct-conversion quadrature radar systems. IEEE Microwave Wireless Compon Lett 2007b;17:403–405.

Park B-K, Yamada S, Lubecke VM, Boric-Lubecke O. Single-channel receiver limitations in Doppler radar measurements of periodic motion. IEEE Radio and Wireless Symposium, San Diego, CA, USA; 2006. p 99–102.

Razavi B. Design considerations for direct-conversion receivers. IEEE Trans Circuits Syst II 1997;44:428–435.

Redman-White W, Leenaerts DMW. 1/f noise in passive CMOS mixers for low and zero IF integrated receivers. Proceedings of the 27th European Solid-State Circuits Conference, 2001 (ESSCIRC 2001); 2001 Sept 18–20. p 41–44.

Svitec R, Raman S. DC offsets in direct-conversion receivers: characterization and implications. IEEE Microwave Mag 2005;6:76–86.

Umstattd R. Operating and evaluating quadrature modulators for personal communication systems. National Semiconductor Application Note 899; 1993.

Vergara AM, Boric-Lubecke O, Lubecke VM. DC information preservation for cardiopulmonary monitor utilizing CW Doppler radar. 30th Annual Conference of the IEEE Engineering in Medicine and Biology Society; 2008 Aug. p 1246–1249.

Xiao Y, Lin J, Boric-Lubecke O, Lubecke VM. Frequency tuning technique for remote detection of heartbeat and respiration using low-power double-sideband transmission in Ka-band. IEEE Trans Microwave Theory Tech 2006;54(5):2023–2032.

Yamada S, Boric-Lubecke O, Lubecke VM. Cancellation techniques for LO leakage and DC offset in direct conversion systems. 2008 IEEE MTT-S International Microwave Symposium Digest; 2008 Jun. p 1191–1194.

Zhu Z, Huang X. Adaptive compensation of gain/phase imbalances and DC-offsets using constant modulus algorithm. IEEE International Conference on Acoustics, Speech, and Signal Processing, 2004, Proceedings (ICASSP '04); 2004 May 17–21. Vol. 4.

Zhu Z, Huang X, Bias analysis of a gain/phase/DC-offset estimation technique for direct frequency conversion modulators. IEEE International Conference on Acoustics, Speech, and Signal Processing; 2005. Vol. 3, p 18–23.

6

SOURCES OF NOISE AND SIGNAL-TO-NOISE RATIO

Amy D. Droitcour[1], Olga Boric-Lubecke[2], and Shuhei Yamada[2]

[1]*Wave 80 Biosciences, Inc., San Francisco, California, United States*
[2]*Department of Electrical Engineering, University of Hawaii at Manoa, Honolulu, Hawaii, United States*

The signal-to-noise ratio (SNR) of the continuous-wave (CW) Doppler radar system for physiological monitoring is derived in this chapter. This derivation can be used to assess the theoretical limits of the radar system and to determine the factors that affect the limits so that design decisions can be made appropriately. The radar equation is used to estimate the received power, taking into account the range to the target, the transmitted power, the radar cross section (RCS), the antenna gain, the wavelength, and the range. The amplitude of the signal at baseband depends on the received power, the mixer's conversion loss, and the amount of phase modulation on the received signal. The amount of the signal modulated is determined by the amount of physiological motion in the direction of the radar transceiver. Noise sources include radio frequency (RF) phase noise from the oscillator, environmental thermal noise, and baseband $1/f$ noise of the mixer and of the baseband signal-conditioning circuits. The SNR depends on the amount of noise at the mixer output from each of these sources. The variation of SNR with range, RCS, and the amount of physiological motion is analyzed. The effects of near-field operation on the antenna gain are also described.

Doppler Radar Physiological Sensing, First Edition.
Edited by Olga Boric-Lubecke, Victor M. Lubecke, Amy D. Droitcour, Byung-Kwon Park, and Aditya Singh.

6.1 SIGNAL POWER, RADAR EQUATION, AND RADAR CROSS SECTION

6.1.1 Radar Equation

The radar equation, as described in Chapter 2, is used to estimate the received signal power in a radar system, helping to determine the system's theoretical limits. The estimated received power is based on the transmitted power, the range to the target, and the properties of the transmit antenna, the target, and the receive antenna. When measuring motion due to heart and respiration with a Doppler radar transceiver, in some cases the residual phase noise will be the limiting factor; otherwise the limiting factor will usually be receiver sensitivity and the received signal power.

To estimate the received signal power, it is necessary to determine how much power is lost and gained at various steps between the transmitter and the receiver. To calculate the received power in a radar system, it is necessary to first calculate the transmitted power in the direction of the target. Antennas are generally measured by their gain, G, which is the ratio of the power radiated toward the center of the target by that antenna to the power radiated in all directions by an isotropic antenna. The power radiated in the desired direction is known as the effective radiated power (ERP) and is equal to the gain multiplied by the transmitted power P_T:

$$\text{ERP} = GP_T \tag{6.1}$$

The power density at a range R from a nonisotropic antenna is the ERP in that direction divided by the surface area of a sphere with radius R, with attenuation due to atmospheric absorption α:

$$P_D = \frac{P_T G_T e^{-\alpha R}}{4\pi R^2} \tag{6.2}$$

The radar target intercepts a portion of the radiated power and reflects it, partially in the direction of the radar receiving antenna. The RCS, σ, is determined by the amount of power incident on the target that is re-radiated toward the antenna. The RCS is not the same as the physical cross section; it depends on the electrical properties of the material and its three-dimensional shape. The signal power reflected from the target is

$$P_{\text{reflected}} = \frac{P_T G_T \sigma e^{-\alpha R}}{4\pi R^2} \tag{6.3}$$

This reflected signal then spreads out in space similar to the transmitted signal. If the receiving antenna is colocated with the transmitting antenna, the power density just before the receiving antenna is the reflected power divided by the surface area of a sphere with radius R:

$$P_{D,\text{receiver}} = \frac{P_T G_T \sigma e^{-2\alpha R}}{(4\pi R^2)^2} \tag{6.4}$$

The receiving antenna is traditionally described by its effective area, $A_{e,R}$, which determines what portion of the radiated energy it can capture. The power received is equal to the power density at the antenna, multiplied by the effective capture area, A_e, of the receiving antenna:

$$P_R = \frac{P_T G_T \sigma A_{e,R} e^{-2\alpha R}}{(4\pi R^2)^2} \tag{6.5}$$

This equation can also be written in terms of the receiving antenna's gain. The relationship between the receiving antenna's gain and effective area is

$$G_R = \frac{4\pi A_{e,R}}{\lambda^2} \tag{6.6}$$

where λ is the wavelength of the RF signal. When this is substituted into Equation 6.5, the resulting expression for the received power is

$$P_R = \frac{P_T G_T \sigma \lambda^2 G_R e^{-2\alpha R}}{(4\pi R^2)^2 (4\pi)} = \frac{P_T G_T G_R \sigma \lambda^2 e^{-2\alpha R}}{(4\pi)^3 R^4} \tag{6.7}$$

This final term is known as the radar equation. When the same antenna is used for transmitting and receiving, the gain is the same for both antennas (Fig. 6.1), and the equation can be simplified to

$$P_R = \frac{P_T G^2 \sigma \lambda^2 e^{-2\alpha R}}{(4\pi)^3 R^4} \tag{6.8}$$

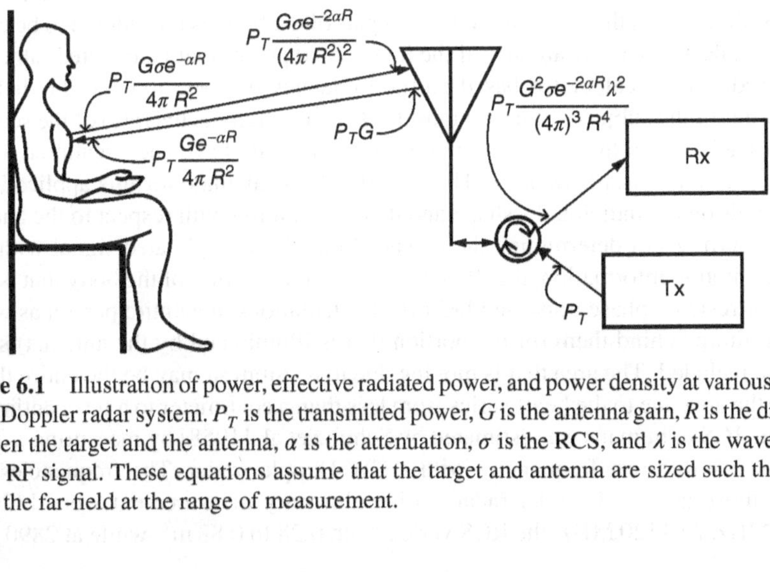

Figure 6.1 Illustration of power, effective radiated power, and power density at various points in the Doppler radar system. P_T is the transmitted power, G is the antenna gain, R is the distance between the target and the antenna, α is the attenuation, σ is the RCS, and λ is the wavelength of the RF signal. These equations assume that the target and antenna are sized such that they are in the far-field at the range of measurement.

For the typical physiological monitoring scenario, the operation is through air indoors and distances are short, such that the attenuation term can be dropped:

$$P_R = \frac{P_T G^2 \sigma \lambda^2}{(4\pi)^3 R^4} \tag{6.9}$$

For radar monitoring of physiological motion, the RCS is the only value that is not clearly defined. This is discussed in more detail in the following section. The radar equation assumes that the target is in the antenna far-field. This is not necessarily true for all cases, especially at very close ranges and for respiratory motion, where the target is larger compared with the range than it is for the heart motion. Effects of near field are described in Chapter 8.

6.1.2 Radar Cross Section

The RCS is a measure of how well the target reflects radar signals in the direction of the radar receiver. It can be described as the ratio of the strength of the reflected signal from the target to the reflected signal from a perfectly smooth and perfectly conducting sphere with a 1 m^2 cross-sectional area [Electronic Warfare and Radar Systems Engineering Handbook, 1992]. It is often described as

$$\sigma = (\text{Projected cross-section}) \times (\text{Reflectivity}) \times (\text{Directivity}) \tag{6.10}$$

When measuring a person, calculation of the projected cross section requires including the whole person as well as the bed or chair on which they are sitting. If the entire person (as well as bed or chair) is not in the beam of the antenna, then only the illuminated portion should be considered. If this is the case, the target is in the near-field. However, for physiological motion measurement, the size of the area in motion determines the RCS; the stationary part of the body is considered to be clutter.

The reflectivity is the amount of the intercepted power that is reflected rather than absorbed. This is calculated based on the frequency of operation and the dielectric properties of the subject's skin and muscle. The directivity is the ratio of the scattered power back toward the antenna to the power that would have back-scattered had the target been an isotropic radiator. This is difficult to calculate for this application, as it depends on the individual's shape and their orientation with respect to the antenna.

For purposes of determining the power of the phase-modulated signal with heart and respiration information, the RCS depends on the fraction of the body that is moving. For residual phase noise and DC offset calculations, the entire person as well as the furniture behind them (or the portion that is illuminated by the antenna) should also be included. The area that is moving due to respiration may be the entire thorax, while the area due to the heart varies from less than a centimeter to a few centimeters.

The RCS of humans was estimated by Schultz et al. [1958] by measuring a 200-lb man at five different frequencies with a CW Doppler radar. The frequencies closest to those used in Doppler radar cardiopulmonary measurement were 1120 and 2890 MHz. At 1120 MHz, the RCS varied from 0.28 to 0.88 m^2, while at 2890 MHz, the RCS varied from 0.20 to 0.72 m^2, depending on the aspect of the subject relative to the antenna and the antenna polarization. With the radar facing the front or back of the subject, at 1120 MHz, the RCS was 0.72 m^2 for horizontal polarization and 0.88 m^2 for vertical polarization. At 2890 MHz, when facing the front, the RCS was 0.41 m^2 for horizontal polarization and 0.50 m^2 for vertical polarization, and when facing the back the RCS was 0.61 m^2 for horizontal polarization and 0.72 m^2 for vertical polarization. Wu [1989] computed the RCS of a human model using a muscle hemisphere head, a cylindrical neck, and a conical torso. He found that with horizontal polarization, the RCS was $15\,\text{dB} < (\text{RCS})/\lambda^2 < 28\,\text{dB}$ and with vertical polarization $0\,\text{dB} < (\text{RCS})/\lambda^2 < 20\,\text{dB}$, where λ represents the wavelength. The variations again occur with respect to the angle to the body. Facing the front of the body, Wu found $(\text{RCS})/\lambda^2 = 14\,\text{dB}$ for both horizontal and vertical polarizations. At 2.4 GHz, this would indicate an RCS area of 0.39 m^2 closely agreeing with Schultz et al. [1958].

Although predictions and measurements of the Doppler signal from walking humans have been made [Geisheimer et al., 2002; van Dorp and Groen, 2003], they do not show measured values of the RCSs of each body part. Individuals can vary greatly in size and, therefore, they will also vary greatly in RCS. In addition, when determining the RCS for measurements of heart and respiration, it is necessary to determine the area of the body moving with pulse and breathing in the direction of the radar, which can also vary greatly from person to person.

6.1.3 Reflection and Absorption

The electrical properties of the biological tissue affect how much of the signal is reflected and transmitted, both at the skin–air interface and at interfaces between different tissues within the body. Of the radiation that enters the body, the electrical properties determine how much of it is attenuated per unit distance, how much is transmitted to the next layer, and how much is reflected back toward the skin surface. Biological tissue is nonmagnetic; therefore, its permeability, μ, is nearly identical to that of free space. The dielectric constant, representing the material's permittivity, ε, and the conductivity, σ, are the two electrical properties that primarily define the electrical characteristics of the biological tissue, as discussed in detail in Chapter 4.

6.1.4 Phase-to-Amplitude Conversion

To determine the signal power at baseband, the received RF signal power, the receiver loss or gain, the mixer conversion loss or gain, and the amount of phase modulation must all be considered. The RF signal power can be determined with the radar equation, as described in the previous section. The conversion of the phase-modulated signal to a baseband signal follows calculations used in communication phase-modulation link equations [Lathi, 1998].

The signal at the local oscillator (LO) input to the mixer is

$$L(t) = A_{\text{LO}} \cos(2\pi f t) \tag{6.11}$$

and the signal at the RF input to the mixer is

$$R(t) = A_{\mathrm{RF}}\sqrt{G_{\mathrm{Rx}}}\cos(2\pi ft + \psi(t) + \theta) \tag{6.12}$$

where $\psi(t)$ is the phase modulation of the signal and θ is a constant relative phase shift between the two signals. A_{RF} and A_{LO} are the amplitudes of the signal and LO, and G_{Rx} is the gain or loss between the antenna and the mixer's RF input. When the LO and RF signals are mixed, after low-pass filtering, the output is

$$B(t) = \sqrt{G_{\mathrm{CL}}G_{\mathrm{Rx}}}A_{\mathrm{RF}}\cos(\psi(t) + \theta) \tag{6.13}$$

where G_{CL} is the conversion gain of the mixer (power gain), representing the ratio of the IF output power to the RF input power when signals are mixed.

The signal power at baseband is

$$S_B = \frac{\overline{B^2(t)}}{Z} = \frac{G_{\mathrm{CL}}G_{\mathrm{Rx}}A_{\mathrm{RF}}^2\overline{(\cos(\psi(t) + \theta))^2}}{Z} \tag{6.14}$$

To determine baseband signal power, the received power in Equation 6.14 must be converted into the voltage amplitude at the RF input to the mixer, A_{RF}. The signal power at the input of the receiver is equal to the mean-squared received voltage divided by the input impedance, Z:

$$P_R = \frac{\overline{R^2(t)}}{Z} \tag{6.15}$$

Since the RF signal is a phase-modulated sinusoid,

$$\overline{R^2(t)} = \frac{A_{\mathrm{RF}}^2}{2} \tag{6.16}$$

and

$$A_{\mathrm{RF}}^2 = 2P_R Z \tag{6.17}$$

Plugging in Equation 6.9 for P_R, the squared amplitude is

$$A_{\mathrm{RF}}^2 = \frac{2P_T G^2 \sigma \lambda^2 Z}{(4\pi)^3 R^4} \tag{6.18}$$

Therefore, the baseband signal power is

$$S_B = \frac{2P_T G_{\mathrm{CL}} G_{\mathrm{Rx}} G^2 \sigma \lambda^2 \overline{(\cos(\psi(t) + \theta))^2}}{(4\pi)^3 R^4} \tag{6.19}$$

For Doppler radar cardiopulmonary monitoring,

$$\psi(t) = \frac{4\pi}{\lambda} x(t) \tag{6.20}$$

where $x(t)$ is the physiological motion in the direction of the antenna, so the power at the output of the mixer is

$$S_B = \frac{2P_T G_{\mathrm{CL}} G_{\mathrm{Rx}} G^2 \sigma \lambda^2 \overline{\left(\cos\left(\frac{4\pi}{\lambda} x(t) + \theta\right)\right)^2}}{(4\pi)^3 R^4} \tag{6.21}$$

If the value of θ is such that the signal is at an optimal phase demodulation point, the small-angle approximation applies and the baseband output is

$$B(t) \approx \sqrt{G_{\mathrm{CL}} G_{\mathrm{Rx}}} A_{\mathrm{RF}} \frac{4\pi}{\lambda} x(t) \tag{6.22}$$

and the signal power at baseband is

$$S_B = \frac{2P_T G_{\mathrm{CL}} G_{\mathrm{Rx}} G^2 \sigma \lambda^2 \left(\cos\left(\frac{4\pi}{\lambda} x(t) + \theta\right)\right)^2}{(4\pi)^3 R^4} \tag{6.23}$$

6.2 OSCILLATOR PHASE NOISE, RANGE CORRELATION AND RESIDUAL PHASE NOISE

6.2.1 Oscillator Phase Noise

The ideal oscillator for a CW radar would be a perfect sinusoid, with amplitude A and frequency f, so that the signal, $s(t)$ is

$$s(t) = A \sin(2\pi f t) \tag{6.24}$$

However, all real oscillators have noise, in both phase and amplitude, which makes the signal

$$s(t) = (A + a(t)) \sin(2\pi f t + \phi(t)) \tag{6.25}$$

where $a(t)$ is the amplitude noise and $\varphi(t)$ is the phase noise. The amplitude noise does not affect the signal at its zero crossing, and the phase noise does not affect the amplitude at the peaks. Since all practical oscillators have some type of amplitude limiting [Lee and Hajimiri, 2000], the amplitude noise from the oscillator is usually negligible compared with the phase noise so that the signal is effectively

$$s(t) = A \sin(2\pi f t + \phi(t)) \tag{6.26}$$

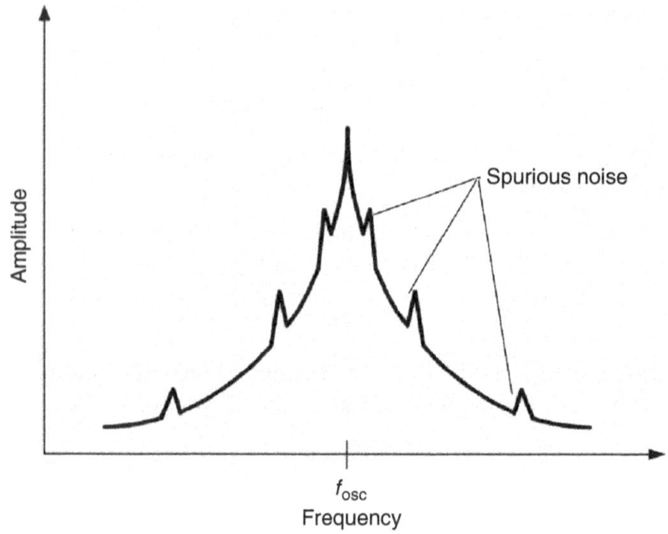

Figure 6.2 RF sideband spectrum, including phase noise and spurious noise. The phase noise spectrum is symmetrical about the oscillation frequency, indicating that phase noise, and not amplitude noise, is dominant in this oscillator. The peaks in the spectrum are spurious noise, indicating modulation by other signals.

Frequency stability is the degree to which an oscillator produces the same frequency over time. All real sources have some variability in frequency. Fluctuations in frequency are due to spurious and phase noise. Spurious noise is caused by signals that modulate the signal frequency, and these appear as discrete components in spectral density plots. Phase noise is random, caused by thermal noise, shot noise, and Flicker noise. A sample spectral density plot is shown in Fig. 6.2. The ideal oscillator's spectrum would be a delta function at f_{osc}.

The signals in Equations 6.24 and 6.26 are compared in the time domain in Fig. 6.3. In this figure, Gaussian white noise with zero mean and a variance of 0.0625 was added to the phase of a 1-Hz sinusoid. Note that although the zero-crossings change, the amplitude of the signal does not vary.

All methods to quantify phase noise measure frequency or phase deviation of the source in either the frequency or the time domain. The most common measurement is spectral density of phase fluctuations per hertz, $S_\phi(f_o)$. This describes the energy distribution as a continuous function in units of radian variance per unit bandwidth:

$$S_\phi(f_o) = \frac{\Delta\phi^2_{\text{RMS}}}{\text{bandwidth used to measure phase deviation}} = \frac{\text{rad}^2}{\text{Hz}} \tag{6.27}$$

The spectral density is given in units of square radians per hertz or in decibels relative to 1 rad^2/Hz.

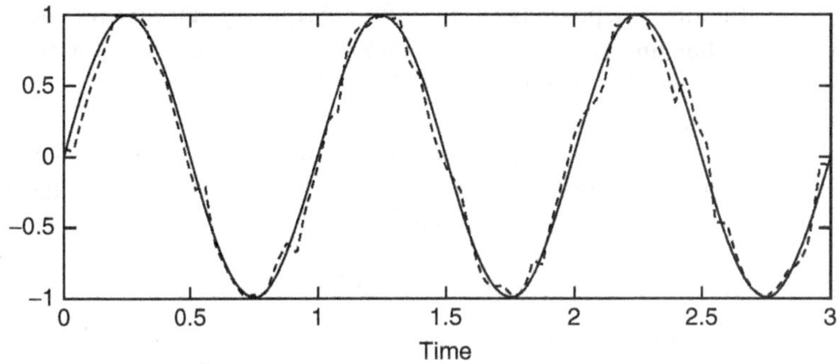

Figure 6.3 Exaggerated depiction of phase noise in the time domain. The solid line is the perfect sinusoid in Equation 6.24 and the dotted line is the sinusoid with phase noise in Equation 6.26.

Another common measure of phase noise is the single-sideband spectrum. This is the ratio of the power at an offset of f_o (Hz) from the carrier to the signal power, a measurement of the noise energy. As shown in Fig. 6.4, oscillator single-sideband phase noise, $L_\phi(f_o)$, is defined as the ratio of the power in a 1-Hz bandwidth at an offset frequency f_o from the carrier frequency to the total carrier power.

$$L_\phi(f_o) = 10 \log \left(\frac{\text{power density (in one sideband)}}{\text{total signal power}} \right) = 10 \log \left(\frac{P_{\text{ssb}}}{P_s} \right) \quad (6.28)$$

It is usually expressed in units decibels below the carrier per hertz (dBc/Hz) at a specific offset. Often round numbers such as 1, 10 kHz, or 1 MHz are used for the offset,

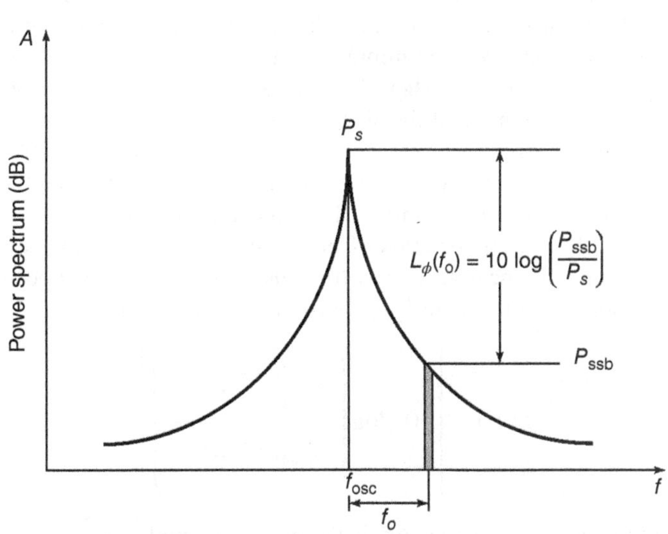

Figure 6.4 Measurement of single-sideband phase noise, $L_\phi(f_o)$.

with the actual number depending on the offset frequency relevant for the application. However, when another frequency is important for the application, the specific frequency will be used; for example, the DCS1800 cellular base station specification is given at 600 kHz.

When the total phase deviation is much less than a radian so that the small-angle approximation applies, the relationship between the spectral density of phase fluctuations and the single-sideband phase noise is

$$S_\phi(f_o) = 2L_\phi(f_o) \tag{6.29}$$

However, at small offset frequencies on free-running noisy oscillators, the phase deviation can be near to or greater than a radian, and then this relation does not apply. In this case, the single-sideband phase noise flattens, while the phase fluctuation spectral density can increase to more than 0 dB/Hz. Phase noise more than 0 dBc/Hz in noisy oscillators indicates that the carrier frequency is wandering over a frequency range, and the spectrum should flatten out when the small-angle approximation no longer applies, indicating a wide spectral line due to frequency variation of the carrier. It is not correct to have a single-sideband phase noise value greater than 0 dBc/Hz, since the noise cannot be greater than the carrier. The carrier can be considered to have wider bandwidth to frequency variation of the carrier.

The variation in the zero crossings is also sometimes referred to as jitter. Jitter is usually defined looking in the time domain at timing accuracy, while phase noise involves looking at the noise spectrum in the frequency domain.

The noise-to-phase transfer function is linear and time-varying. The transfer function is linear because the oscillator phase disturbance is proportional to the resonator's amplitude disturbance. The time-varying nature of the relationship is shown by the response to an impulse at different points in the cycle. If the impulse occurs at a voltage maximum, the timing of zero-crossings (and therefore the phase) is not changed, but if the impulse occurs at any other time, the zero crossings change, and the amount they change depends on when the impulse occurs. Since the phase disturbance due to a noise impulse depends on when the impulse occurs, the noise-to-phase transfer function is time-varying, and the shape of the oscillation waveform affects how sensitive the oscillator's phase is to noise impulses [Lee and Hajimiri, 2000]. The sensitivity of different waveforms to phase noise can be described through the impulse sensitivity function for the waveform, Γ [Lee and Hajimiri, 2000].

Based on this theory, components of noise near integer multiples of the carrier frequency fold into noise near the carrier frequency, as described by Lee and Hajimiri [2000]. White noise generates the $1/f_o{}^2$ portion of the phase noise:

$$L(f_o) = \left[10 \cdot \log\left(\frac{\frac{\overline{i_n^2}}{\Delta f_n}\Gamma^2_{\mathrm{RMS}}}{2q^2_{\max} \cdot (2\pi f_o)^2}\right)\right] \tag{6.30}$$

where $\overline{i_n^2}$ is the mean noise power, Δf_n is the noise bandwidth, $q_{\max}$ is the maximum charge displacement in the resonator, and f_o is the offset frequency from the carrier.

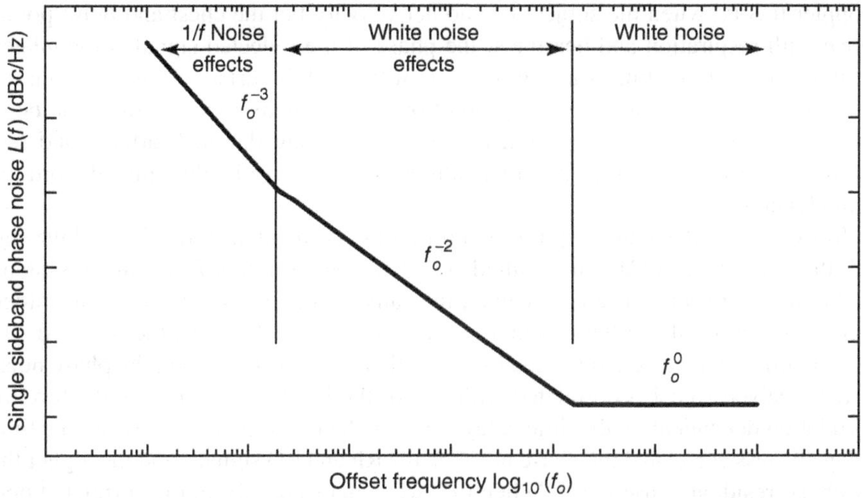

Figure 6.5 Example phase noise spectrum: a typical phase noise spectrum will have a $1/f_o^3$ dependence close to the carrier, a $1/f_o^2$ dependence beyond that, and be flat farther from the carrier.

Because phase noise is proportional to f_o^{-2} multiplied by the noise spectrum, as shown in Equation 6.30, white noise near DC and other integer multiples of the carrier frequency is up-converted to the carrier with $1/f_o^2$ slope, and Equation 6.31 shows that $1/f$ noise near DC gets up-converted to the carrier, weighted by the coefficient c_0, with a $1/f_o^3$ slope:

$$L(f_o) = \left(10 \cdot \log\left(\frac{\frac{\overline{i_n^2}}{\Delta f_n} c_0^2}{8q_{\max}^2 \cdot (2\pi f_o)^2} \cdot \frac{f_{1/f}}{f_o}\right)\right) \tag{6.31}$$

White noise near the carrier remains at the same frequency. This typically leads to an oscillator phase noise spectrum with a $1/f_o^3$ dependence close to the carrier, a $1/f_o^2$ dependence beyond that, and flat at frequencies far from the carrier [Leeson, 1966]. This spectrum is shown in Fig. 6.5.

Because Doppler monitoring of heart and respiration signals focuses on modulation on the order of 1 Hz from the carrier, the $1/f_o^3$ phase noise dependence is the only relevant part of the spectra for this application. The phase noise with the $1/f_o^3$ slope is up-converted baseband $1/f$ noise at the transistor [Lee and Hajimiri, 2000]. Oscillators developed in technologies with poor $1/f$ noise performance will have high close-in phase noise.

6.2.2 Range Correlation and Residual Phase Noise

The Doppler radar physiological monitoring system transmits a CW signal, which is reflected off the subject and then demodulated in the receiver. In accordance with

Doppler theory, when the subject has no net velocity but the chest and pulse points move with respiration and heartbeat, the phase of the reflected signal is modulated proportionally to the time-varying position of the body's surface. Demodulating the phase then gives a signal directly proportional to the body motion. Since the body motion contains information about the chest movement due to heartbeat and respiration, heartbeat and respiration signatures and rates can be determined from the demodulated signal.

Since the heartbeat and respiration information is encoded in the phase of the signal, the phase noise of the transmitted signal can be a limiting factor in the system. In the direct-conversion radar receiver, the same source is used for the transmitted signal and the local oscillator signal in the receiver, which means the received signal is a time-delayed version of the local oscillator signal. Therefore, the phase noise of the received signal is correlated with that of the local oscillator, with the level of correlation dependent on the time delay between the two signals: the greater the time delay, the less correlated the phase noise on the RF and LO signals, and the higher the baseband residual phase noise. When the two signals are mixed, the correlated portion of the phase noise effectively cancels, leaving a residual phase noise spectrum at baseband that is far below the phase noise spectrum at RF. In a radar application, this time delay is the time it takes the signal to travel to the target and back, which is proportional to the target range. Since this time delay is proportional to the target range, the target range determines the level of phase noise reduction provided by the range correlation effect. The dependence of the amount of correlation between the signals on range gave the range correlation effect its name [Budge and Burt, 1993a]. Range correlation theory describes how to calculate the residual phase noise spectrum, and it was first proposed to explain why CW radar systems were not swamped by ground clutter noise [Raven, 1966]. Range correlation is particularly important when measuring the motion due to the heartbeat and respiration since the information is encoded in phase modulations of 0.1–10 Hz, where the phase noise is near its peak [Lee and Hajimiri, 2000]. Range correlation, as described in the following equations, is illustrated in Fig. 6.6.

The radar transmits the signal:

$$T(t) = \cos(2\pi ft + \phi(t)) \tag{6.32}$$

where f is the oscillation frequency, t is the elapsed time, and $\phi(t)$ is the phase noise of the oscillator. Phase noise can be considered as a random fluctuation in the signal's phase. If the transmitted signal is reflected by a target at a nominal distance d_0 that has a time-varying displacement given by $x(t)$, the received signal is approximately

$$R(t) \approx \cos\left[2\pi ft - \frac{4\pi d_0}{\lambda} - \frac{4\pi x(t)}{\lambda} + \phi\left(t - \frac{2d_0}{c}\right) + \theta_0\right] \tag{6.33}$$

where the wavelength is λ and θ_0 is the constant phase shift due to reflection at the body surface. The received signal is similar to the transmitted signal with a time delay determined by the nominal distance to the target, d_0, with its phase modulated by the

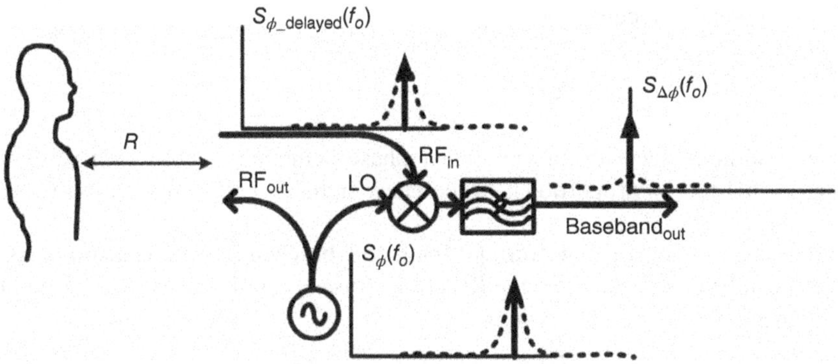

Figure 6.6 Illustration of the range correlation phase noise filtering effect. Since the transmitted signal is derived from the same source as the received signal, the phase noise on the LO, $S_\phi(f_o)$, and the RF input, $S_{\phi\_delayed}(f_o)$, are correlated. When the two signals are mixed, most of the phase noise at baseband is effectively cancelled, leaving only the residual phase noise, $S_{\Delta\phi}(f_o)$.

periodic motion of the target, $x(t)$. The information about the periodic target motion can be demodulated if this signal is multiplied by a local oscillator (LO) signal that is derived from the same source as the transmitted signal. Ignoring amplitude variations, the LO signal is expressed by

$$L(t) = \cos(2\pi ft + \phi(t)) \tag{6.34}$$

The phase fluctuations of the LO due to oscillator phase noise are correlated to those of the received signal.

When the received and LO signals are mixed and the output is low-pass filtered, the resulting baseband signal is

$$B(t) = \cos\left[\theta + \frac{4\pi x(t)}{\lambda} + \Delta\phi(t)\right] \tag{6.35}$$

where

$$\Delta\phi(t) = \phi(t) - \phi\left(t - \frac{2d_0}{c}\right) \tag{6.36}$$

is the residual phase noise and

$$\theta = \frac{4\pi d_0}{\lambda} - \theta_0 \tag{6.37}$$

is the constant phase shift dependent on the nominal distance to the target, d_0. This baseband signal can be demodulated and processed to be

$$B(t) \approx \frac{4\pi x(t)}{\lambda} + \Delta\phi(t) \tag{6.38}$$

with a single-ended receiver, or

$$\theta(t) = \theta + \frac{\pi}{4} + \frac{4\pi x(t)}{\lambda} + \Delta\phi(t) \tag{6.39}$$

with a quadrature receiver and nonlinear phase demodulation. In both cases, the desired signal that is proportional to the chest signal is summed with the residual phase noise.

According to Budge and Burt [1993a], the baseband noise spectral density, $S_{\Delta\phi}(f_o)$, can be calculated from the RF phase noise spectral density, $S_\phi(f_o)$, and the target range, R:

$$S_{\Delta\phi}(f_o) = S_\phi(f_o)\left[4\sin^2\left(2\pi\frac{Rf_o}{c}\right)\right] \tag{6.40}$$

where f_o is offset frequency. At values relevant for radar monitoring of heart and respiration, Rf_o/c will be on the order of 10^{-9}, so the small-angle approximation is valid, and range correlation will cause the baseband noise spectrum to increase proportionally to the square of the target range, R, and the square of the offset frequency, f_o:

$$S_{\Delta\phi}(f_o) \approx S_\phi(f_o)\left[16\pi^2\frac{R^2 f_o^2}{c^2}\right] \tag{6.41}$$

To maintain a constant level of the residual phase noise when increasing the range (which would be necessary to maintain a minimum SNR when residual phase noise is the dominant noise source), the oscillator phase noise must decrease. Therefore, the range requirements and noise level limits for a given application set the required oscillator phase noise specification, which determines the technology requirements.

The close-in RF phase noise spectrum of almost all oscillators has a −30-dB/dec slope [Lee and Hajimiri, 2000; Leeson, 1966]. Range correlation effectively multiplies the phase noise spectrum by that of a filter with a +20-dB/dec slope (because the range correlation effect is proportional to the square of the offset frequency), so the resulting baseband noise spectrum is expected to have a −10-dB/dec slope. For a 50-cm range and an offset frequency of 1 Hz, typical values for heartbeat monitoring, the residual phase noise is 154 dB less than the RF phase noise.

Shrader and Gregers-Hansen [1990] recommend increasing the single-sideband power spectral density of the phase noise value by 6 dB before applying the range correlation filtering effect. This accounts for a 3-dB increase because both sidebands of noise affect clutter residue and another 3-dB increase because the oscillator contributes noise during both transmitting and receiving. The first 3-dB factor would not be present if an intermediate frequency was used, as in a heterodyne receiver. Both of these factors of two are represented in Equation 6.40.

Because there is no carrier at DC, residual phase noise needs to be expressed as a spectral density, in decibels per hertz, rather than as a single-sideband phase noise in decibels below the carrier per hertz. The decibels per hertz units are the ratio of the spectrum in rad^2/Hz to $1\,\text{rad}^2/\text{Hz}$. When the phase deviation due to phase noise is small (with a quiet oscillator or at a high offset frequency), $S_\phi(f_o)$ is 3 dB greater than

$L_\phi(f_o)$ since it includes both sidebands. When the phase deviation is high, the phase noise is causing the signal frequency to vary over a bandwidth greater than 1 Hz, and the phase noise spectral density can be greater than 0 dB. For calculating the level of residual phase noise, the spectral density should be used.

Since the residual phase noise appears as additive noise on the baseband signal when a radar system is used for physiological monitoring, as shown in Equation 6.38, the phase noise reduction due to the range correlation effect is particularly important. If two different oscillators with uncorrelated phase noise were used for transmitting and receiving, it would be impossible to detect the small phase variations created by heart motion, unless the phase noise level was extremely low in both oscillators.

Range correlation has a much less significant effect on amplitude noise. The range correlation effect on amplitude noise is described in Budge and Burt [1993a] as follows:

$$S_{AA}(f_o) = S_A(f_o)\left[4\cos^2\left(\frac{2\pi R f_o}{c}\right)\right] + 2R_A(t_d)\delta(f_o) \tag{6.42}$$

where $S_A(f_o)$ is the spectrum of the amplitude noise and term in brackets accounts for the effects of range delay. R_A is the autocorrelation of the amplitude noise, and for Gaussian white amplitude noise, $R_A(t_d)$ is much less than one, and the second term is negligible. Since, as described earlier, Rf_o/c is very small, the small-angle approximation applies, and Equation 6.55 can be approximated as

$$S_{AA}(f_o) \approx 4S_A(f_o) \tag{6.43}$$

For small Rf_o/c, and Gaussian white amplitude noise, range correlation results in an amplitude gain of 6 dB [Budge and Burt, 1993b].

6.3 CONTRIBUTIONS OF VARIOUS NOISE SOURCES

There are three main sources of noise for the physiological signals monitored by Doppler radar: residual phase noise, down-converted RF additive white Gaussian noise (AWGN) from the front end of the receiver, and baseband $1/f$ noise. These three noise sources are combined at baseband. Each of these sources at baseband must be calculated as the noise power due to each of these sources. These calculations are made at the output of the mixer, where the signal has been converted from a phase-modulated signal to a baseband amplitude signal and all the noise sources are additive.

6.3.1 Phase Noise

Since the close-in phase noise has a −30 dB/dec slope, the phase noise can be defined by the phase noise at an arbitrary frequency f_1:

$$S\phi(f_o) = \frac{S_\phi(f_1)}{f_1^{-3}} \cdot f_o^{-3} \tag{6.44}$$

This is most easily defined at the 1 Hz intercept, $S_\phi(1)$:

$$S_\phi(f_o) = \frac{S_\phi(1)}{(1\,\text{Hz})^{-3}} f_o^{-3} = S_\phi(1) \cdot \left(\frac{f_o}{1\,\text{Hz}}\right)^{-3} \tag{6.45}$$

A model of residual phase noise can be found by using this result and the range correlation equation:

$$\begin{aligned} S_{\Delta\phi}(f_o) &= 2\left(S_\phi(1)\left(\frac{f_o}{1\,\text{Hz}}\right)^{-3}\right) \cdot \left[16\left(\pi^2 \cdot \frac{\left(R + \frac{ct_d}{2}\right)^2 f_o^2}{c^2}\right)\right] \\ &= 32\pi^2 \left(\frac{\left(R + \frac{ct_d}{2}\right)^2}{c^2}\right) (1\,\text{Hz})^3 S_\phi(1) f_o^{-1} \end{aligned} \tag{6.46}$$

The signal with phase noise at the transceiver's RF input is

$$R_{\text{PN}}(t) = A_{\text{RF,PN}} \cos(2\pi f t + \phi(t - T)) \tag{6.47}$$

where $A_{\text{RF,PN}}$ is the amplitude, f is the carrier frequency, and $\phi(t - T)$ is the phase noise of the signal. T indicates the time elapsed from when the signal left the transceiver to when it is received. The signal has a received power of

$$P_{R,\text{PN}} = \frac{P_T G^2 \sigma_c \lambda^2}{(4\pi)^3 R^4} \tag{6.48}$$

This is calculated from the radar equation 6.10, with σ_c the RCS of the clutter that reflects the signal with phase noise. The received phase noise power is equal to the mean square of its voltage divided by the input impedance, Z:

$$P_{R,\text{PN}} = \frac{\overline{R_{\text{PN}}{}^2(t)}}{Z} \tag{6.49}$$

Therefore, the squared amplitude is

$$A_{\text{RF,PN}}^2 = 2P_{R,\text{PN}} Z = \frac{2P_T G^2 \sigma_c Z \lambda^2}{(4\pi)^3 R^4} \tag{6.50}$$

Since phase noise is a phase modulation, the baseband power can be calculated with the phase modulation link equation as it was for the phase-modulated signal, but the phase term at baseband is replaced with the residual phase noise term $\Delta\phi(t)$, because the RF and LO phase noise are combined when the signals are mixed. The baseband residual phase noise voltage is

$$B_{\mathrm{RPN}}(t) = \sqrt{G_{\mathrm{CL}} G_{\mathrm{Rx}}} A_{\mathrm{RF,PN}} \cos(\Delta\phi(t)) \tag{6.51}$$

and the baseband residual phase noise power is

$$N_{\mathrm{RPN},B} = \frac{\overline{B^2_{\mathrm{RPN}}(t)}}{Z} = \frac{G_{\mathrm{CL}} G_{\mathrm{Rx}} A^2_{\mathrm{RF,PN}} \overline{(\cos(\Delta\phi(t)))^2}}{Z} \tag{6.52}$$

$$N_{\mathrm{RPN},B} = \frac{\overline{B^2_{\mathrm{RPN}}(t)}}{Z} = \frac{2P_T G_{\mathrm{CL}} G_{\mathrm{Rx}} G^2 \sigma_c \lambda^2 \overline{(\cos(\Delta\phi(t)))^2}}{(4\pi)^3 R^4} \tag{6.53}$$

and applying the small-angle approximation,

$$N_{\mathrm{RPN},B} = \frac{\overline{B^2_{\mathrm{RPN}}(t)}}{Z} = \frac{2P_T G_{\mathrm{CL}} G_{\mathrm{Rx}} G^2 \sigma_c \lambda^2 \overline{(\Delta\phi(t))^2}}{(4\pi)^3 R^4}$$

The mean squared residual phase noise in the time domain is the integral of the spectrum over the received frequencies:

$$\Delta\phi^2_{\mathrm{RMS}} = \int_{f_{\min}}^{f_{\max}} S_{\Delta\varphi}(f_o) df_o \tag{6.54}$$

where $f_{\max}$ is the highest frequency and $f_{\min}$ is the lowest frequency passed through the filters. Using the expression for $S_{\Delta\phi}(f_o)$ in Equation 6.45, we can express the mean squared residual phase noise as follows:

$$\begin{aligned} \Delta\phi^2_{\mathrm{RMS}} &= 32\pi^2 (1\,\mathrm{Hz})^3 S_\phi(1) \frac{\left(R + \frac{ct_d}{2}\right)^2}{c^2} \int_{f_{\min}}^{f_{\max}} f_o^{-1} df_o \\ &= 32\pi^2 (1\,\mathrm{Hz})^3 S_\phi(1) \frac{\left(R + \frac{ct_d}{2}\right)^2}{c^2} \ln \left[\frac{f_{\max}}{f_{\min}}\right] \end{aligned} \tag{6.55}$$

This expression allows the baseband mean-squared phase noise to be calculated from the RF phase noise, the range to the target, the transceiver time delay, and the selected filtering frequencies. The RMS noise is

$$\Delta\phi_{\mathrm{RMS}} = 4\sqrt{2}\pi \left(\frac{\left(R + \frac{ct_d}{2}\right)}{c} \right) \sqrt{S_\phi(1) \ln \left(\frac{f_{\max}}{f_{\min}}\right)} \tag{6.56}$$

We can substitute this value into 6.66 to get an estimate of the residual phase noise with range, as follows:

$$N_{\mathrm{RPN},B} = \frac{P_T \sigma_c G^2_{\mathrm{antenna}} G_{\mathrm{Rx}} G_{\mathrm{CL}}}{\pi f^2} S_\phi(1) \frac{\left(R + \frac{ct_d}{2}\right)}{R^4} \ln\left(\frac{f_{\max}}{f_{\min}}\right) \tag{6.57}$$

This value can be checked by assuming that with perfect phase demodulation, the residual phase noise SNR will be the same as the ratio of the phase variation of the signal to that of the noise, multiplied by the ratio of the signal RCS to that of the clutter:

$$\left(\frac{S_0}{N_{\mathrm{RPN}}}\right) = \left(\frac{\phi_{\mathrm{RMS,heart}}}{\Delta\phi_{\mathrm{RMS}}}\right)^2 \cdot \frac{\sigma}{\sigma_c} \tag{6.58}$$

or

$$\left(\frac{S_0}{N_{\mathrm{RPN}}}\right) = \frac{\overline{\sigma x^2(t)} f^2}{2R^2 \sigma_c S_\phi(1) \ln\left(\frac{f_{\max}}{f_{\min}}\right)} \tag{6.59}$$

If the signal power is scaled to be that calculated in Equation 6.33 the noise power is

$$N_{\mathrm{RPN},B} = \frac{P_T \sigma_c G^2_{\mathrm{antenna}} G_{\mathrm{Rx}} G_{\mathrm{CL}}}{\pi f^2 R^2} S_\phi(1) \ln\left(\frac{f_{\max}}{f_{\min}}\right) \tag{6.60}$$

which matches the calculation in Equation 6.57 except for the term for the transceiver time delay that was not included in this estimate.

6.3.2 Baseband 1/*f* Noise

In this application, 1/*f* noise from the mixer and from the baseband signal-conditioning circuitry dominates the baseband noise spectrum. The 1/*f* baseband receiver noise can be approximated as

$$N_{1/f,B} = \int_{f_{\min}}^{f_{\max}} P_{1/f}(1) f^{-1}\, df = P_{1/f}(1) \ln\left(\frac{f_{\max}}{f_{\min}}\right) \tag{6.61}$$

where $P_{1/f}(1)$ is the noise power in a 1-Hz bandwidth centered at 1 Hz.

6.3.3 RF Additive White Gaussian Noise

Since the information in Doppler radar cardiorespiratory sensing is encoded as a phase modulation, the RF SNR is not the same as the signal's SNR after it has been demodulated to baseband. The amplitude noise at RF affects the phase of the signal based on the percentage of phase modulation in addition to the RF SNR. The RF noise at the input is

$$N_{\mathrm{RF,in}} = NB \tag{6.62}$$

where N is the white channel noise power spectral density and B is the receiver bandwidth.

Because the SNR is being calculated after the mixer, the noise figure, receiver gain, and mixer conversion loss need to be included in the equations. The noise figure can be expressed as the ratio of the input SNR to the output SNR, or as the ratio for the noise output from the actual receiver to the noise output from an ideal receiver:

$$NF = \frac{\left(\frac{S}{N}\right)_{\mathrm{in}}}{\left(\frac{S}{N}\right)_{\mathrm{out}}} = \frac{N_{\mathrm{RF,out}}}{G_{\mathrm{Rx}} G_{\mathrm{CL}} N_{\mathrm{RF,in}}} \tag{6.63}$$

where G_{Rx} is the gain of the receiver and G_{CL} is the mixer's conversion loss. The noise figure expresses the amount of noise added to the signal by the receiver. Therefore, the signal at the mixer output is

$$S_B = G_{\mathrm{Rx}} G_{\mathrm{CL}} (S_0)_{\mathrm{in}} \tag{6.64}$$

while the noise after the mixer is

$$N_{\mathrm{RF},B} = G_{\mathrm{Rx}} G_{\mathrm{CL}} (NF) N_{\mathrm{RF,in}} = 2 G_{\mathrm{CL}} G_{\mathrm{Rx}} (NF)(NB) \tag{6.65}$$

There is a factor of 2 because the thermal noise in the two sidebands in uncorrelated, so the noise power adds.

The dominant RF noise at the input to the receiver is thermal noise; thermal noise is zero-mean, has a Gaussian distribution, and does not vary with frequency. This is additive to the RF signal. The thermal noise power is expressed by

$$P_{N,\ \mathrm{thermal}} = 4kTB \tag{6.66}$$

where k is Boltzmann's constant, T is the absolute temperature, and B is the bandwidth. Therefore, N can be substituted with $4kT$ in most cases. Therefore, the total RF noise converted to baseband is

$$N_{\mathrm{RF},B} = 8 G_{\mathrm{CL}} G_{\mathrm{Rx}} (NF)(kTB) \tag{6.67}$$

6.4 SIGNAL-TO-NOISE RATIO

The three main sources of noise, residual phase noise, down-converted RF AWGN, and baseband $1/f$ noise, are combined at the mixer output after they have been converted to their values in baseband. Because the noise from the three sources is uncorrelated, the noise powers simply add. Therefore, the SNR for the system is

$$\frac{S_B}{N_B} = \frac{S_0}{N_{1/f,B} + N_{\mathrm{RF},B} + N_{\mathrm{RPN},B}} \tag{6.68}$$

This can be expanded to

$$\left(\frac{S_B}{N_B}\right) = \frac{\frac{P_T G^2 G_{\mathrm{Rx}} G_{\mathrm{CL}} \sigma}{2\pi R^4} \cdot \overline{x^2(t)}}{P_{1/f}(1) \ln\left(\frac{f_{\max}}{f_{\min}}\right) + 2G_R G_{\mathrm{CL}}(NF)(NB) + \frac{P_T \sigma_c G^2 G_R C_{\mathrm{CL}}}{\pi f^2} S_\phi(1) \ln\left(\frac{f_{\max}}{f_{\min}}\right) \cdot \left(\frac{\left(R + \frac{ct_d}{2}\right)^2}{R^4}\right)} \tag{6.69}$$

This is equivalent to

$$\left(\frac{S_B}{N_B}\right) = \frac{P_T G_{\mathrm{ant}}^2 G_{\mathrm{Rx}} G_{\mathrm{CL}} \sigma \overline{x^2(t)}}{2\pi(N_{1/f,B} + N_{\mathrm{RF},B})R^4 + 2\left(\frac{P_T \sigma_{\mathrm{ciu}} G_{\mathrm{ant}}^2 G_{\mathrm{Rx}} G_{\mathrm{CL}}}{r^2} S_\phi(1) \ln\left(\frac{f_{\max}}{f_{\min}}\right)\right)\left(R + \frac{ct_d}{2}\right)^2} \tag{6.70}$$

When residual phase noise is dominant, the SNR will be proportional to $(R + 0.5ct_d)^{-2}$, and when either the baseband noise or the RF AWGN is dominant, the SNR will be proportional to R^{-4}. If one noise source is not dominant for all ranges, the residual phase noise will be dominant close to the target, and the baseband or RF noise will be dominant further from the target. The equation also indicates that the SNR should be linear with the RCS of the target, and target RCS should not affect the dominant type of noise. The RCS and the amount of motion for both heart and respiration are expected to vary from subject to subject, and likely also with orientation with respect to the antenna.

In this analysis, it was assumed that the signal is at the optimal phase demodulation point. This gives the best-case signal-to-noise calculation for a single-channel receiver. The output at the mixer will be filtered and amplified before it is digitized. For the quadrature receiver, if the signal is determined by choosing between the *I* and *Q* signals, the signal power and residual phase noise power would be cut in half. This would not affect the SNR if residual phase noise is dominant, but does if either RF amplitude noise or baseband $1/f$ noise is the dominant noise source. If the *I* and *Q* signals are combined, the baseband noise from the filtering and amplifying stages is added before the combination takes place. If residual phase noise is dominant, the SNR of the combined of *I* and *Q* signals will be similar to that of the single-channel receiver at the optimal phase demodulation point. If one of the other noise sources is dominant, the SNR would be a factor of 2 worse.

For radar monitoring of heartbeat and respiration, the RCS, the mean-squared motion (MSM), and the clutter cross section are the only values in the SNR expression that change with different subjects and environments. The desired RCS–MSM product is actually the integral of the motion over the area of the body, multiplied by factors for the body reflectivity, the directivity of the reflected signal, and the amount of motion that is in the direction of the antenna. The peak-to-peak chest motion due to respiration in adults ranges from 4 to 12 mm [De Groote et al., 1997; Kondo et al., 1997], while the peak-to-peak motion due to the heartbeat is about 0.5 mm [Ramachandran and Singh, 1989].

6.5 VALIDATION OF RANGE CORRELATION

The range correlation effect was validated by Droitcour et al. [2004] with direct conversion radar transceivers oscillators with different levels of phase noise. The effect of range correlation on baseband residual phase noise for different offset frequencies and time delays was estimated using Equation 6.40 and measured phase-noise data. The range-correlation theory was verified by varying the delay between the transmitter and the receiver and measuring the baseband noise spectrum at the output and comparing it with the predicted values. The setup of this experiment is shown in Fig. 6.7. The RF output of the radar was connected to the phase-shifter input through a 30-cm SMA cable and a 10-dB attenuator. The 10-dB attenuator was used to reduce voltage controlled oscillator (VCO) loading by the phase shifter. An SMA cable connected the phase-shifter output to the RF input of the chip. The length of this cable was varied to change the time delay between the RF and LO signals at the mixer. The baseband

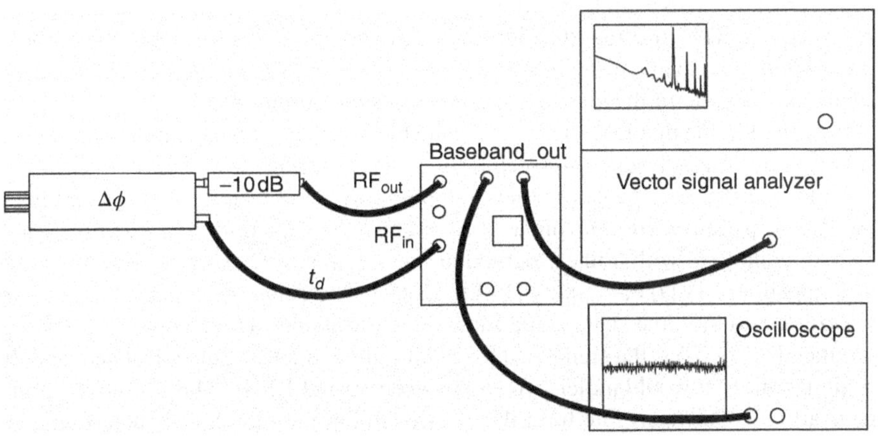

Figure 6.7 Setup for the range-correlation verification experiment. The baseband noise spectrum was measured with the VSA. Cables of various lengths were connected in the place of the cable marked t to change the time delay between the RF and LO signals. © 2004 IEEE, Reprinted, with permission, from Droitcour et al. [2004].

output of the chip was measured with a vector signal analyzer. The baseband noise spectrum was measured and converted to a phase-noise equivalent by calculating the ratio of the measured noise power to the power a 30-kHz IF signal would have with the same RF and LO power. This was then converted into spectral density of phase fluctuation by multiplying by 2 [Vendelin et al., 1990]. The time delay and loss through the cables, attenuator, and phase shifter were measured with an RF network analyzer, and the loss was taken into account when calculating the equivalent IF power.

The measured phase noise and the 30-dB dec slope line used to predict the baseband noise are shown in Fig. 6.8(a) for offset frequencies from 1 Hz to 1 kHz. The predicted and measured phase fluctuation spectral densities are plotted in Fig. 6.8(b) for delays of 6.2, 12.6, and 28.0 ns, and offset frequencies from 1 Hz to 1 kHz. On average, the measured values were within 5 dB of the predicted values. The baseband phase noise was reduced by 148 to 136 dB at 1 Hz for the time delays from 6.2 to 28.0 ns, which correspond to ranges from 0.93 to 4.2 m.

The measured baseband noise spectral density was in the same range as predicted based on the previously measured phase noise and range-correlation theory. The measured baseband noise increased as the time delay increased and had approximately a 10-dB/dec slope, as was predicted. The phase noise was not measured at the same time as the baseband noise spectrum, and this may be another cause for some of the discrepancy between the predictions and measured results.

6.6 HUMAN TESTING VALIDATION

In Droitcour et al. [2009] the SNR theory was validated with human testing on 22 subjects with a CMOS radar and discrete analog baseband processing that included measurements of SNR and detection of heart and respiration rates. The digital signal processing (DSP) filters the signals to remove noise and to isolate the Doppler heart signal from the respiration signal, combines signals from the I and Q channels, determines the rate of the signal, and smoothens the output rate.

In this work, the first DSP step is to isolate the heart signal from the combined heart and respiration signals with a 400-tap Kaiser high-pass filter with β of 6.5 and a cutoff of 0.6 Hz. The heart signals are then low-pass filtered to remove out-of-band noise with a 20-tap Kaiser filter having a β of 6.5 and a 20-Hz cutoff. The I and Q signals are then combined with a linear demodulation method, referred to as singular value decomposition (SVD) combining [Jollife, 2002]. The delay introduced by the filters is corrected so the time scale is the same for all channels. The Doppler heart rate is calculated every 0.5 s; the signal in a 8-s Hamming window is autocorrelated, and the local maxima that would indicate a rate between 30 and 120 BPM is used to calculate the heart rate. The heart rate from the electrocardiogram reference is determined by extracting the R waves using a wavelet-based algorithm [Li et al., 1995] and inverting the mean of the interbeat interval (in seconds) in an 8-s window and multiplying by 60 to obtain breaths per minute. The rates from both the ECG and the Doppler system are then smoothed with an exponential filter having an α value of 0.93.

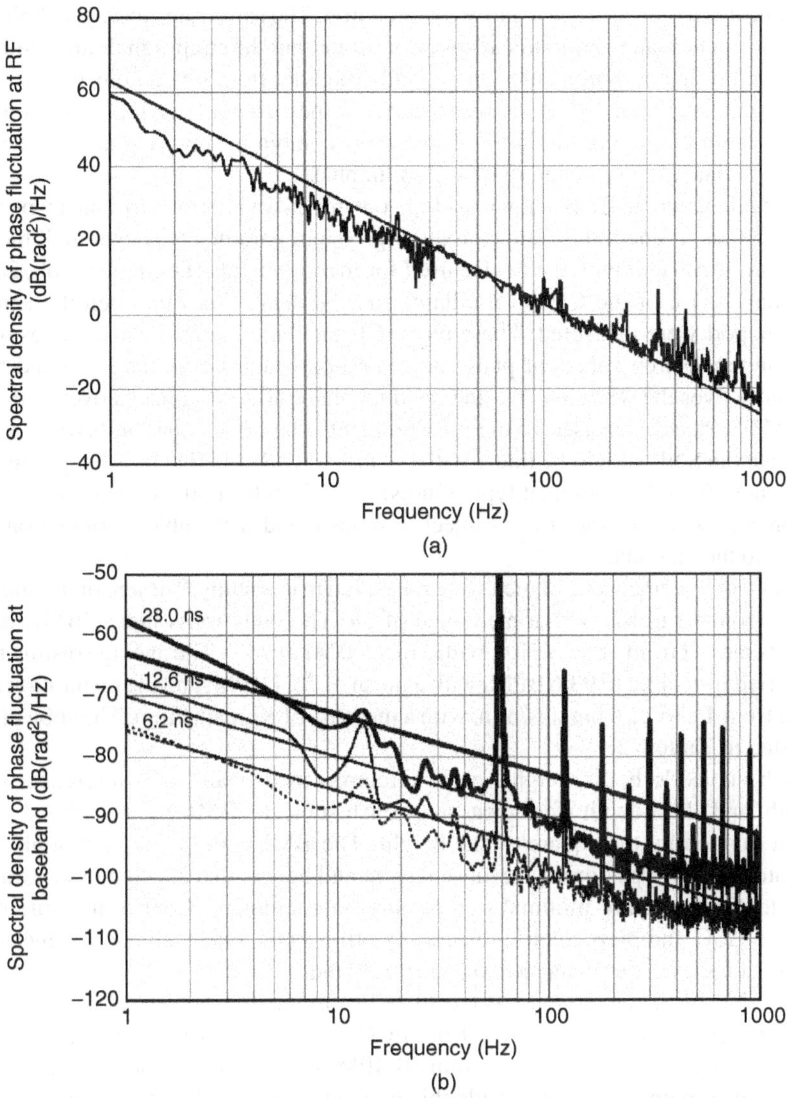

Figure 6.8 (a) Measured RF phase noise with −30 dB/dec fit line used to predict baseband noise and (b) measured and predicted baseband residual phase noise with time delays of 28.0, 12.6, and 6.2 ns. © 2004 IEEE, Reprinted, with permission, from Droitcour et al. [2004].

The respiration is extracted from the combined heart and respiration signal. The respiration signal is low-pass filtered with a 50-tap 1.5-Hz cutoff filter to remove out-of-band noise. Then, the I and Q signals are combined with a linear demodulation method referred to as SVD combining [Jollife, 2002]. The respiration rate is found by autocorrelating the signal in an 18-s Hamming window and finding the local maxima that indicates a rate between 4 and 30 breaths/min. The same rate-finding technique is used for the chest and abdominal respiration straps, but the strap signals are combined with equal-ratio combining, are low-pass filtered after they are combined rather than before, and have their DC offset removed since they do not have DC offset removal in the analog signal conditioning. Equal-ratio combining involves adding the two sources of data after ensuring that they are in phase.

The calculated SNR is shown in Fig. 6.9 for heart motion (a) and respiratory motion (b) as a function of range, assuming system parameters shown in Table 6.1, which are consistent with the device used for measurements [Droitcour et al., 2009]. The phase noise of the CMOS oscillator and the delay times between the antenna and data port were measured. These plots indicate that with calculated values of RF noise and measured values of phase noise and baseband noise, the baseband noise dominates over the RF noise, but the residual phase noise dominates over both with the CMOS transceiver used in the measurement. The SNR equation 6.70 also indicates that the SNR should be linear with the target RCS, and that the changes in RCS should not affect the dominant type of noise. The RCS for both heart and respiratory motion is expected to vary from subject to subject, and with subject orientation with respect to the antenna.

A total of 7 women and 15 men were measured in this study. The age of the subjects ranged from 19 to 67, with a mean age of 34. The body mass index (BMI) of the subjects ranged from 18.3 to 31.4, with a mean BMI of 24.3. The average resting heart rate varied from 43.2 to 93.6 BPM, with a mean of 70.4 BPM, and the respiration rates varied from 4.8 to 21.0 breaths/min, with a mean of 12.8 breaths/min. The subject data are listed in Table 6.2.

Each subject's body weight, chest circumference, waist circumference, chest breadth, and chest depth were measured. Each subject's BMI was calculated as the weight divided by the square of the height. The BMI is the most commonly used estimate of body type, largely because weight and height are easy to measure, highly consistent, and require minimal expenditure on equipment. Control measurements included ECG and piezoelectric respiratory effort belts. Each subject was measured for 90 s at each of four distances: 0.5, 1.0, 1.5, and 2.0 m.

The SNR of the heart signal is calculated from the power spectral density of the Doppler signal. The average rate of the ECG signal is determined to be the center rate of the signal, and the power within 10 BPM of the center is considered to be the signal power, with all power outside this window considered to be the noise power. The same technique is used for the respiration, but the signal is the power within 6 breaths/min of the rate from the belts, and if the rate is below 6 breaths/min, the minimum rate is 0.1 breaths/min and the maximum rate is 12.1 breaths/min. The respiration signal's power is corrected for the amount of the signal that is removed by the baseband DC-blocking filter by dividing the SNR by the gain of the filter.

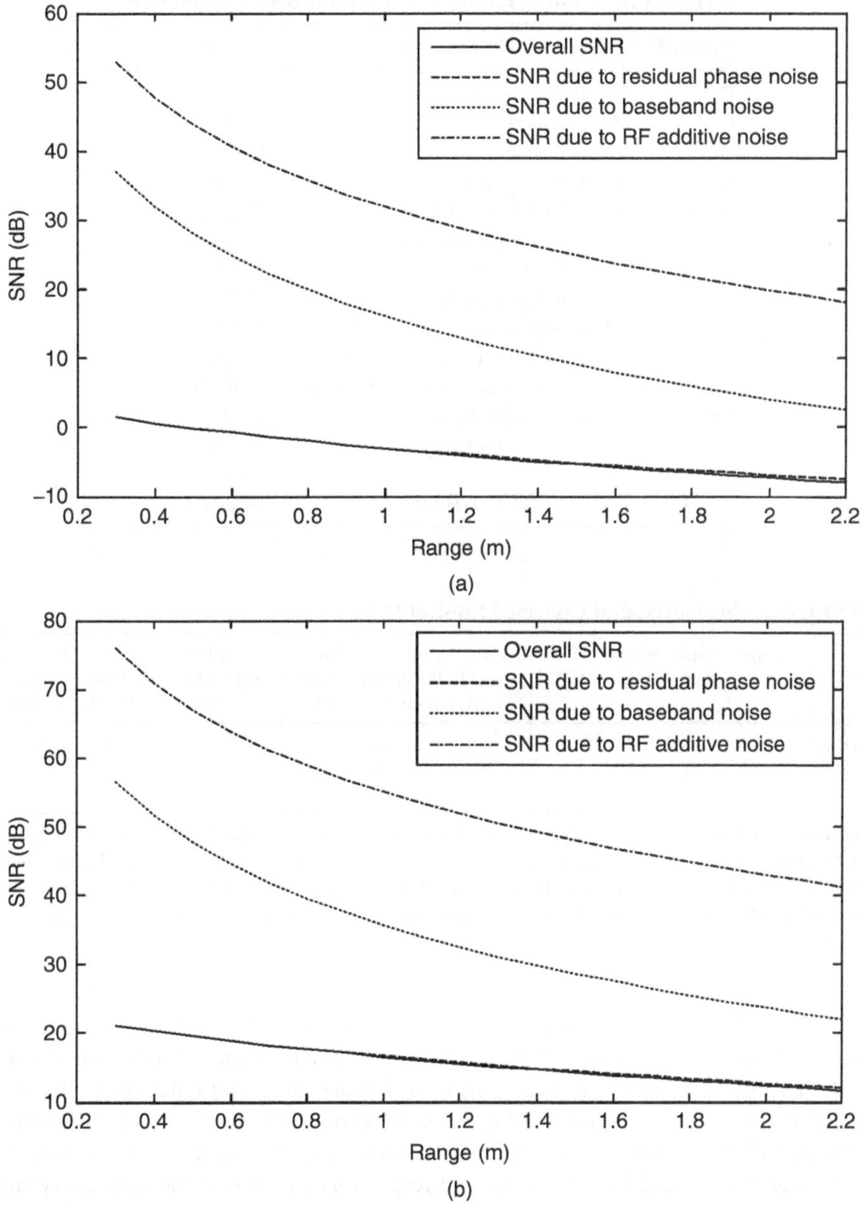

Figure 6.9 Predicted signal-to-noise ratio for (a) heart and (b) respiration with each noise source and all noise sources, using the parameters in Table 6.1. © 2009 IEEE, Reprinted, with permission, from Droitcour et al. [2009].

TABLE 6.1 System Parameters Used for SNR Calculation

Symbol	Description	Value
P_T	Transmit power	0 dBm
G	Antenna gain	6 dBi
G_{Rx}	Receiver gain	6 dB
G_{CL}	Conversion gain	−3 dB
$\sigma x^2(t)$	RCS–MSM product (heart)	2.25 mm^4
$\sigma x^2(t)$	RCS–MSM product (respiration)	500 mm^4
$P_{1/f}(1)$	$1/f$ noise power at 1 Hz	−130 dBm/Hz
f_{max}	Maximum frequency – heart	10 Hz
f_{max}	Maximum frequency – respiration	10 Hz
f_{min}	Minimum frequency – heart	0.6 Hz
f_{min}	Minimum frequency – respiration	0.01 Hz
NF	Receiver noise figure	6 dB
T	RF noise temperature	300 K
t_d	Delay	5 ns
$S_\phi(1)$	Phase noise at 1-Hz intercept	64 dB/Hz

TABLE 6.2 Measured and Collected Subject Data

	Age	Height (cm)	Weight (kg)	BMI (kg/m^2)	Chest Breadth (cm)	Chest Depth (cm)	Waist Circumference (cm)	Chest Circumference (cm)	Heart Rate (BPM)	Respiration Rate (rpm)
Average	34.0	174.2	74.6	24.3	27.5	18.7	80.8	86.1	70.4	12.8
Standard deviation	11.2	9.5	17.4	4.2	3.9	2.9	18.6	12.5	11.4	4.0

The subject's age was reported by the subject. The BMI is the subject's BMI, calculated as weight per height square, with the weight in kilogram and the height in meter. The subject's chest breadth, chest depth, waist circumference, and chest circumference were all measured at exhale. The heart rate and respiration rate are the subject's average heart and respiration rates measured in beats per minute and respirations per minute.

The Bland–Altman analysis technique for method comparison involves plotting the difference between the two methods' measurement values against the average of the two measurements, a Tukey mean difference plot, and calculating the 95% confidence upper and low limits of agreement between the two methods [Bland and Altman, 1986]. In this case, Doppler radar results were compared with those obtained from references, ECG for heart rate, and respiratory effort belt for respiratory rate. The mean difference is then an estimate of the average bias of one method relative to the other. Assuming the measurement error has a Gaussian distribution about the bias, the 95% confidence intervals can be calculated as the bias ± 1.96 times the standard

TABLE 6.3 Bland–Altman Data for Heart Rate and Respiratory Rate Measurements at Each Range

	Range (m)	95% Confidence Interval	Mean	Standard Deviation
Heart rate	0.5	+7.04/ − 7.00	0.02	3.58
	1.0	+11.18/ − 9.59	0.80	5.30
	1.5	+19.74/ − 16.74	1.50	9.31
	2.0	+25.9/ − 16.15	4.88	10.73
Respiratory rate	0.5	+4.75/ − 4.01	0.37	2.23
	1.0	+4.32/14.78	−0.23	2.32
	1.5	+4.44/ − 5.06	−0.31	2.42
	2.0	+7.12/ − 10.53	−1.70	4.50

deviation of the differences. The bias is calculated as the mean of the difference between the two measurements. The Bland–Altman statistics are calculated for 60 s from each measurement, starting after 22 s, so that the filters and exponential average have time to settle.

The mean and standard deviation of the difference between the heart or respiration rates and the 95% confidence interval at each range found with the Doppler system and the control are shown in Table 6.3.

The signatures and rates collected from subject 4062 at the ranges of 0.5 and 1.5 m are shown in Figs. 6.10 and 6.11, respectively. For each range, the 60-s traces that were used for Bland–Altman rate comparison are shown. These traces are the combined *I* and *Q* heart and respiration signals from the Doppler radar, the ECG signal, and the combined chest and abdomen respiratory effort straps.

The SNR was calculated for the heart and respiration traces for each subject for each range. The average SNRs at each range for heart and respiration are shown in Table 6.4 and Fig. 6.12.

The correlation coefficient was calculated between each measured subject parameter and the calculated SNR at each range. The heart SNR did not show statistically significant correlation with any of the measured parameters, but the respiration SNR had significant correlations with chest circumference, waist circumference, chest depth, and height–waist circumference product, at 0.5, 1.0, and 1.5 m. The correlation coefficient and its associated p value are shown for each of these parameters in Table 6.5. The heart and respiration SNR versus range data shown in Fig. 6.12 match much more closely with the residual phase noise model than the other noise model, validating the prediction that residual phase noise is the dominant noise source.

If the measured variables are input into the predicted SNR equation, the RCS–MSM product can be calculated for the heart and respiration, and these values

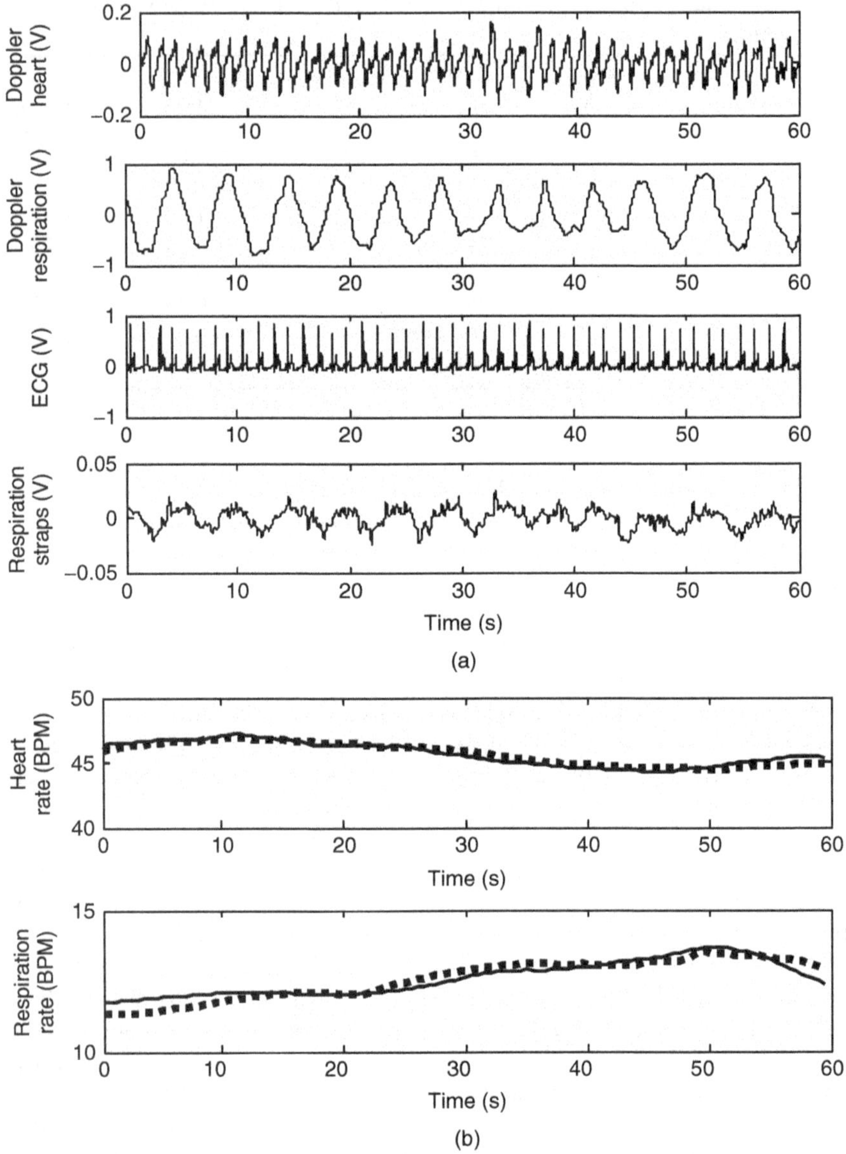

Figure 6.10 Data from subject 4062 at 0.5 m. (a) The top trace is the combined heart signal from the radar; the second trace is the combined respiration signal from the radar, the third trace is the ECG, and the fourth trace is the combined respiration signal from the straps. (b) Heart and respiratory rates calculated from the Doppler radar and the reference. © 2009 IEEE, Reprinted, with permission, from Droitcour et al. [2009].

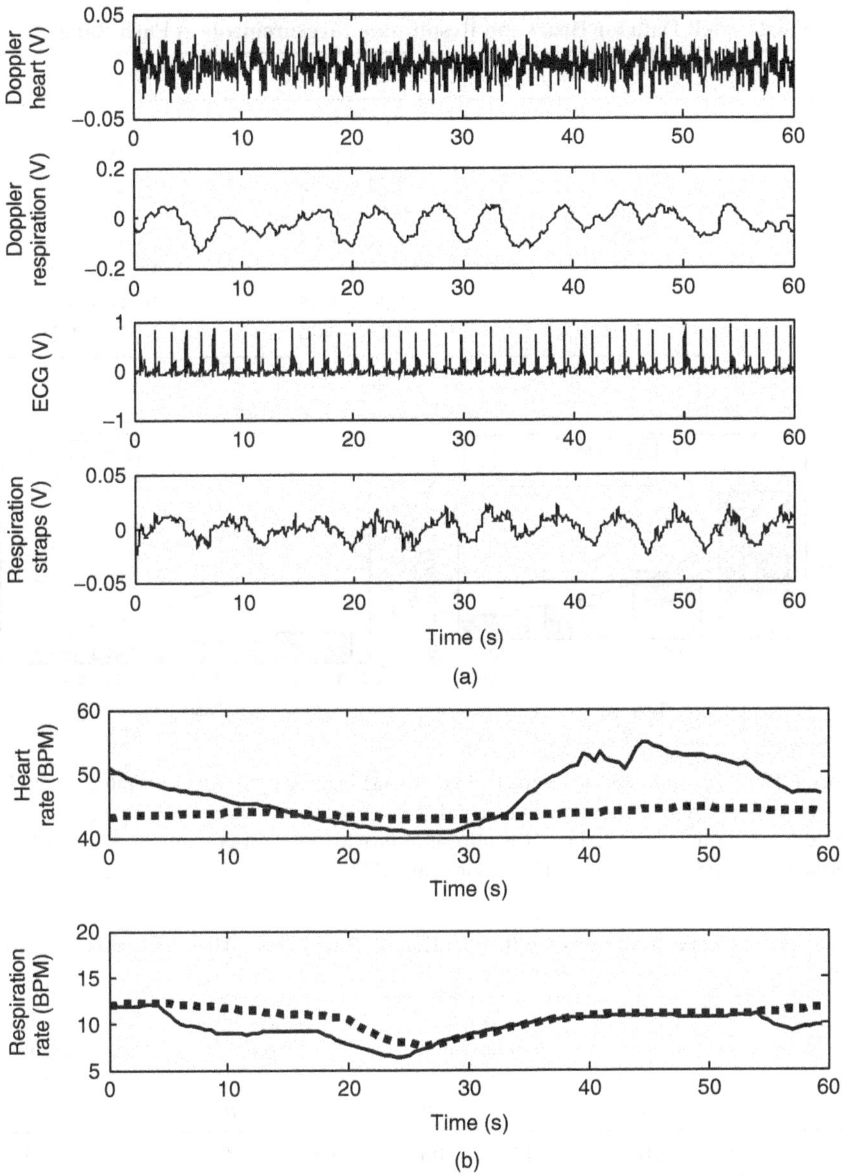

Figure 6.11 Data from subject 4062 at 1.5 m. (a) The top trace is the combined heart signal from the radar; the second trace is the combined respiration signal from the radar, the third trace is the ECG, and the fourth trace is the combined respiration signal from the straps. (b) Heart and respiratory rates calculated from the Doppler radar and the reference. © 2009 IEEE, Reprinted, with permission, from Droitcour et al. [2009].

TABLE 6.4 SNR Data for Heart and Respiration Measurements at Each Range

	Range (m)	Average SNR	Average SNR (dB)
Heart	0.5	1.07 ± 0.83	0.29 ±
	1.0	0.49 ± 0.22	−3.10 ±
	1.5	0.24 ± 0.18	−6.20 ±
	2.0	0.17 ± 0.06	−7.70 ±
Respiration	0.5	31.61 ± 10.13	15.00 ± 1.08
	1.0	19.16 ± 5.49	12.82 ± 1.05
	1.5	11.00 ± 1.95	10.41 ± 0.77
	2.0	10.95 ± 3.80	10.39 ± 1.23

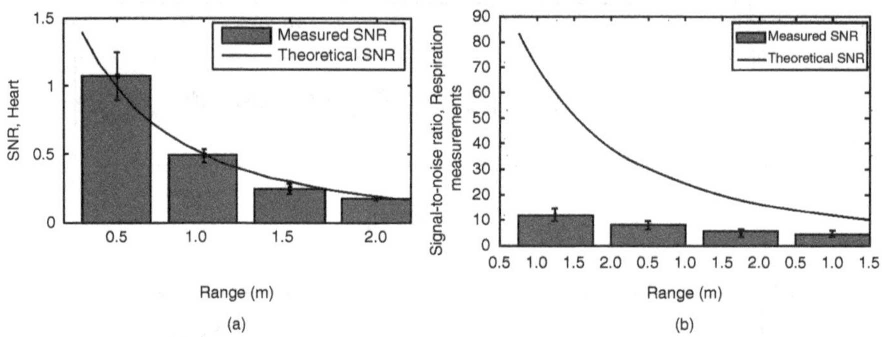

Figure 6.12 Measured and theoretical SNR for (a) heartbeat and (b) respiratory rate. The theoretical radar-cross-section–mean-squared ratio product was set at a value of 250 mm^4 for respiratory motion and 2.25 mm^4 for heartbeat to provide the best possible fit.

TABLE 6.5 Correlation Between Respiration SNR and Body Measurements at Each Range

Range (m)	Chest Circumference		Waist Circumference		Chest Depth		Height–Waist Circumference Product	
	r	*p*	*r*	*p*	*r*	*p*	*r*	*p*
0.5	0.36	0.10	0.40	0.06	0.40	0.07	0.36	0.10
1.0	0.53	0.01	0.43	0.05	0.55	0.01	0.42	0.05
1.5	0.48	0.02	0.44	0.04	0.67	0.00	0.43	0.05
2.0	0.28	0.21	0.23	0.30	0.23	0.30	0.27	0.21

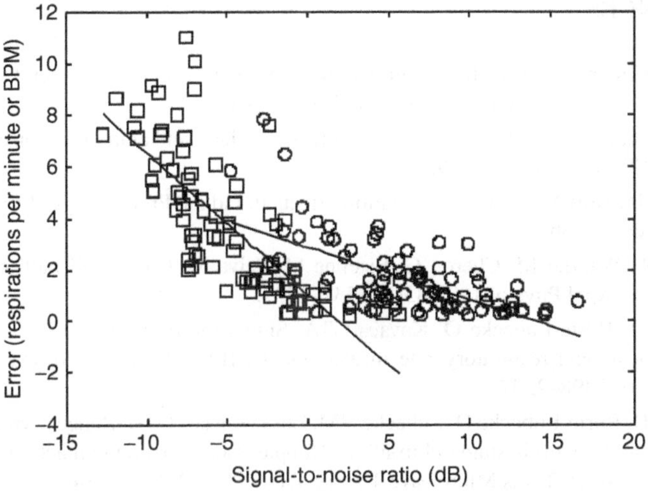

Figure 6.13 Scattergram of error versus signal-to-noise ratio for heart and respiratory rate measurement with the Doppler radar. The error is defined as the standard deviation of the difference between the radar-based measurement and the reference, and the signal-to-noise ratio is measured as described in this chapter. A linear regression is performed on the data; the model for the heart rate is $E = 1.03 - 0.55^{*}$ SNR with R^2 of 0.59. The model for respiration is $E = 2.86 - 0.20^{*}$ SNR with R^2 of 0.42. © 2009 IEEE, Reprinted, with permission, from Droitcour et al. [2009].

can be used in future assessments of similar systems. Based on system parameters from Table 6.1, RCS–MSM product is calculated to be $250\,\text{mm}^4$ for respiratory motion, and $2.25\,\text{mm}^4$ for heart motion. Thus, the respiratory RCS–MSM product was found to be 110 times that of the heart signal.

The error in rate was plotted versus the SNR for all measurements in Fig. 6.13. When the SNR is plotted in decibels, there is an approximately linear relationship, so the accuracy is proportional to the log of the SNR. The model for heart accounts for 59% of the variation in heart, and the model for respiration accounts for 42% of the variation in respiration. This indicates that the SNR does affect the ability to detect heart and respiration rates with the methods used by Droitcour et al. [2009]. The accuracy was sometimes very good with an SNR as low as −1 dB, but it was not consistently good until the SNR was greater than 10 dB. This indicates that improvements in the SNR will improve the accuracy for rate-finding. This also indicates that with better signal processing, the accuracy could be improved for signals with SNR as low as −1 dB.

REFERENCES

Bland JM, Altman DG. Statistical methods for assessing agreement between two methods of clinical measurement. Lancet 1986;1(8476):307–310.

Budge MC, Jr., Burt MP. Range correlation effects on phase and amplitude noise. Proceedings of the IEEE Southeastcon; 1993a.

Budge MC, Jr., Burt MP. Range correlation effects in radars. Proceedings of the IEEE Radar Conference; 1993b.

De Groote A, Wantier M, Cheron G, Estenne M, Paiva M. Chest wall motion during tidal breathing. J Appl Physiol 1997;83(5):1531–1537.

Droitcour AD, Boric-Lubecke O, Kovacs GTA. Signal-to-noise ratio in Doppler radar system for heart and respiratory rate measurements. IEEE Trans Microwave Theory Tech 2009;57(10):2498–2507.

Droitcour AD, Boric-Lubecke O, Lubecke VM, Lin J, Kovacs GTA. Range correlation and I/Q performance benefits in single-chip silicon Doppler radars for non-contact cardiopulmonary monitoring. IEEE Trans Microwave Theory Tech 2004;52(3):838–848.

Electronic Warfare and Radar Systems Engineering Handbook, Avionics Department of the Naval Air Warfare Center Weapons Division in 1992, document number TP 8347.

Geisheimer JL, Greneker E, Marshall W. A high-resolution Doppler model of human gait. Proc SPIE: Radar Sensor Technol Data Visualization 2002;4744:8–18.

Gentilli GB, Tesi V, Linari M, Marsili M. A versatile microwave plethysmograph for the monitoring of physiological parameters. IEEE Trans Biomed Eng 2002;49(10):1204–1210.

Jollife IT. Principal Component Analysis. Secaucus, NJ: Springer-Verlag New York, Inc.; 2002.

Kondo T, Uhlig T, Pemberton P, Sly PD. Laser monitoring of chest wall displacement. Eur Respir J 1997;10:1865–1869.

Lathi BP. Modern Digital and Analog Communication Systems. New York: Oxford University Press; 1998.

Lee TH, Hajimiri A. Oscillator phase noise: a tutorial. IEEE J Solid State Circuits 2000;35(3):326–336.

Leeson DB. A simple model of feedback oscillator noise spectrum. Proc IEEE 1966;54:329–330.

Li C, Zheng C, Tai C. Detection of ECG characteristic points using wavelet transforms. IEEE Trans Biomed Eng 1995;42(1):21–28.

Ramachandran G, Singh M. Three-dimensional reconstruction of cardiac displacement patterns on the chest wall during the P, QRS, and T-segments of the ECG by laser speckle interferometry. Med Biol Eng Comput 1989;27(5):525–530.

Raven RS. Requirements on master oscillators for coherent radar. Proc IEEE 1966;54(2):237–243.

Schultz FV, Burgener RC, King S. Measurement of the radar cross section of a man. Proc IRE 1958;46:476–481.

Shrader WW, Gregers-Hansen V. MTI radar. In: Skolnik MI, editor. Radar Handbook. 2nd ed. San Francisco: McGraw-Hill, Inc.; 1990.

van Dorp P, Groen FCA. Human walking estimation with radar. IEEE Proc Radar Sonar Navig 2003;150(5):356–365.

Vendelin GD, Pavio AM, Rohde UL, Rohde EL. Microwave Circuit Design Using Linear and Nonlinear Techniques. New York: Wiley; 1990.

Wu T. Radar cross section of arbitrarily shaped bodies of revolution. Proc IEEE 1989;77(5):735–740.

7

DOPPLER RADAR PHYSIOLOGICAL ASSESSMENTS

John Kiriazi[1], Olga Boric-Lubecke[2], Shuhei Yamada[2], Victor M. Lubecke[2], and Wansuree Massagram[3]

[1]*QCT RF Systems, Qualcomm Inc., San Diego, California, United States*
[2]*Department of Electrical Engineering, University of Hawaii at Manoa, Honolulu, Hawaii, United States*
[3]*Department of Computer Science and Information Technology, Naresuan University, Phitsanulok, Thailand*

The Doppler radar detects all motion in the radar field of view, through detection of phase variations in the received signal. For a relatively still, seated, or recumbent subject, this includes random fidgeting motion and quasi-periodic positional variations of the chest surface due to cardiopulmonary activity. Periodic chest motion will map an arc in the complex plane, with magnitude of received radio frequency (RF) power proportional to arc radius, and amplitude of motion proportional to the angle spanned by the arc. Phase demodulation provides the output proportional to chest displacement, and this information can be further analyzed to extract respiratory and heart rates, analyze the shape of respiratory signals, assess heart rate variability (HRV) parameters, and estimate displacement amplitude and related respiratory volume. Moreover, the magnitude of received RF power can be analyzed to determine cardiopulmonary radar cross section (RCS) and further determine subject orientation.

Doppler Radar Physiological Sensing, First Edition.
Edited by Olga Boric-Lubecke, Victor M. Lubecke, Amy D. Droitcour, Byung-Kwon Park, and Aditya Singh.

7.1 ACTIGRAPHY

The challenge in physiological monitoring via Doppler radar is to effectively isolate the subject's random fidgeting physiological motion. This motion will result in irregular arc shapes in the complex plane, due to sudden changes in reference position and motion amplitude. These changes can be tracked and analyzed to detect the periods of significant motion; however, it is problematic to effectively separate cardiopulmonary motion from fidgeting motion. On the other hand, it is possible to detect and eliminate these periods of significant motion and extract cardiopulmonary parameters only during periods with no significant motion. The percentage of measurement interval containing significant motion may be used as a measure of subject rest/activity cycle, determining the degree of restlessness, for example, actigraphy, which is associated with sleep disorder studies.

The Doppler radar physiological measurements were conducted according to the Committee on Human Studies (CHS) protocol number CHS14884, approved by the CHS of University of Hawaii at Manoa. The data were taken from subjects in seated and supine positions. The subjects wore normal clothing and were instructed to breath normally during the measurements. The challenges in designing the test setup for this study were simultaneous measurement of reference and radar signals, human safety, and signal integrity. Figure 7.1 shows the block diagram of the measurement setup. The reference respiration signals – ECG, finger pulse, respiration signals from piezoelectric sensors on a subject's upper and lower torso, and air volume from spirometer – were captured through the BIOPAC data acquisition system (DAQ).

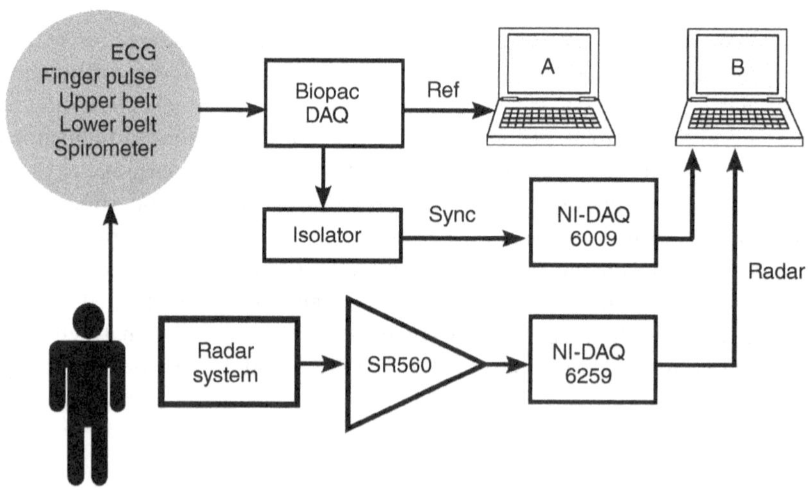

Figure 7.1 A block diagram of test protocol setup. The reference signals, ECG, finger pulse, upper chest belt, lower chest belt, and spirometer, were completely isolated from the radar system. The sync signal provided the marker for when the BIOPAC reference system starts to line up the data from the two systems.

The cardiopulmonary motion signals from Doppler radar were captured through the National Instrument DAQ. The two DAQ systems were synchronized with a marker signal from the BIOPAC system via an optical isolator and a National Instrument DAQ card (National Instruments NI-DAQ PCI-6009). The first pulse was sent out when BIOPAC DAQ started capturing data. The pulse train had different width to minimize the synchronization error.

The Doppler radar system, at 2.4 GHz with 0-dB m power level at the antenna connector, used these following commercially available components: one transmitting antenna (Antenna Specialist ASPPT2988), four receiving antennas (Antenna Specialist ASPPT2988), eight zero-degree power splitters (Mini-Circuits ZFSC-2-2500), four 90° power splitters (Narda 4033C), and eight mixers (Mini-Circuits ZFM-4212). The baseband output signals were amplified and filtered with low-noise amplifiers (LNAs; Stanford Research Systems SR560) and then digitized with an onboard analog-to digital converter (ADC) of a National Instruments DAQ card (National Instruments NI-DAQ PCI-6259). The software to collect and process the data was written in MATLAB.

The 2.4 GHz AC-coupled system was used in this study for low cost and simplicity. Linear demodulation was chosen to recover the phase variations, since it is robust with respect to distortion due to AC coupling and noise.

After linear demodulation, the threshold-based peak detection was used to identify the exact locations of respiration peaks, as well as detect the period of motion. Motion detection was based on sudden changes in amplitude of motion and reference position. Duration of motion was measured, and statistical analysis performed on respiratory peaks detected when there was no significant motion.

The results from 17 healthy volunteers are presented. In both seated and supine positions, the antennas were 1-m away from the front torso of the subjects. The 1 m distance was chosen for mounting simplicity for supine measurements. Data were taken for 30 min in the seated position, and 10 min in supine position. Figure 7.2 shows an example of 30 min data from the Doppler radar output after linear demodulation, taken for a seated subject. The spikes indicate fidgeting or a shift in body position; dotted gray lines indicates fidgeting; solid dark gray line shows respiration; gray cross markers indicates peak inhalation; solid light gray trace represents partial indication of motion artifacts.

The duration of motion artifact period caused by subject movement could be considered as the activity period. The results showed the reality of how much data could be accurately extracted with the current system setup and algorithm. The 30-min data were then divided into three segments of 10-min data: the first 10 min, the second 10 min, and the last 10 min, to evaluate which segment contains more activity periods in general.

The results of 17 subjects from 30-min seated measurement (with their corresponding 10-min segments) and 10-min supine measurement are shown in Table 7.1. The data from supine position contained less activity periods than seated position with only exception of subject 2201 and 2203.

Figure 7.3 shows the percentage of radar signal without motion artifacts, which illustrates the behavior of human subjects. Almost equal portions of the subjects

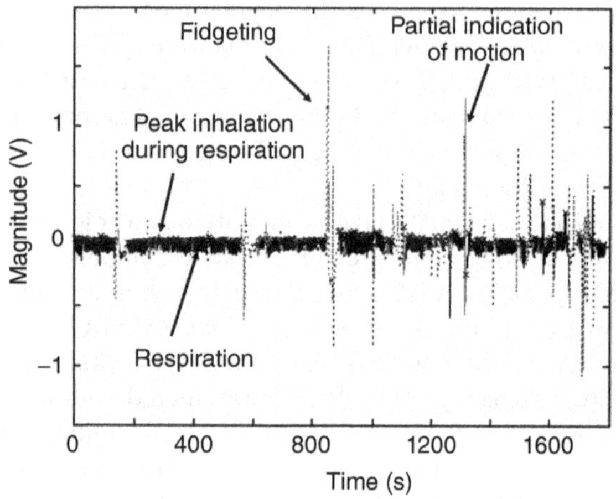

Figure 7.2 An example of 30-min data from the Doppler radar output after linear demodulation, taken for a seated subject. Gray cross markers indicate detected peaks, and dotted gray line shows detected periods of motion. © 2011 IEEE. Reprinted, with permission, from Massagram et al. [2011].

TABLE 7.1 Percentage of Data Without Motion Artifacts

Subject	Seated				Supine
	30-min	First 10-min	Second 10-min	Third 10-min	10-min
2201	75.32	82.79	70.54	70.86	35.57
2202	44.79	54.55	39.89	41.46	98.05
2203	59.70	47.53	67.63	64.46	46.32
2204	75.72	62.66	87.16	77.09	92.15
2205	88.09	63.66	100.00	100.00	97.81
2206	38.57	24.50	34.52	57.73	87.61
2209	93.00	92.06	92.06	94.44	97.38
2210	95.55	95.37	98.84	92.08	97.14
2301	93.55	100.00	98.41	82.55	100.00
2302	65.78	70.54	64.81	62.34	90.74
2303	65.61	64.74	61.34	70.43	91.99
2304	39.69	51.57	19.37	51.06	85.25
2306	67.40	74.51	79.21	48.97	100.00
2401	97.85	99.90	97.85	95.62	100.00
2402	99.19	100.00	97.52	100.00	99.40
2403	78.71	35.02	100.00	100.00	100.00
2404	51.75	54.62	52.40	49.64	98.46
Average	72.37	69.06	74.21	74.04	89.28

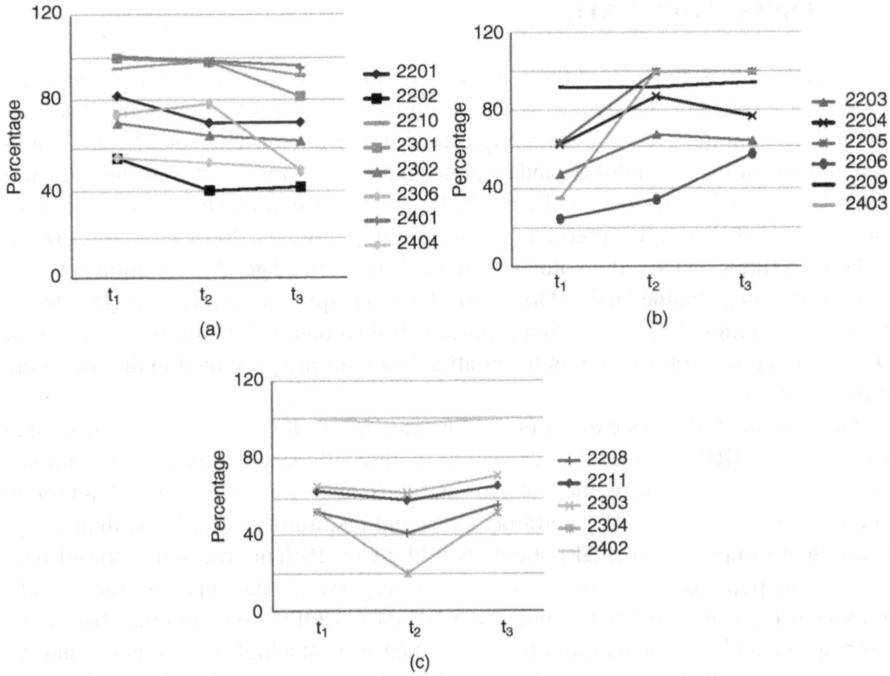

Figure 7.3 The behavior of human subjects varied from (a) remained still at the beginning of the measurement then started fidgeting, (b) started fidgeting at first then remained still in the rest of the measurement, and (c) remained still at first, started to fidget, and then back to still during the measurement. © 2011 IEEE. Reprinted, with permission, from Massagram et al. [2011].

(a) remained still at the beginning, (b) started fidgeting at first then remained still, and (c) remained still at the beginning and end but fidgeted in the middle of the measurement.

The information about the subjects' "active" versus "inactive" periods can be determined. Both Table 7.1 and Fig. 7.3 show the percentage of time subject is quiescent. Evidently, most subjects were not able to remain still for the entire duration of measurement. The behavior of subjects varied and can be categorized into three groups as mentioned. These three behavioral groups contained almost the same amount of subjects. On the average, each subject could remain still at approximately more than 72% of the data length for seated and 89% for supine positions. Moreover, in 15 out of 17 subjects, there was no significant motion for more than 85% of the measurement interval in supine positions. The experimental results in this study show that Doppler radar could be used as a means of determining how often the subject shifts position while sleeping. This technique may be particularly suitable for sleep apnea monitoring, infant sudden infant death syndrome (SIDS) monitoring, fatigue monitoring, in-hospital monitoring, and home health care.

7.2 RESPIRATORY RATE

Respiratory rate is considered the next critical vital sign and yet often goes underestimated or ignored primarily due to shortcomings of the currently used measurement methods. Respiratory rate provides important information on a person's health condition and physiological stability, and an abnormal respiratory rate is a strong indicator that a health crisis is imminent [Buist, 2004]. In fact, a sudden change in respiratory rate is one of the strongest predictors of mortality [Hodgetts, 2002; Cretikos, 2006].

Human studies on healthy volunteers have demonstrated good correlation between respiratory rates obtained using Doppler radar and respiratory effort belts [Droitcour, 2006; Massagram, 2008]. The first reported clinical data validating the accuracy of Doppler radar respiratory rate on hospitalized patients are presented in the following text/paragraphs.

The human study was carried out at Queen's Medical Center under institutional review board (IRB) number RA-2008-061 on clinically stable patients. The patients included in this study were not selected randomly; they were selected with an intent to cover a broader range of respiratory rates and respiratory waveforms than a representative sample of hospital patients would cover. Patients receiving opioid pain medication, patients recovering from thoracic surgery, and patients with lung conditions such as chronic obstructive pulmonary disease (COPD), pneumonia, obstructive sleep apnea, and pulmonary embolism were measured at a higher than representative frequency. Twenty-four patients were evaluated in this study. In pilot studies performed under the same IRB approval, the difference in respiratory rates between the Doppler radar respiratory rate and references had a standard deviation of approximately 1.3. To achieve a standard error of limits of agreement of approximately 0.5 breaths/min, the desired sample size was determined to be between 20 and 25 patients:

$$n = 3\,\frac{(\text{expected standard deviation of difference})^2}{(\text{desired limits of agreement})^2} = 3\frac{1.3^2}{0.5^2} = 20.3$$

Twenty-four subjects were included in this study, including 15 males and 9 females. Their age ranged from 43 to 91 years, with a mean age of 70 years. Body mass index ranged from severely underweight (BMI = 14.0) to morbidly obese (BMI = 48.1), with a mean BMI of 29.7. The demographic information is summarized in Table 7.2. For one subject, clinical information beyond age and sex were not collected. Of the remaining 23 subjects, 5 had undergone surgery during their current hospital admission, 3 of which were open-heart surgeries. Six patients were receiving pain medications and four of them were receiving opioid analgesics. Four patients were receiving supplemental oxygen during the measurements.

Patients were monitored using radar while their vital signs were being monitored by other equipment. Low-power 2.4 GHz Doppler radar with proprietary hardware and software was used in the study, facing the device toward the patient's thorax at a distance of about 1 m. The radio power emitted by this Doppler radar device is well below that of many consumer and hospital wireless electronic devices, so

TABLE 7.2 Patient Demographic Information

	Age ($n = 24$)	Sex ($n = 24$)	BMI ($n = 23$)	Respiratory Rate ($n = 24$)
Mean	69	15 males 9 females	29.7	18
Standard deviation	15		8.3	4
Max	91		48.1	26
Min	43		14.0	11

the radio power does not pose any significant safety risk. After measuring for the user-selected interval, the Doppler radar device processes the data to determine the quality of the signal, and if the signal is of adequate quality to provide an accurate rate, it displays the patient's respiratory rate on the screen. All patients had the reference measurements: the Welch Allyn Propaq provided respiratory rate via thoracic impedance measurement, and the Embla system provided respiratory rate through inductive plethysmographic measurement of respiratory effort. Several of the patients also had continuous pulse oximetry and ECG monitored by another device. A respiratory rate was also obtained by counting respiratory excursions for the same duration as the Doppler radar measurement interval, simultaneously with the Doppler radar measurement. The counting of chest excursions involved counting the number of peak inhalations in the specified time interval, as timed with a stopwatch, and multiplying by the appropriate number to calculate breaths per minute.

Once powered and connected to the patient, the Welch Allyn Propaq Encore model 242 continuously updates and displays a respiratory rate if the RESP function is enabled. The Welch Allyn Propaq requires affixing electrodes on the patient's skin, attaching lead wires to the electrodes, and plugging the ECG leads into the Propaq 200-series unit. It measures respiratory effort by running a small AC current between the electrodes and monitoring the change in impedance as the patient breathes. Cardiogenic artifact is removed from the impedance waveform, and it is analyzed to determine a respiratory rate. This rate is displayed on the local screen. For this study, the rate displayed at the end of the Doppler radar measurement was recorded for comparison.

The Embla Embletta GOLD system with XactTrace belts and Somnologica software is a body-worn system, which continuously records respiratory signals. Once an XactTrace abdomen belt and an XactTrace thorax belt are connected to the patient, the recording is initiated by pressing the "start" button. In this configuration, the Embletta system records the respiratory effort waveforms in the Embletta unit. After the measurement is complete and the belts are detached from the Embletta unit, the data are transferred to a PC running the Somnologica software. The Somnologica software analyzes the signal and provides a respiratory rate. The XactTrace belts, when used with the Embletta system, use inductance pneumography to obtain a respiratory effort. These chest belts include an embedded wire coil; the respiratory effort signal is obtained by sending an AC signal through the wire in the chest belts and measuring the change in the chest belts' inductance, as the shape of the patient's chest changes with breathing. This system cannot be configured to provide a respiratory rate in real time. For the measurements in this study, the time stamp button on the Embletta unit was pressed at the beginning and end of each spot check respiratory measurement. The rate used for this analysis was the rate presented by the system at the time stamp at the end of the measurement interval.

For a respiratory rate spot check, literature on the repeatability and interobserver variability in visual assessment provides indication of the clinically relevant range for agreement. Lim et al. [2002] found a repeatability coefficient of 4.1 breaths/min for respiratory rate measurements made sequentially by the same observer, and a repeatability coefficient of 5.7 breaths/min for sequential respiratory measurements made by different observers, and a repeatability coefficient of 4.3 breaths/min for simultaneous measurements made by different observers, with all the measurements performed on adults. Based on these data, the 95% limits of agreement for a respiratory rate spot check should be less than ±4 to ±6 breaths/min.

The primary data analysis method used was Bland–Altman analysis: identification of the 95% limits of agreements. It is expected that 95% of differences in measurements made simultaneously with the two analyzed methods would lie within these limits. The 95% limits of agreement are calculated as the bias (the mean difference between each method) ±2 standard deviations of the difference between the measurements from each method. The data and the 95% limits are plotted in the Bland–Altman plots for each method comparison, and the bias, standard deviation, and 95% limits are shown in the table for the comparison of each method. The difference between the methods is also shown as the root mean square of the difference between measurements with each method. Finally, a linear regression is performed, and the equation of the regression line and the correlation coefficient are plotted. The agreement between Doppler radar and the three references is summarized in Table 7.3. As shown in Table 7.3, the 95% limits of agreement between the Kai

TABLE 7.3 Summary of Agreement of Doppler Radar with Reference Measurements

	Doppler Radar and Welch Allyn Propaq Encore	Doppler Radar and Embla Embletta System	Doppler Radar and Visual Assessment
Bias (mean of difference between measurements)	−0.5	−1.31	−0.81
Standard deviation of difference between measurements	1.8	1.6	1.1
95% confidence limit: high	3.0	1.8	1.4
95% confidence limit: low	−4.0	−4.5	−3.1
RMS difference	1.8	2.0	1.4
Linear regression equation	$y = 0.81x + 2.65$	$y = 0.92x + 0.14$	$y = 0.97 - 0.29$
Correlation coefficient	$R^2 = 0.89$	$R^2 = 0.89$	$R^2 = 0.94$

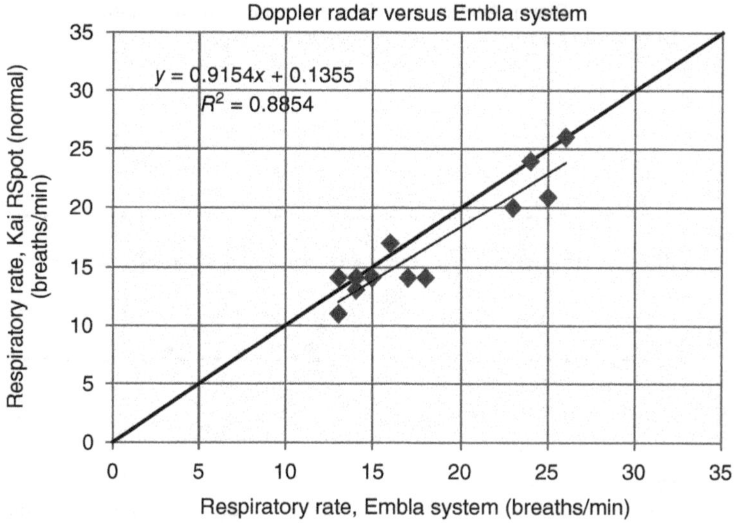

Figure 7.4 The linear regression of the respiratory rate provided by the Doppler radar system and that provided by the Embla system. © 2009 IEEE. Reprinted, with permission, from Droitcour et al. [2009].

RSpot and all three reference measurements fall within ±5 breaths/min. Correlation coefficient between the Doppler radar respiratory rate and that obtained by all three references is at least 0.89. Standard deviation of difference between measurements and the root mean square of the difference are both less than 2 breaths/min.

Figure 7.4 shows the linear regression of the respiratory rate provided by the Doppler radar system and that provided by the Embla system, showing strong correlation between the two measurements. Figure 7.5 shows the Bland–Altman plot of the difference versus the mean of measurement of respiratory rates provided by the Doppler radar and by the Embla system. As indicated in Table 7.3, the 95% confidence intervals fall between +1.8 and −4.5 breaths/min.

7.3 TIDAL VOLUME

Tidal volume is the volume of air an individual is normally breathing in and out in one cycle. Example studies by Kondo et al. [1997] and Siebens [2008] have separately demonstrated the linear relationship between volume and chest wall displacement during unobstructed breathing. The current practices to measure respiration rates and lung volumes are measurement of airflow and respiratory effort/movement. Direct measurement of airflow typically uses face masks or mouthpiece, which can be obtrusive and change the subject's respiration. Indirect measurement of airflow, such as thermocouple or capnography, has less adverse affects, but still requires the placement of sensors in front of the nose and/or month. Respiratory effort/movement

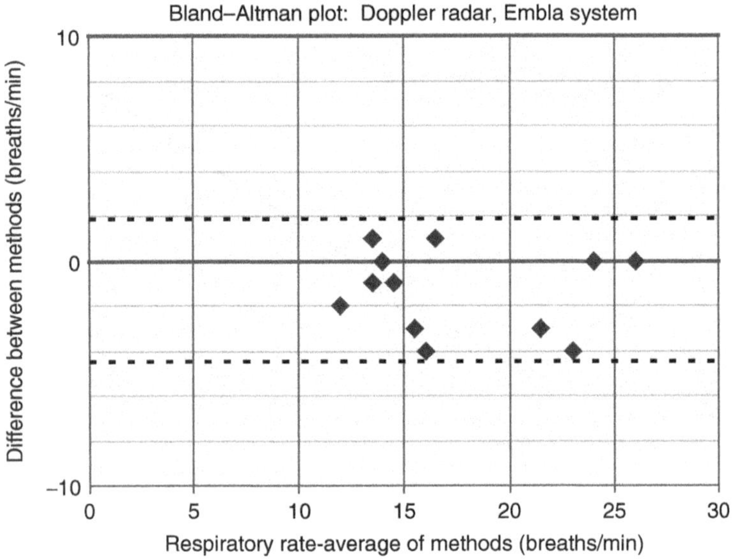

Figure 7.5 Bland–Altman plot: the difference versus the mean of measurement of respiratory rates provided by the Doppler radar and by the Embla system. © 2009 IEEE. Reprinted, with permission, from Droitcour et al. [2009].

measurement typically requires direct contact with the patient through various chest bands that may impede unrestricted chest motion.

Considering the linear relationship between the chest wall displacement and air volume changes, the output of the Doppler radar system should be proportional to the tidal volume during normal unobstructed breathing. Unlike the contact methods, Doppler radar offers the benefit of unobtrusive noncontact chest wall displacement measurement.

The microwave signal has a large average clutter DC offset due to the reflection caused by the environment and the subject's stationary parts. In order to remove the comparatively large baseline shifts and allow for sensitive measurements, the detected signal must be AC coupled. However, the AC-coupling circuit changes the shape of the actual microwave signal due to the high-pass filter used to remove DC (Fig. 7.6).

Since the shape and amplitude of the DC-coupling signal are linearly proportional to the shape and amplitude of the volume displacement, it is thus necessary to transform the recorded AC-coupled signal using the transfer function of the AC-coupling circuit to obtain the actual shape of the waveform. The AC-coupling circuit in the preamplifier LNA uses an analog two-pole high-pass filter with a roll-off frequency of 0.03 Hz. Thus, the DC-corrected signal can be recovered by integrating the signal twice. Each integral has to take the RC time constant into account. The RC was determined from the measured step-response of the LNA and found to be 49.75 ms (the measured roll-off frequency is approximately 0.0201 Hz instead of 0.03 Hz). Figure 7.6 shows an example of actual signal, AC-coupled

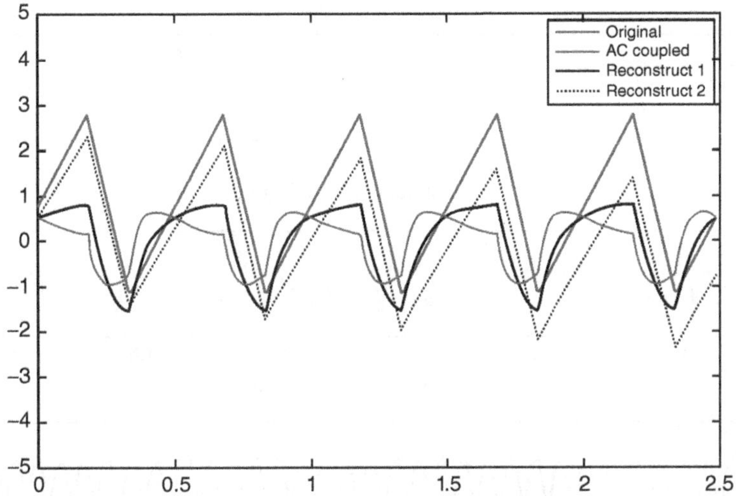

Figure 7.6 Example of DC-corrected operation. The signal, reconstruct 1, represents the output from the first integration, and the signal, reconstruct 2, represents the output from the second integration.

signal, and the DC-corrected signal by transforming the AC-coupled signal. Notice that the baseline of the DC-corrected signal shifts due to the nature of integration. This can be corrected by removing the mean and performing a polynomial fit to the signal.

The measurements were performed as described in Section 7.1 with spirometer used as a reference. The respiratory signals are shown in Fig. 7.7, and the instantaneous respiration rates are shown in Fig. 7.8. The output from radar and lower chest belt corresponds well with the output from the spirometer. The upper chest belt, however, did not produce a clean and well-correlated signal.

Tidal volume represents the volume of the air moved during normal breathing and is calculated from the peak-to-peak amplitude of the volume displacement. Figure 7.9 shows the tidal volume corresponding to the volume displacement. The radar signal is well within $\pm 5\%$ of the spirometer. The statistical analysis in Fig. 7.10 shows that the tidal volume of the radar correlates very well with the tidal volume from the spirometer. The Pearson product-moment correlation coefficient (PMCC) is 0.95378. The Bland–Altman analysis shows that the mean difference of the two tidal volumes is less than 10 mL, with the standard deviation of the difference of 20 mL. For the Gaussian distribution, the 95% confidence interval is -32 to 50 mL range. All methods worked well for finding the instantaneous respiration rate, as shown in Fig. 7.10. The experiments performed on 10 human subjects showed similar outcomes.

The respiratory signal output of Doppler radar system shows linear relation to the volume displacement when compared with conventional airflow and respiratory effort/movement measurements. Calibration of displacement to airflow before

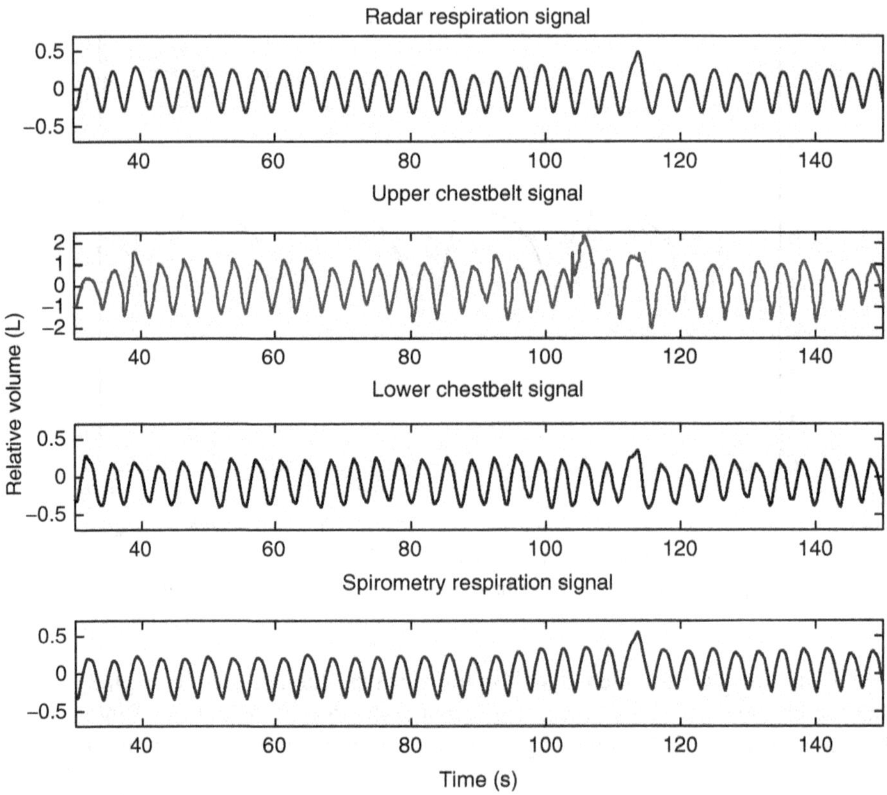

Figure 7.7 Relative volume displacement of the radar, upper chest belt, lower chest belt, and spirometer respiration signals. © 2009 IEEE. Reprinted, with permission, from Massagram et al. [2009].

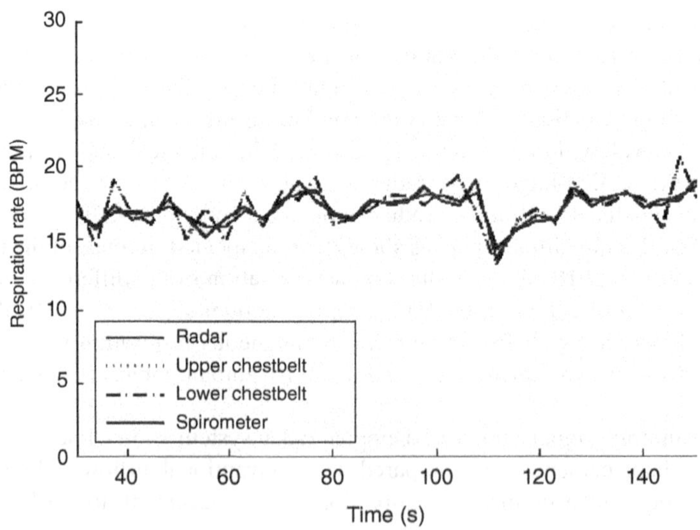

Figure 7.8 Instantaneous respiration rates for all signals. © 2009 IEEE. Reprinted, with permission, from Massagram et al. [2009].

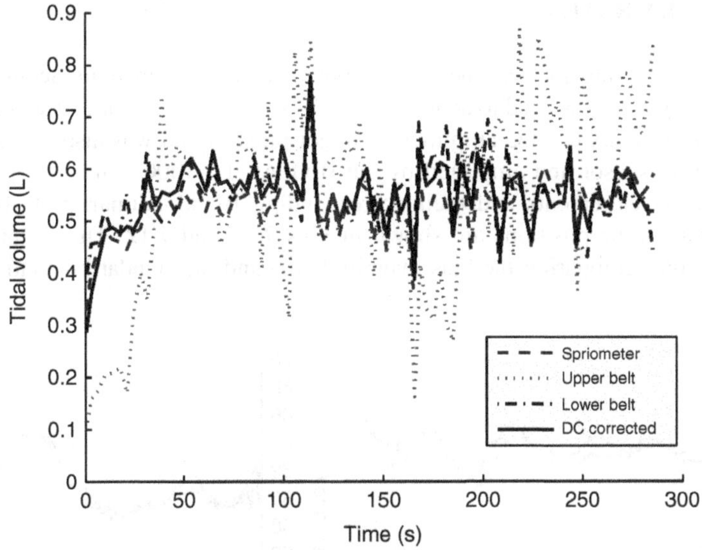

Figure 7.9 Tidal volumes of the radar, upper chest belt, lower chest belt, and spirometer respiration signal. © 2009 IEEE. Reprinted, with permission, from Massagram et al. [2009].

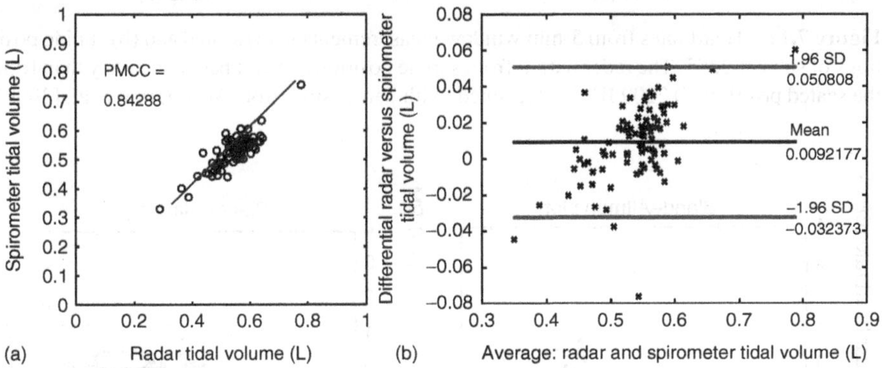

Figure 7.10 Statistical analysis of the tidal volume: (a) correlation plot and (b) Bland–Altman analysis. © 2009 IEEE. Reprinted, with permission, from Massagram et al. [2009].

subject measurements and accurate chest wall position information enable mean differences of less than 10 mL, with the standard deviation of the difference of 20 mL between radar and reference measurements. Unlike the current practices of respiration rate and lung volume measurements via airflow and measurement of respiratory effort/movement, Doppler radar offers the benefit of reliable unobtrusive noncontact measurement.

7.4 HEART RATES

Data for heart rate measurement were obtained as explained in Section 7.1. A three-lead system was used to acquire the ECG signal for reference. It was amplified by the BIOPAC ECG100C amplifier. A finger pulse signal was also captured via a piezoelectric sensor and amplified by the BIOPAC DA100C. The examples of the heart rate measurements and their corresponding Bland–Altman plots from both seated and supine positions are shown in Figs 7.11 and 7.12. The Bland–Altman statistical plot comparing the bias magnitude ($\overline{d}$) and the standard deviation (s_d) of

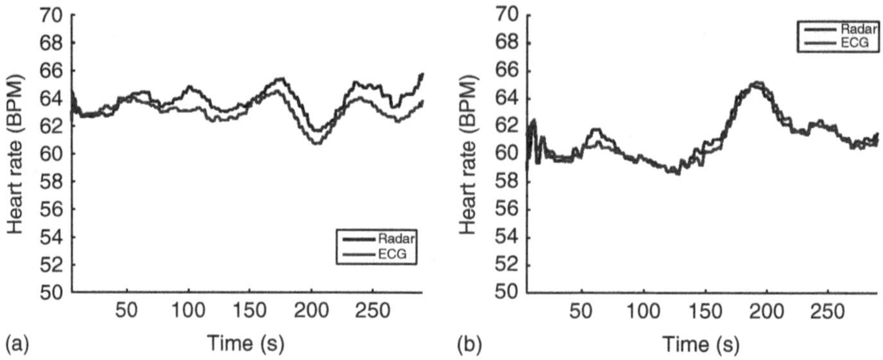

Figure 7.11 Heart rates from 5-min window measurements in (a) seated and (b) supine positions of subject 2205. The radar output from supine position shows a better accuracy than from the seated position. © 2009 IEEE. Reprinted, with permission, from Massagram et al. [2009].

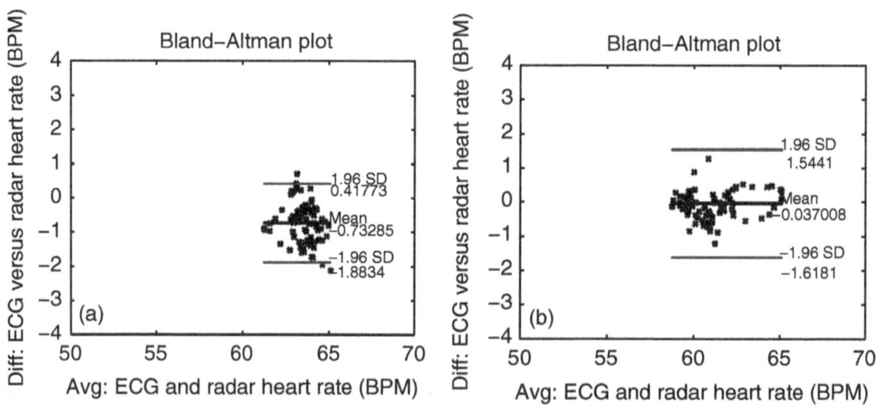

Figure 7.12 Bland–Altman plots from 5-min window measurements in (a) seated and (b) supine positions of subject 2205. All data points in supine measurement were with 95% confidence intervals unlike for seated measurement. The Bland–Altman bias magnitude in seated position is much greater than in supine position (seated −0.733 mL vs supine −0.037 mL). The BA plot in (a) displays a cloud shape, a sign of bad correlation, whereas (b) is almost a line.

TABLE 7.4 Average Heart Rate and Bland–Altman Analysis for Seated Position Measurements

Subject	Average Heart Rate (BPM)		Bland–Altman Analysis	
	ECG	Radar	$\bar{d}$ (BPM)	s_d (BPM)
2201	63.08	63.47	0.3897	3.9134
2202	67.58	70.27	2.6899	5.2314
2203	61.06	63.16	2.1	7.1696
2204	64.39	68.19	3.7963	2.6283
2205	63.88	63.15	−0.7328	0.587
2206	69.79	72.80	3.0085	4.4347
2208	75.03	72.80	−2.2264	5.9458
2209	70.45	70.60	0.1561	2.7381
2210	63.31	63.18	−0.129	2.0617
2211	56.64	56.43	−0.2096	1.9183
2301	76.72	79.31	2.59	3.0827
2302	68.67	71.08	2.4165	6.4347
2303	62.98	76.26	13.2801	8.5438
2304	73.95	78.03	4.0752	7.5066
2306	63.96	58.66	−5.2989	3.1783
2401	68.04	68.32	0.2812	1.1528
2402	58.99	58.28	−0.7123	0.9903
2403	56.01	57.18	1.1662	1.8073
2404	64.63	60.95	−3.6866	8.8792

the heart rates found in the Doppler radar and ECG signals are shown in Tables 7.4 and 7.5 for the seated and supine measurements, respectively.

The radar measurements obtained from subject 2303 in both seated and supine position did not correlate well with the references. One hypothesis is that the breathing spectrum of this subject interfered with his heart rate. The breathing of subject 2303 was not periodic, unlike the breathing of subject 2205 as shown in Fig. 7.13. Figure 7.14 shows the difference between the breathing power spectrum density plots of subject 2303 versus 2205. The breathing frequency of subject 2205 is observed clearly at 0.217 Hz while the same cannot be said for subject 2303. The breathing spectrum of subject 2303 spread wider and could overlap the heart period.

The Bland–Altman analysis in Tables 7.4 and 7.5 shows that Doppler radar heart rate signal is more accurate when the subject is lying down in supine position.

7.5 HEART RATE VARIABILITY

HRV refers to the beat-to-beat alterations in heart rate. This indicator of the activity of autonomic regulation of circulatory function reflects how well the cardiovascular system works. It has received a tremendous amount of attention and has been used in countless studies to assess the effects of autonomic nervous system and

TABLE 7.5 Average Heart Rate and Bland–Altman Analysis for Supine Position Measurements

Subject	Average Heart Rate (BPM)		Bland–Altman Analysis	
	ECG	Radar	$\bar{d}$ (BPM)	s_d (BPM)
2201	60.75	62.57	1.8151	5.5053
2202	70.72	71.45	0.7384	0.9575
2203	58.92	71.19	12.2721	8.123
2204	58.26	57.44	−0.816	1.3442
2205	61.17	61.13	−0.0411	0.7973
2206	66.77	70.06	3.2985	2.8418
2209	60.02	61.80	1.7749	2.0546
2210	64.32	64.17	−0.1468	0.42
2301	83.93	89.85	5.9256	8.2259
2302	66.65	67.97	1.3207	2.4439
2303	60.30	76.54	16.2396	7.0326
2304	72.87	74.74	1.8718	1.3737
2305	67.32	67.61	0.2909	1.5212
2306	62.81	62.66	−0.1517	1.5581
2401	58.59	58.54	−0.0479	0.9809
2402	55.61	63.89	8.283	7.3905
2403	55.07	55.97	0.9026	3.1365
2404	59.89	59.59	−0.3002	1.192

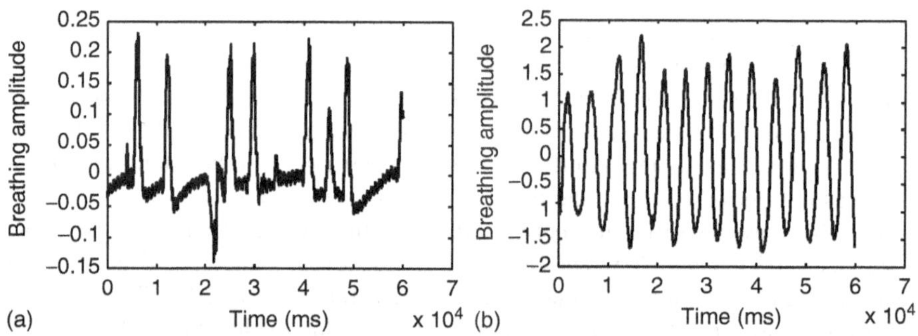

Figure 7.13 Breathing signals (a) subject 2303 with irregular period and nonsinusoidal pattern and (b) subject 2205 with regular period and sinusoidal pattern.

cardiovascular activity. Continuous beat-to-beat interval or instantaneous heart rate measurements are the source information for HRV analysis. The beat-to-beat interval expresses the time duration between the heart beat, normally within milliseconds of accuracy. The heart rate is the number of heart beat per unit of time, normally expressed as beats per minute (BPM). The electrocardiograph (ECG) is traditionally considered the standard way to measure the beat-to-beat intervals. The other

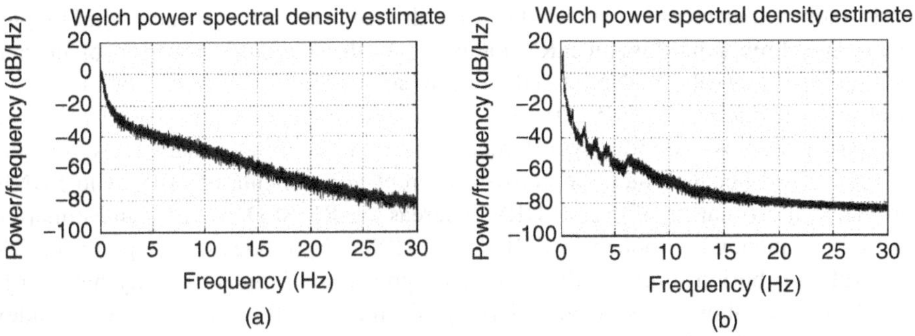

Figure 7.14 PSD of the breathing signals (a) subject 2303 with wilder spectrum spread and (b) subject 2205 well-defined peak at 0.217 Hz and narrower spectrum spread.

approach to measure beat-to-beat intervals is the measurement of pulse waves. The pulse wave methods, such as (1) a photoplethysmograph (PPG) or (2) a piezoresistor sensor, are less invasive and simpler than ECG, yet all the methods mentioned here require patients to be tethered to the sensing devices. Doppler radar detection of respiratory and heart rates has been known for more than three decades [Lin, 1975, 1979]. The challenge for using Doppler radar to acquire the beat-to-beat intervals for HRV analysis is whether it could provide acceptable accuracy for a period of time needed (from 2 to 5 min, and up to 24 h). The remote sensing of HRV could prove to be a powerful tool for health-care monitoring and medical studies. Such unobtrusive sensing could benefit patients with conditions that may be altered or worsened by contact sensors, such as when monitoring for sleep disorders, in SIDS prevention, and in providing burn victim care. The challenge for using Doppler radar to acquire the beat-to-beat intervals for HRV analysis is whether the detection method used is robust enough to provide acceptable accuracy in comparison against other methods.

A task force of the European Society of Cardiology and the North American Society of Pacing and Electrophysiology published standards of measurement, interpretation, and use of HRV in 1996 [Task force, 1996]. The task force specified many different HRV metrics for both short-term records (5 min) and long-term records (24 h). Many other measures of HRV have been proposed and investigated throughout the years, nevertheless those specified by the task force have been the most widely applied. Methods of obtaining the HRV analysis maybe divided into four main groups: (1) time domain methods, (2) frequency domain methods, (3) mathematic modeling methods, and (4) nonlinear methods. The most common HRV measurement methods are the time and frequency domain methods. In this study, the time domain methods were applied. The basis of HRV calculation in time domain is either the heart rate at any point in time or the intervals between successive complexes. In a continuous electrocardiographic (ECG) record, each QRS complex is detected and the intervals between adjacent QRS complexes (normal-to-normal or N–N, also called beat-to-beat or R–R), or the instantaneous heart rate is determined.

In these experiments to assess the feasibility of HRV measurement via Doppler radar, the three time-domain HRV indexes: (1) the standard deviation of normal beat-to-beat intervals (SDNN), (2) the root mean square of differences of successive beat-to-beat intervals (RMSSD), and (3) the triangular index (HRV TI) were compared and observed for any differences. These three time-domain measurements were selected because they emphasize different components of HRV. The SDNN provides an estimation of overall HRV, whereas the RMSSD provides an estimation of high-frequency components of HRV. The HRV TI provides the shape of the RR interval distribution; uniform distributions representing large variability have large values and distributions with single large peaks have small values. Each HRV index is described in details as follows.

SDNN: Simple time-domain variables can be calculated include the mean NN interval, the mean heart rate (HR), the difference between night and day HR, and so on. More complex statistical time-domain measures can be calculated from a series of instantaneous HR or cycle intervals. These measures may be divided into two classes: (1) those derived from measurements of the NN intervals or instantaneous HR and (2) those derived from the difference between NN intervals. These variables may be derived from the analysis of the total ECG recording or calculated using smaller segments of the recording periods. The latter one allows comparison of HRV to be made during varying activities, for example, rest and sleep. The following are the mathematical interpretations derived from the task force's description. Each time the N beat occurs is denoted as $t(n)$ for $n \in \{1, \ldots, N\}$. The interval between beats is denoted as $\delta(n) = t(n) - t(n-1)$. The time of occurrence of each interval $\delta(n)$ is $t(n)$. The SDNN is determined as

$$\text{SDNN} = \sqrt{\frac{1}{N-2}\sum_{n=2}^{N}(\delta(n) - \overline{\delta})^2} \tag{7.1}$$

where $\overline{\delta}$ is the average NN interval,

$$\overline{\delta} = \frac{1}{N-1}\sum_{n=2}^{N}\delta(n) \tag{7.2}$$

The scaling factor is $N-2$ because there are $N-1$ intervals in the record and one degree of freedom is used to estimate the mean NN interval. The SDNN is considered to be the simplest variable to be calculated. Since the SDNN is the square root of variance and is mathematically equal to the total power of spectral analysis (TP in frequency domain), it reflects all the cyclic components responsible for variability in the period of recording. It is important to note that the SDNN is length dependent since the total variance of HRV increases with the length analyzed recording. The standardized durations of recording for SDNN analysis are the short-term 5 min and the nominal long-term 24 h.

RMSSD: The RMSSD between adjacent intervals is determined as

$$\mathrm{RMSSD} = \sqrt{\frac{1}{N-2}\sum_{n=3}^{N}[\delta(n)-\delta(n-1)]^2} \tag{7.3}$$

The RMSSD is related to the high-frequency (HF) energy of HRV in the band from 0.15 to 0.4 Hz.

HRV-TI: The series of NN intervals can also be converted into a geometric pattern using three general approaches: (1) a basic measurement of the geometric pattern (e.g., the width of the distribution histogram at the specific level) is converted into the measure of HRV, (2) the geometric pattern is interpolated by a mathematically defined shape (e.g., approximation of the distribution histogram by a triangle, or approximation of the differential histogram by an exponential curve) and then the parameters of this mathematical shape are used, and (3) the geometric shape is classified into several pattern-based categories that represent different classes of HRV (e.g., elliptic, linear, and triangular shapes or Lorenz plots). These methods require the NN interval sequence to be measured on or converted into a discrete scale to permit the construction of smoothed histogram. Mostly, the bins are approximately 8 ms long ($1/2^7$ bit $= 1/128$ s $= 7.8125$ ms corresponding to the precision of current commercial equipment), which are not too fine or too coarse.

HRV triangular index is a measure of the shape of the NN interval distribution. In general, uniform distributions representing large variability have large values and distributions with single large peaks have small values. The metric is defined in terms of a histogram of the NN intervals. $b(t_i)$ represents the number of intervals in the ith bin centered at t_i. HRV TI is defined as

$$\mathrm{HRV\ TI} = \frac{\sum_{i=1}^{N_b} b(t_i)}{\max_i b(t_i)} = \frac{N-1}{\max_i b(t_i)} = \frac{N-1}{Y} \tag{7.4}$$

The geometrical differential index is the difference between the widths of the histogram of differences between adjacent RR intervals measured at selected heights (e.g., 1000 and 10,000 samples).

Since the RR intervals or instantaneous heart rates are the two main parameters for HRV analysis, inaccurate measurement of both metrics will have a direct effect in HRV index calculation. The measurements obtained via protocol number CHS14884 showed that the heart rates from Doppler radar can be measured fairly accurately. The data of 5-min duration with minimum motion artifacts were extracted from the subjects. The heart signal output from the linear-demodulation Doppler radar was then compared with the ECG reference signal for heart rate and HRV indexes. This section shows whether the accuracy in Doppler radar measurement of heart signal is feasible for HRV time-domain analysis of SDNN, RMSSD, and HRV TI.

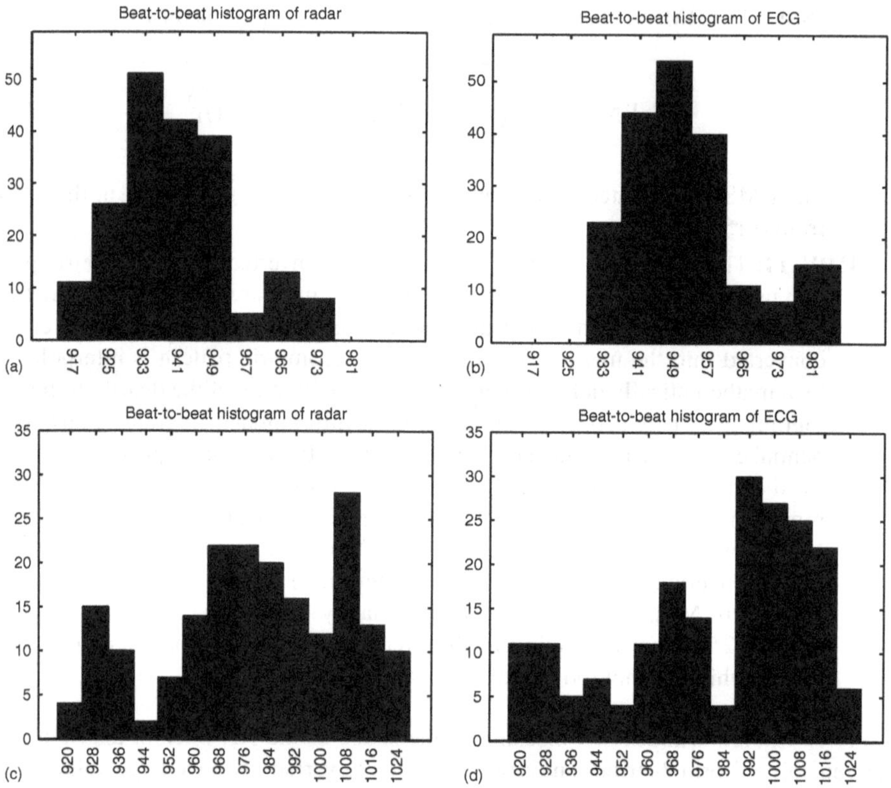

Figure 7.15 RR interval histogram plots of subject 2205 in (a, b) seated positions and (c, d) supine positions, which have wider spread of distribution in RR intervals. © 2009 IEEE. Reprinted, with permission, from Massagram et al. [2009].

The examples of the histograms of RR intervals extracted from Doppler radar and ECG signals are shown in Fig. 7.15. A histogram with 8-ms bin is used in order to derive the HRV TI. The heart rates in supine position varied more than in seated position. Tables 7.6 and 7.7 summarize the HRV indexes from Doppler radar and ECG in seated and supine positions, respectively.

A high degree of accuracy in instantaneous heart rate could be achieved for still subjects during a short period of time (30–60 s) from a Doppler radar system. However, HRV analysis requires at less 2–5 min of accurate RR interval measurements. The results in this section show that it is possible, especially in supine position, to extract an accurate HRV index from Doppler radar measurement.

7.6 RESPIRATORY SINUS ARRHYTHMIA

Respiratory sinus arrhythmia (RSA) has been shown in many empirical studies to represent a sensitive noninvasive index of parasympathetic cardiac control. The literal

TABLE 7.6 HRV Indexes from Doppler Radar and ECG Signals of Subjects in Seated Positions

Subject	SDNN		RMSSD		HRV TI			
	Radar (ms)	ECG (ms)	Radar (ms)	ECG (ms)	Index		Maximum Bin (ms)	
					Radar	ECG	Radar	ECG
2201	54.14	26.08	368.45	49.76	7.50	4.53	942	950
2202	75.23	17.41	601.24	49.70	6.96	4.88	852	852
2203	82.57	24.75	715.96	90.24	8.86	4.15	988	988
2204	25.41	25.44	105.68	53.75	5.91	6.72	934	878
2205	13.48	13.86	31.52	29.06	3.82	3.61	933	949
2206	32.80	13.93	258.78	24.22	8.48	4.53	823	815
2208	38.25	37.86	102.65	111.39	4.64	7.80	757	805
2209	25.35	26.22	48.89	38.87	5.91	7.22	835	819
2210	24.93	25.16	72.28	72.00	7.50	6.29	941	957
2211	28.71	17.92	266.76	57.11	4.64	3.48	1071	1079
2301	22.46	19.59	70.27	36.47	5.42	5.27	767	767
2302	55.14	20.11	482.36	47.59	3.82	3.98	873	841
2303	118.83	28.15	881.59	61.08	11.47	6.96	975	791
2304	62.98	14.78	447.38	28.97	4.53	2.03	770	778
2306	33.83	33.14	139.76	53.93	7.80	6.72	932	1012
2401	18.90	18.93	71.92	65.94	4.33	4.43	908	908
2402	22.44	20.18	58.49	35.88	6.72	5.42	1031	1055
2403	25.74	34.65	71.17	83.72	5.27	8.13	1063	1039
2404	88.43	17.67	569.50	39.81	6.96	5.91	971	995

definition of RSA is a sinus rhythm coincident with breathing: acceleration during inhalation and deceleration during exhalation. RSA phenomenon is observed through the measurement of heart rate and respiration. The ability of Doppler radar to simultaneously obtain both physiological measurements remotely could prove to be useful for future RSA research.

RSA is the fluctuations with the phase of respiration – cardioacceleration during inhalation and cardiodeceleration during exhalation. It is predominantly mediated by respiratory gating of parasympathetic efferent activity to the heart: vagal efferent traffic to the sinus node occurs primarily in phase with expiration and is absent or attenuated during inspiration. RSA is believed to be a major component of HRV. It represents one frequency range of interest (0.15–0.3 Hz) in the entire spectrum of HRV. Many studies have used RSA to indicate the level of mental stress, cardiac aging, denervation and reinnervation after heart transplant, autonomic cardiac control, and the effect of drugs on the sympathetic and parasympathetic systems.

Unlike the guidelines established by the HRV task force [Task force, 1996], there has been no general agreement among investigators as to the standardized quantification techniques for assessing the RSA. The three common quantification procedures

TABLE 7.7 HRV Indexes from Doppler Radar and ECG Signals of Subjects in Supine Positions

Subject	SDNN		RMSSD		HRV TI			
	Radar (ms)	ECG (ms)	Radar (ms)	ECG (ms)	Index		Maximum Bin (ms)	
					Radar	ECG	Radar	ECG
2201	78.88	44.03	553.41	104.71	9.75	7.50	928	936
2202	18.26	22.36	30.66	37.63	3.82	5.00	863	855
2203	62.85	61.08	542.04	158.97	13.00	9.75	1080	832
2204	27.25	29.51	86.47	40.21	5.91	5.57	1049	1049
2205	27.96	29.52	40.90	41.11	6.96	6.50	1008	992
2206	11.65	19.14	70.26	32.19	2.57	3.98	886	862
2209	22.28	18.43	80.52	34.33	6.29	3.25	1011	955
2210	36.86	36.00	58.26	52.17	4.76	6.29	950	950
2301	44.96	12.72	135.38	17.62	3.00	3.68	683	675
2302	24.47	15.84	67.37	22.94	5.91	4.64	882	866
2303	111.30	30.23	968.89	43.44	12.19	5.91	1074	794
2304	14.07	9.82	55.78	21.50	3.48	3.42	820	796
2305	11.48	12.63	34.74	53.38	2.35	4.43	882	890
2306	47.05	41.40	128.72	53.50	5.27	6.72	929	929
2401	18.23	18.40	61.75	75.00	4.15	3.98	1029	1021
2402	67.37	91.60	189.00	166.22	8.13	12.19	1152	1032
2403	31.38	38.55	95.90	91.59	6.29	6.29	1125	1125
2404	12.52	18.33	67.84	39.28	4.31	4.56	1007	999

for estimating RSA are vagal tone ($\widehat{V}$), spectral analysis, and peak–valley, which are explained in detail as follows:

Vagal tone estimation: This method is applicable both time and frequency domain approaches. It eliminates complex aperiodic trends in time series and separates RSA from other frequency components of HRV. It is accomplished by moving a polynomial equation of variable length stepwise through the data, and subsequently estimating the variance of the remaining, filtered time series of points within the presumed respiratory band.

Spectral analysis estimation: The heart period time series is decomposed into separate components with mutually exclusive bandwidths. This method partitioned RSA within the frequency band characterized by the respiration rate of an individual and separated from the other slower rhythmic components. The decomposition approach makes used of FFT and is often referred to as spectral analysis. For this method to be reliably applied, it is generally considered necessary that the specific time series being analyzed meet a requirement of weak stationary. In other words, the mean level and variability of the signal be relatively constant across the assessment period. Studies have shown that spectral estimation of RSA is sensitive to pharmacological vagal manipulations [Akselrod et al., 1981, 1985; Pomeranz et al., 1985].

Peak-to-valley estimation: RSA was quantified in the time domain and measured as the difference in millisecond between the shortest R–R interval accompanying inspiration and the longest interval accompanying expiration. The differences were then averaged across the number of breaths occurred during a measurement period. This method and other similar ones are known as the "peak-to-valley" quantification. Various studies [Fouad et al., 1984; Grossman et al., 1990a; Katona et al., 1977] provide substantial evidence that the peak-to-valley method is highly sensitive to variations in tonic parasympathetic cardiac control. This method uses both respiratory and heart period measurement data. First, the heart-period time series were synchronized with inspiration and expiration periods. The RSA magnitude was computed on a breath-by-breath basis. Inspiration period was used as a window for identifying the minimum heart period for a breath, and expiration period was used as a window for finding the maximum heart period for that breath. Both inspiration and expiration windows were extended forward 750 ms to accommodate the phase shifts occurring between respiration and heart period that are respiratory-rate dependent. In order to ensure that heart period minima and maxima were associated with the phase of respiration, only minima immediately preceded by a longer interval and maxima immediately preceded by a shorter interval were considered valid. In other words, a minimum interval at the beginning of inspiration immediately preceded by an even shorter interval in the previous expiratory phase would not be considered valid. When either of these criteria was violated, the RSA score for a specific breath was scored as zero. This zero was averaged into the analyses – indicating that no observable respiratory modulation of heart period occurred for that breath. When there were valid minimal and maximal heart periods for a breath, RSA was calculated as maximum minus minimum heart period in milliseconds. The mean RSA value for a period was the sum of the individual RSA scores for each breath divided by the number of breaths occurring during the measurement period.

In this study, RSA peak–valley was estimated using the respiratory signal from Doppler radar output and the heart period from ECG. The results were then compared with the RSA amplitude extracted from the upper-torso and lower-torso piezosensor chest belts.

The 5-min duration data with minimum motion artifacts were extracted from the measurement in Section 7.1. The heart period data in milliseconds were obtained from the QRS peak detection algorithm. The statistical correlations of the three RSA values were assessed using PMCC and Bland–Altman analysis. Figure 7.16 is the graphical representation of RSA for subject 2205 in both measurement positions. The RSA peak–valley amplitudes are shown in Fig. 7.17 for the same subject. The results were compiled into Table 7.8 for seated subjects and Table 7.9 for supine subjects.

The peak–valley estimation of RSA amplitude utilizes both physiological signals from the subjects. The technique could be easily applied for a quasi-real-time usage

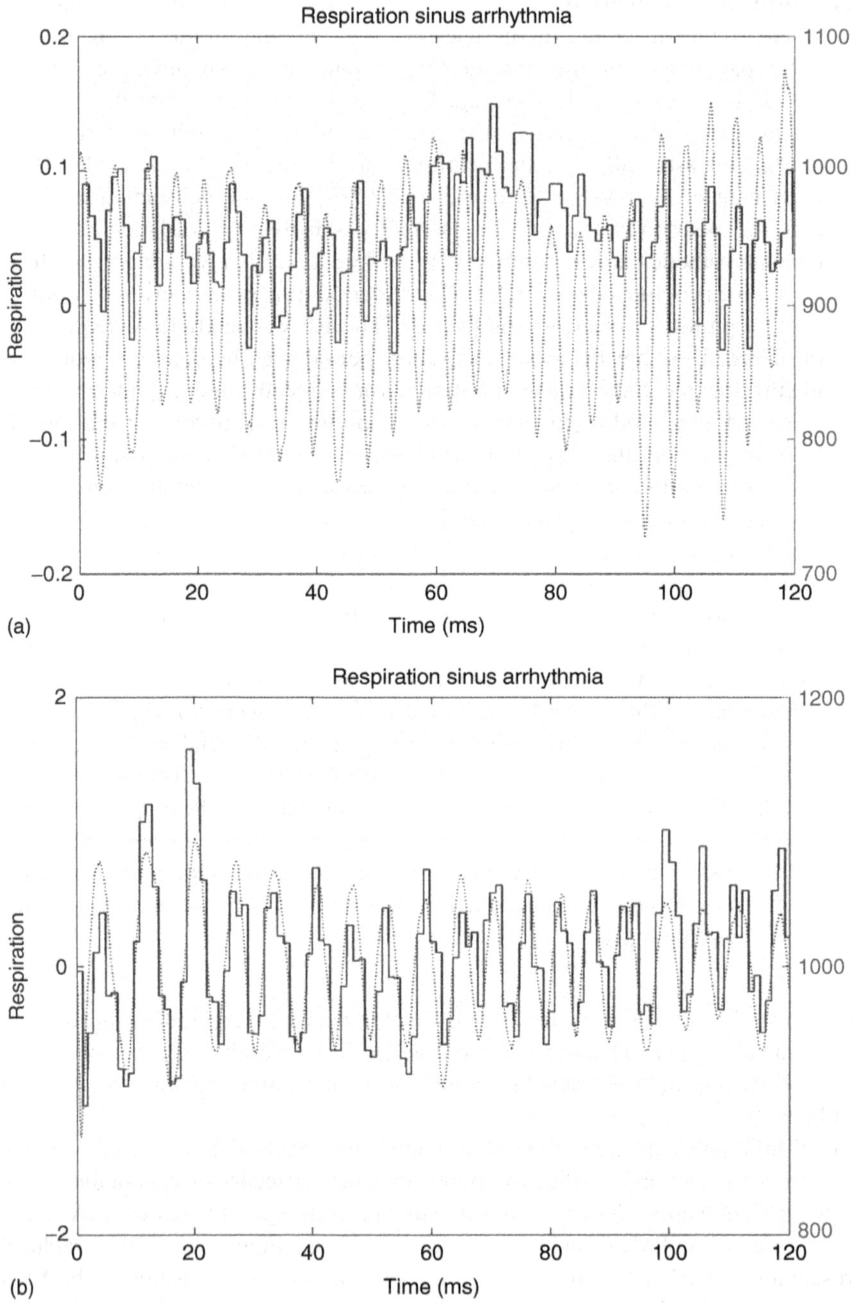

Figure 7.16 Two-axis plots representing the respiratory signal and the heart period for (a) seated position and (b) supine position, for subject 2205. © 2009 IEEE. Reprinted, with permission, from Massagram et al. [2009].

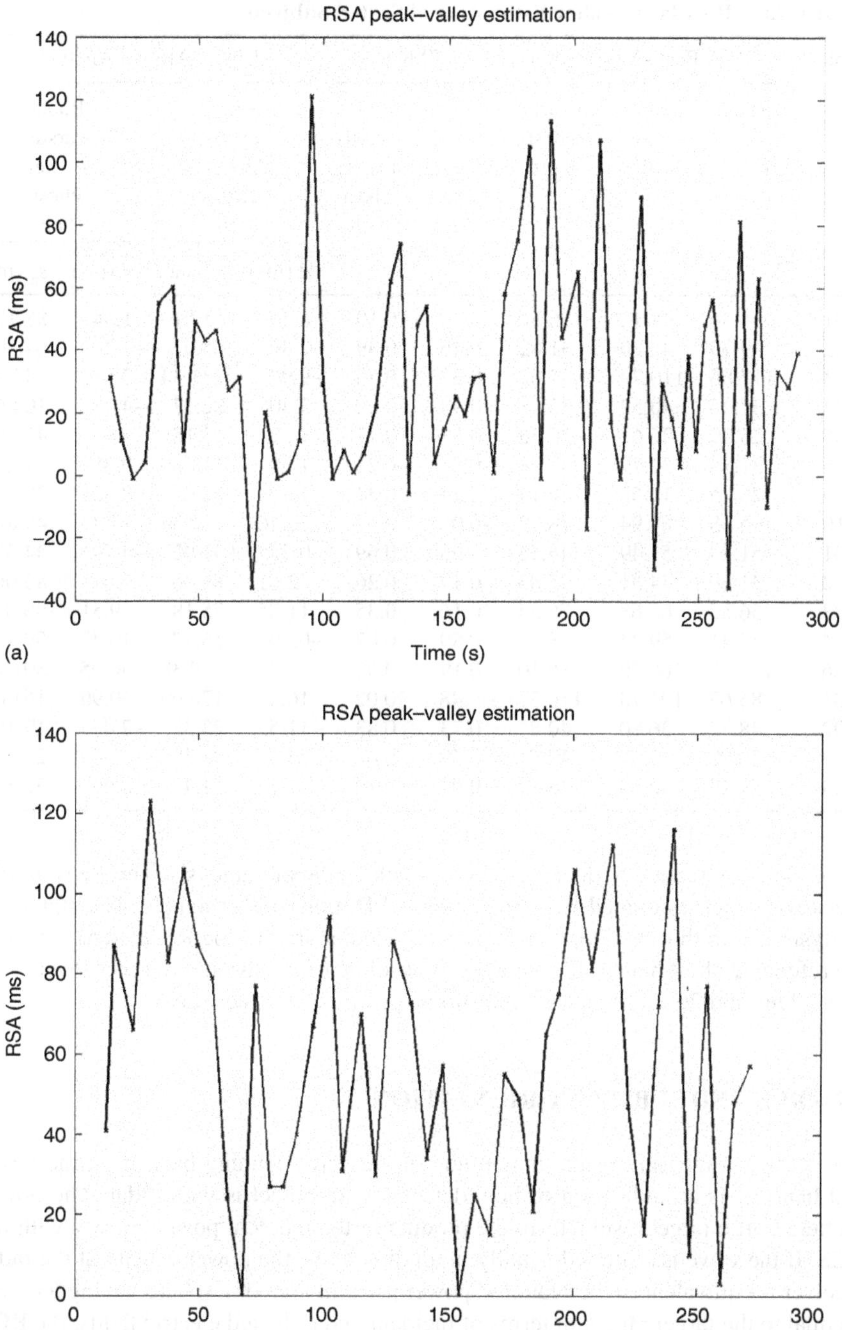

Figure 7.17 RSA peak–valley amplitude estimation for (a) seated position and (b) supine position, for subject 2205. © 2009 IEEE. Reprinted, with permission, from Massagram et al. [2009].

TABLE 7.8 RSA Peak–Valley Estimation of Seated Subjects

Subject	RSA Peak–Valley (ms)			PMCC		Bland–Altman Analysis			
	Radar	Upper chest belt	Lower chest belt	Radar versus Upper chest belt	Radar versus Lower chest belt	Radar versus Upper chest belt		Radar versus Lower chest belt	
						$\overline{d}$ (ms)	s_d (ms)	$\overline{d}$ (ms)	s_d (ms)
2201	54.36	48.12	56.30	0.38	0.19	−6.25	71.81	1.94	88.19
2202	39.79	42.20	31.82	0.46	0.39	2.89	65.07	−7.76	64.50
2203	117.85	110.79	87.03	0.32	0.70	−9.97	231.34	−30.72	142.28
2204	37.45	40.52	32.48	0.98	−0.20	3.40	22.17	−4.73	104.68
2205	25.29	26.61	28.32	0.68	0.77	1.32	23.18	3.03	41.49
2206	31.83	32.79	30.29	−0.21	−0.21	1.68	110.30	−0.38	110.67
2209	26.45	13.55	26.14	0.34	0.34	−12.91	32.09	−0.32	29.74
2210	65.48	87.94	78.92	−0.08	0.87	22.46	158.54	13.44	49.86
2301	51.24	51.49	46.44	0.63	0.69	0.25	47.92	−4.79	44.35
2302	31.69	44.31	25.83	0.19	0.26	12.61	86.26	−6.61	83.06
2303	56.83	68.68	76.34	0.16	0.35	11.85	75.18	19.51	88.04
2304	57.41	59.24	48.61	0.89	0.87	−0.89	55.77	−10.47	59.15
2306	132.58	166.20	86.10	0.19	0.02	35.81	309.38	−46.05	301.47
2401	85.67	101.73	116.57	−0.48	−0.02	16.06	170.09	30.90	161.08
2402	48.33	36.80	40.87	0.73	0.83	−11.53	23.32	−7.47	19.10
2403	36.92	41.57	42.51	0.58	0.60	4.65	26.58	5.60	25.66
2404	39.40	42.59	48.90	0.05	0.16	1.72	41.43	8.90	32.52

since the computation is done based on breath-by-breath analysis. The RSA amplitudes were then estimated from the collected Doppler radar signals. The statistical analysis shows that the Doppler RSA were comparable to the RSA extracted from the reference chest belts. The averages of absolute of the mean differences were less than 12 ms and 20 ms for seated and supine positions, respectively.

7.7 RCS AND SUBJECT ORIENTATION

The RCS is a measure of the magnitude of the wave echoing back from the target and hence it is an indication of how detectable an object is with radar. The power scattered off a target results from the product of the incident power density with the RCS. If the wave is scattered equally in all directions, the power density at the radar receiver is equivalent to the scattered power per unit area of a sphere having a radius R equal to the target range. In terms of incident and reflected electric fields, the RCS is expressed as

$$\sigma = \lim_{R\to\infty} 4\pi R^2 \frac{|E_r|^2}{|E_i|^2} \tag{7.5}$$

TABLE 7.9 RSA Peak–Valley Estimation of Supine Subjects

Subject	RSA Peak–Valley (ms)			PMCC		Bland–Altman Analysis			
	Radar	Upper chest belt	Lower chest belt	Radar versus Upper chest belt	Radar versus Lower chest belt	Radar versus Upper chest belt		Radar versus Lower chest belt	
						$\bar{d}$ (ms)	s_d (ms)	$\bar{d}$ (ms)	s_d (ms)
2201	162.23	209.60	173.70	0.37	0.61	47.37	317.94	11.47	317.57
2202	43.56	66.18	49.37	0.08	0.50	22.62	146.49	6.94	149.69
2203	237.84	232.65	210.29	0.17	−0.29	68.79	322.82	36.11	379.39
2204	23.62	31.89	24.50	0.36	0.17	8.27	29.35	1.89	25.93
2205	112.71	85.19	114.71	0.96	0.98	−27.52	55.36	2.00	42.09
2206	19.57	16.28	22.11	0.60	0.67	−2.97	22.80	2.83	21.40
2209	35.61	57.03	39.41	0.52	0.48	22.48	76.52	4.55	70.20
2210	86.67	61.81	57.47	0.21	0.32	−24.86	167.23	−29.19	159.49
2301	23.43	20.57	20.97	0.79	0.82	−2.87	20.88	−2.47	20.94
2302	30.63	29.60	29.90	0.40	0.87	−1.03	38.51	−0.73	19.46
2303	46.46	51.88	62.51	0.55	0.12	5.71	101.77	16.56	92.13
2304	117.31	141.69	126.66	0.10	0.08	1.54	405.49	−1.27	392.90
2306	144.21	124.78	132.06	−0.30	0.63	−18.76	423.45	−9.73	265.87
2401	209.78	165.81	133.61	0.51	0.96	−43.97	250.36	−75.38	72.22
2402	83.03	67.33	50.43	0.78	0.31	−15.70	52.46	−32.60	93.17
2403	94.72	105.06	98.53	−0.08	0.23	15.04	315.14	8.13	249.32
2404	30.16	21.08	22.43	0.31	0.38	−9.08	43.84	−7.74	41.54

where E_i and E_r are the incident and reflected fields, respectively. The limit of R tending to infinity aims to consider far-field targets where the incident radiation is a uniform plane wave, leading to the IEEE definition of RCS [Jay, 1984]. By definition, the RCS of a target is a fictional area that intercepts the incident wave that, if scattered uniformly, produces an echo power at the receiver equal to that produced by the real target. The RCS depends on the physical characteristics of the target namely size, shape, and material of its surface. In human vital signs monitoring, the RCS of interest is that of the surface of the torso moving due to the cardiopulmonary activity. It is referred to as the cardiopulmonary effective radar cross section (ERCS). Measuring this quantity is possible with a continuous-wave Doppler radar since signals returning from the moving surface of the torso can be isolated from those returning from stationary parts of the body and clutter.

Analytical solutions for RCSs are possible for targets with simple geometrical shapes. This is especially true at high frequencies where specular reflections are assumed. Results obtained give an insight into the dependence of RCS on target geometry and operating frequency. They also facilitate the understanding of RCSs of human subjects.

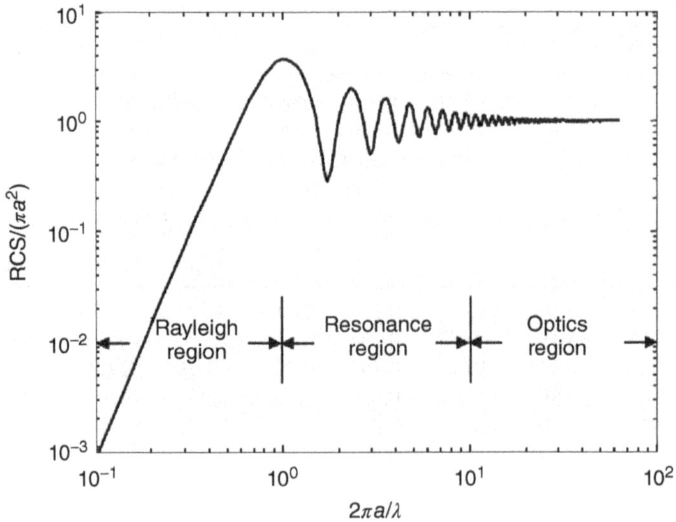

Figure 7.18 Radar cross section of metallic sphere of radius a as a function of wavelength.

The RCS of a perfectly conducting sphere is obtained from Mie series solution of wave scattering [Knott et al., 1993]. In the backscatter direction, a plot of this solution versus frequency is shown in Fig. 7.18, similar to the representation of Skolnik [2001], where a is the radius and λ is the wavelength. The y-axis is the RCS normalized with respect to the physical cross-sectional area, while x-axis is the circumference with respect to wavelength. The variation of the RCS with frequency is classified into three different zones. At relatively low frequencies, the RCS increases with frequency due to Rayleigh scattering. At intermediate frequencies, the wavelength begins to be exceeded by the circumference of the sphere, and the RCS is the result of the superposition of a specular reflection from the front and a wave creeping on the back and returning to the front, as shown in Fig. 7.19. The creeping wave is due to multiple diffractions on the curved surface of the sphere. The two components may combine in phase or out of phase depending on the frequency. This leads to fluctuation of RCS in this zone that is called resonance region. As the frequency increases, the magnitude of the creeping wave diminishes and the specular return

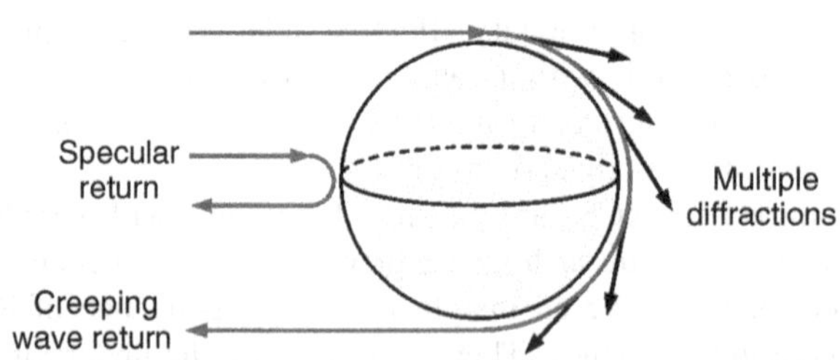

Figure 7.19 Two types of waves scattering off a metallic sphere.

becomes dominant. This corresponds to the optical region where the incoming wave illuminates the sphere with a bright spot at the specular point. In this case, the RCS of the perfectly conducting sphere becomes equal to the physical cross-sectional area, $\sigma = \pi a^2$.

In the optical region, the theory of specular scattering [Knott et al., 1993] states that the backscatter RCS of any object can be expressed as

$$\sigma = 4\pi \frac{(A_{\text{eff}})^2}{\lambda^2} \tag{7.6}$$

where A_{eff} is the effective area of incidence from which reflected components add in phase. For curved surfaces, it is considered as the area at the specular point where the phase variation is within 22.5° or $\lambda/16$. For a perfectly conducting finite cylinder with a radius a and a height b, the solution of RCS is

$$\sigma = \frac{2\pi \cdot a \cdot b^2}{\lambda} \tag{7.7}$$

For a perfectly conducting plate, the reflected components are all in phase and the effective area equals to the physical one. Therefore, the RCS is

$$\sigma = \frac{4\pi \cdot {A_{\text{ph}}}^2}{\lambda^2} \tag{7.8}$$

where A_{ph} is the physical surface area of the rectangular sheet. This relation can be generalized for any flat surface normal to the wave since specular reflections are independent of the plate geometry.

The comparison between the solution for a cylinder to that for a flat plate shows that the curvature of the cylindrical surface results in a reduction in its effective area with respect to its projected rectangle. This demonstrates that two surfaces having the same projected cross section but different curvatures will show different RCSs where the larger belongs to the less curved surface, that is, larger radius of curvature. The variation of RCS with size and curvature of the target surface is the basis for detecting orientation of a human subject.

Testing of human cardiopulmonary characteristics has potential for application in clinics and hospitals where patients are usually lying on beds. In a recumbent position, the orientation of the subject is determined by the sleeping position, which can be classified into three main positions: supine, prone, and fetal-like position. To study the feasibility of detecting the three orientations or sleeping positions, the geometry of the average human torso is simplified into a half-cylinder model, as shown in Fig. 7.20, where the front body corresponds to the curved surface and the back to the flat one. For a perfectly conducting body in the optical region, the ratio of the RCS of the back to that of the front is calculated using the optical region formulas for a rectangular plate and a cylinder:

$$\frac{\sigma_{\text{back}}}{\sigma_{\text{front}}} = \frac{4\pi(2ab)^2/\lambda^2}{2\pi ab^2/\lambda} = \frac{8a}{\lambda} \tag{7.9}$$

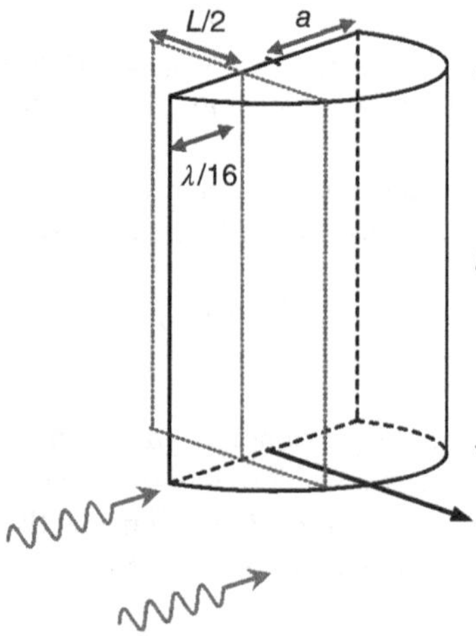

Figure 7.20 A metallic half-cylinder.

The resulting ratio is a function of wavelength. Considering the average human chest breadth to be 30 cm, this ratio equals to 9.6 at 2.4 GHz, and it increases to 23.2 at 5.8 GHz. The choice of these two frequencies is based on their ubiquitous use in common communications equipment and the possibility to realize a Doppler system at a very low cost in these bands. Precedent with similar ISM-band equipment in hospitals, they also reduce the risk of interference or any compliance issue.

For the side of the torso, a wave that is incident normally on the side of a half-cylinder will see an effective area having a width b and a length $L/2$, where L is the effective length of a full cylinder. In the optical region, the radius a is much larger than the wavelength and L is approximated to

$$L \approx \sqrt{\frac{a\lambda}{2}} \tag{7.10}$$

The ratio of RCS of the side with respect to the front is then calculated as

$$\frac{\sigma_{\text{side}}}{\sigma_{\text{front}}} = \frac{\pi a b^2/2\lambda}{2\pi a b^2/\lambda} = \frac{1}{4} \tag{7.11}$$

This indicates that the radar cross of the side is expected to be a quarter that of the front. Although this ratio is frequency independent, frequency variations are expected in practical measurements due to any deviation from the optical region assumption and due to the actual reflectivity of the body surface. A higher complexity is also introduced by the composite movement of the torso consisting of two moving objects: the thorax and abdomen. The relative displacement between the two parts affects the phase by which the reflected components add and hence the measured RCS. Nonetheless, the simplified model indicates that in the gigahertz range, there is a relatively large discrepancy between the RCS of the three sides of the torso, which is a key parameter to identify the orientation of a subject with respect to the transceiver.

The RCS of an arbitrary target can be measured with CW radar by comparing the received power to the transmitted one and applying Friis relation for wave propagation. For a moving target, the Doppler radar developed for detecting heartbeat and respiration rates can be extended to measure the RCS of the moving portion of the torso, defined as the cardiopulmonary ERCS. Ideally, in a quadrature receiver architecture, the baseband signal at the receiver traces an arc on the complex *I–Q* plot such that the radius is proportional to the signal amplitude while the scanned angle to the displacement magnitude. However, the arc is usually distorted and offset from the center of the coordinates due to the signal DC content. This requires a system physical layer with DC-cancellation feature while in the digital processing stage, demodulation algorithms are needed to center the arc at the origin and calculate the radius. In an algorithm proposed, the baseband signal is first divided into segments, each corresponding to one full respiration cycle. The center estimation algorithm [Park et al., submitted for publication] is then applied to each segment of data and a radius is calculated. The segment with maximum radius is selected and the final value is obtained using a circle fitting algorithm. The fitting method deployed is the one proposed by Kasa [1976], and it is the fastest compared with the methods of Pratt [1987] and Taubin [1991].

The receiver signal power is related to the power of the transmitter and the ERCS of the subject through the different propagation and conversion losses in the system. While the contribution of the path loss for a given range can be calculated from Friis relation, the loss from the several hardware parts can be estimated from the manufacturer data sheet and by tracking the wave path through the system. However, a system calibration is possible by measuring a moving object with a known RCS and obtaining the coefficients for extracting the RCS from other moving objects.

In a Doppler system setup for calibration purposes, the losses or gains through the systems must be classified into two categories: a fixed gain factor that remains unchanged in the actual system used for human testing, and a set of variable gain parameters that enable flexibility in the practical setup. These parameters can be variable attenuation for power control, cable extensions, or variable target range. On the other hand, the choice of the calibration target is critical for the accuracy of the measurement because it affects the accuracy of the alignment, the complexity of the mechanical supports, and the validity of the far-field conditions within the limited space of the measurement room. For these reasons, spherical metallic targets are the most suitable for calibration. The actual RCS of the spherical targets is calculated from Mie series solution for perfectly conducting spheres [Wu, 1989]. This relation is valid in all regions of operation. However, the optical region is preferred in practical measurements because specular reflections from metallic spheres are concentrated on a bright spot at the tip of the sphere. This would minimize the effects of the mechanical supports attached to the target.

To measure the RCS in the backscatter direction, the target is placed at a distance along the line-of-site of the transmitting antenna and in the direction of the main radiation lobe. The range is set such that the target is in the far-field zone where the wave radiation and planarity conditions are satisfied. Since the Doppler radar system detects only moving objects in one dimension, the target must be put into a linear periodic motion in the axial direction and programmed with the desired frequency and displacement magnitude.

The CW Doppler radar used for human testing is characterized by a constant gain term that is calculated from the input and measured parameters through the following relation:

$$\Re = \frac{R^4}{\sigma_{\text{cal}}} \cdot \frac{1}{P_{\text{in}}} \left(\frac{A}{G_{\text{LNA}}} \right)^2 \tag{7.12}$$

where $\Re$ includes all fixed losses in the systems, R is the range, P_{in} is the signal generator power, σ_{cal} is the RCS of the calibration target, G_{LNA} is the gain of the LNA, and A is the measured arc radius. The purpose of calibration is to estimate the gain factor $\Re$ through measurements of known targets. Once the system is calibrated, this constant can be used to calculate the ERCS by rearranging the equation in the following form:

$$\sigma_{\text{eff}} = \frac{R^4}{\Re} \cdot \frac{\text{IL}_{\text{add}}}{P_{\text{in}}} \left(\frac{A}{G_{\text{LNA}}} \right)^2 \tag{7.13}$$

where IL_{add} represents any additional losses inserted in the transmitter path and those that were not present in the calibrated system. When using more than one target to calibrate the same system, a sanity check can be performed by comparing the ratio of the measured arcs radii with that calculated using the nominal RCS of the targets:

$$\frac{A_1}{A_2} = \sqrt{\frac{\sigma_1}{\sigma_2}} \tag{7.14}$$

For human cardiopulmonary sensing, the measurement of the ERCS of the torso aims to identify the sleeping position of the subject whether it is supine, prone, or side. The transceiver antennas are suspended above the subject, preferably through the ceiling to avoid interference from any vertical support. The three main sleeping positions are shown in Fig. 7.21. In supine, the subject lies on the back with the straight arms at the side, such that the Doppler radar monitors the front of the body. In prone, the subject lies on the front with the arms at the side. Soldier arms are preferred over the falling position to avoid the tendency of the subject to push up during the test. In the side position, the subject lies on the left side while the right side is facing the antennas and the hands under the head. The subject's knees are allowed to bend to maintain balance, similar to the fetal sleeping position.

To achieve RCS characteristics similar to those of the half-cylinder model, in the Doppler system design, it is important to use an operating frequency that allows near optical scattering. In addition, the target range must satisfy the conditions of far-field measurements so that the derived RCS equations and system calibration are valid.

In addition to the radar measurements, piezoelectric sensors are used on the chest, abdomen, and fingers for reference. A total of three tests are to be performed for

Figure 7.21 Body shape in three sleeping positions: (a) supine, (b) prone, and (c) side.

the subject at a single frequency and a single target range. Before starting each of the actual measurements, an initialization test is required to estimate the initial value of the residual DC component. The latter is a function of the target range as well as the frequency. In actual test, the first 10–12 s of the recorded data correspond to the DC estimation and cancellation settling time, and must be discarded. The useful data vector is broken into segments; each corresponds to a full respiration cycle. The respiration cycles are identified by tracking every three zero-crossings of a respiration reference signal or a baseband signal that is not at null point. The center estimation algorithm is then applied to each data segment and the corresponding arc radius A is obtained. The result is a vector of length equal to the number of segments. A heuristic is followed to determine the value of the amplitude A at each recumbent position. It consists of selecting the segment of data that has the maximum arc radius generated from the center estimation algorithm. The arc is then relocated by applying circle fitting and a new radius is calculated. This heuristic provides a common criterion to compare different positions and is especially useful in the case of distorted arcs. The torso displacement magnitude is also calculated from the angle scanned by the arc resulting from this heuristic.

This technique for measuring the cardiopulmonary ERCS has been applied on a healthy subject using two operating frequencies – 2.4 and 5.8 GHz [Kiriazi et al., 2012]. The center tracked arcs obtained at 2-m range are shown in Fig. 7.22. These results showed that the cardiopulmonary characteristics change with sleeping position in terms of both the ERCS and the torso displacement magnitude.

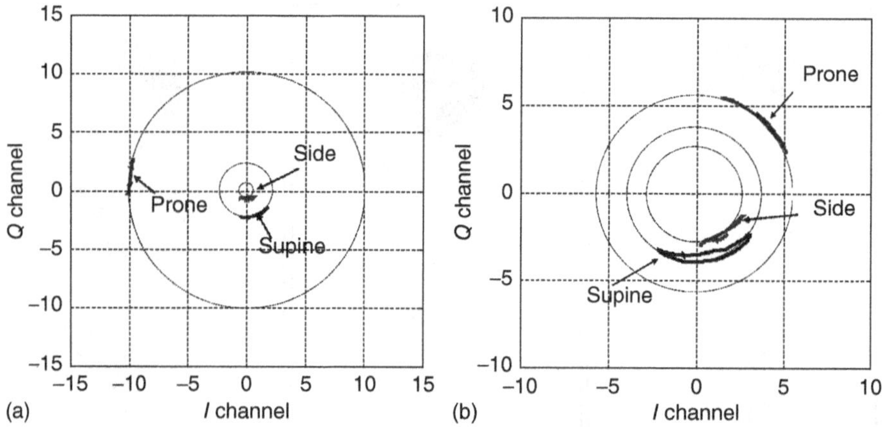

Figure 7.22 Center-tracked arcs for the subject in the supine, prone, and side positions at 2-m range with (a) 2.4 GHz and (b) 5.8 GHz carriers. © 2011 IEEE. Reprinted, with permission, from Kiriazi et al. [2012].

REFERENCES

Akselrod S, Gordon D, Mahwed JB, Snidman NC, Shannon DC, Cohen RJ. Hemodynamic regulation: investigation by spectral analysis. Am J Physiol 1985;249:H867–H875.

Akselrod S, Gordon D, Ubel FA, Shannon DC, Barger AC, Cohen RJ. Power spectral analysis of heart rate fluctuation: a quantitative probe of beat-to-beat cardiovascular control. Science 1981;213:220–222.

Buist MS. Association between clinically abnormal observations and subsequent in-hospital mortality: a prospective study. Resuscitation 2004;62:137–141.

Cretikos M. An Evaluation of Activation and Implementation of the Medical Emergency Team System. Sydney: The University of New South Wales; 2006.

Droitcour AD. Non-contact measurement of heart and respiration rates with a single-chip microwave Doppler radar [PhD dissertation]. Stanford University; 2006.

Droitcour AD, et al. Non-contact respiratory rate measurement validation for hospitalized patients. Annual International Conference of the IEEE Engineering in Medicine and Biology Society, 2009. EMBC 2009; 2009. p 4812–4815.

Fouad FM, Tarazi RC, Ferrario CM, Fighaly S, Alicandro C. Assessment of parasympathetic control of heart rate by a noninvasive method. Am J Physiol 1984;246:H838–H842.

Grossman P, Stemmler G, Meinhardt E. Paced respiratory sinus arrhythmia as index of cardiac parasympathetic tone during varying behavioral tasks. Psychophysiology 1990;27:404–416.

Hodgetts TJ. The identification of risk factors for cardiac arrest and formulation of activation criteria to alert a medical emergency team. Resuscitation 2002;54:125–131.

Jay F(editor in chief). IEEE Standard Dictionary of Electrical and Electronics Terms. 3rd edANSI/IEEE Std 100-1984. New York: IEEE Press; 1984.

Kasa I. A curve fitting procedure and its error analysis. IEEE Trans Instrum Meas 1976; 25:8–14.

Katona PG, Lipson D, Dauchot PJ. Opposing central and peripheral effects of atropine on parasympathetic cardiac control. Am J Physiol 1977;232:H146–H151.

Kiriazi J, Boric-Lubecke O, Lubecke V. Dual-frequency technique for assessment of cardiopulmonary effective RCS and displacement. IEEE Sens J 2012;12(3):574–582.

Knott EF, Shaeffer JF, Tuley MT. Radar Cross Section. 2nd ed. Norwood, MA: Artech House; 1993.

Kondo T, Uhlig T, Pemberton P, Sly PD. Laser monitoring of chest wall displacement. Eur Respir J 1997;10:1865–1869.

Lim WS, Carty SM, Macfarlane JT, Anthony RE, Christian J, Dakin KS, Dennis PM. Respiratory rate measurement in adults – how reliable is it? Resp Med 2002;96:30–33.

Lin JC. Non-invasive microwave measurement of respiration. Proc IEEE 1975;63:1530.

Lin JC. Microwave apexcardiography. IEEE Trans MTT 1979;27:618–620.

Massagram W. A study of feasibility in long-term cardiopulmonary monitoring via Doppler radar [PhD dissertation]. University of Hawaii; 2008.

Massagram W, Lubecke VM, Boric-Lubecke O. Feasibility assessment of Doppler radar long-term physiological measurements. 2011 Annual International Conference of the IEEE Engineering in Medicine and Biology Society, EMBC; 2011. p 1544–1547.

Massagram W, Lubecke VM, Host-Madsen A, Boric-Lubecke O. Assessment of Heart Rate Variability and Respiratory Sinus Arrhythmia via Doppler Radar. IEEE Trans Microwave Theory Tech 2009;57(10):2542–2549.

Park BK, Lubecke V, Boric-Lubecke O, Host Madsen A. Center tracking quadrature demodulation for a Doppler radar motion detector, IEEE/MTT-S Internation Microwave Symposium; 2007. p 1323–1326.

Pomeranz B, Macaulay RJ, Caudill MA, Kutz I, Adam D, Gordon D, Kilborn KM, Barger AC, Shannon DC, Cohen RJ, Benson H. Assessment of autonomic function in humans by heart rate spectral analysis. Am J Physiol 1985;17:H151–H153.

Pratt V. Direct least-squares fitting of algebraic surfaces. Comput Graph 1987;21:145–152.

Siebens A. 2008. Human respiration: the respiratory pump and its performance. Encyclopaedia Britannica. http://www.britannica.com/eb/article-66150/human-respiration#537226.hook.

Skolnik ML. Introduction to Radar Systems. 3rd ed. New York: McGraw-Hill Inc.; 2001.

Task Force of the European Society of Cardiology and the North American Society of Pacing and Electrophysiology. Heart rate variability: standards of measurement, physiological interpretation and clinical use. Circulation 1996;93:1043–1065.

Taubin G. Estimation of planar curves, surfaces and nonplanar space curves defined by implicit equations, with applications to edge and range image segmentation. IEEE Trans Pattern Anal Mach Intell 1991;13:1115–1138.

Wu T. Radar cross section of arbitrarily shaped bodies of revolution. Proc IEEE 1989; 77(5):735–740.

8

ADVANCED PERFORMANCE ARCHITECTURES

ADITYA SINGH[1], ALY FATHY[2], ISAR MOSTAFANEZHAD[3], JENSHAN LIN[4], OLGA BORIC-LUBECKE[5], SHUHEI YAMADA[5], VICTOR M. LUBECKE[5], AND YAZHOU WANG[6]

[1]*University of Hawaii Neuro-science and MRI research Program, John A. Burns School of Medicine, Honolulu, Hawaii, United States*
[2]*Department of Electrical Engineering and Computer Science, University of Tennessee, Knoxville, Tennessee, United States*
[3]*Department of Physics and Astronomy, University of Hawaii at Manoa, Honolulu, Hawaii, United States*
[4]*Department of Electrical and Computer Engineering, University of Florida, Gainesville, Florida, United States*
[5]*Department of Electrical Engineering, University of Hawaii at Manoa, Honolulu, Hawaii, United States*
[6]*Boston Design Center, RF Micro Devices, Inc., Billerica, Massachusetts, United States*

Continuous-wave (CW) quadrature homodyne receivers have been extensively used for wireless life sign monitoring applications due to range correlation and noise performance benefits. Such systems were used to obtain the data presented in the previous chapters. One limitation of such systems is a large DC offset at a down-conversion mixer output, which can saturate the baseband amplifiers and limit the dynamic range. AC coupling or baseband DC cancellation techniques can help improve system performance but they introduce new challenges. In addition, performance of quadrature receivers is limited by the amplitude and phase imbalance between in-phase and quadrature channels. These challenges can be overcome by using a single-channel architecture with phase or frequency tuning to overcome the null point issue, or by using a low-intermediate frequency (IF) receiver. Another challenge in CW Doppler

Doppler Radar Physiological Sensing, First Edition.
Edited by Olga Boric-Lubecke, Victor M. Lubecke, Amy D. Droitcour, Byung-Kwon Park, and Aditya Singh.

systems is separation of motion interference from wanted signals. For example, if the subject is fidgeting or walking, or if another subject appears in the field of view, CW Doppler radar will detect the sum of all motions along observation axes. In that case, extraction of physiological signals becomes very challenging either due to dynamic range limitations, or to overlapping spectra of wanted and unwanted signals. While several techniques have been demonstrated to be effective for separation of motion interference and physiological signals, this is still an active research area, and innovative solutions will be required to enable more widespread use of physiological Doppler radar. Finally, a fundamental limitation of the CW radar is inability to discriminate target range. As discussed in Chapter 2, frequency-modulated continuous-wave (FMCW) and pulsed radar can be used to detect both target range and Doppler shift. Ultra-wideband (UWB) is a special case of pulsed radar, which provides excellent range resolution due to the short pulse duration. FMCW and UWB radar have both been used for the detection of physiological signals, and these systems offer promising capability of simultaneous subject location. In this chapter, advance performance architectures that overcome issues of DC offset, spectrum folding, motion interference, and range detection are discussed.

8.1 DC OFFSET AND SPECTRUM FOLDING

The main disadvantage of CW radar results from its nature of constant transmission and reception. Due to the leakage of the transmitted signal to the receiver, the CW radar system receives a large signal at the transmit frequency that has not reflected off the target. In addition, stationary clutter reflections contribute to the received signal at the transmit frequency due to leakage. These unwanted signals result in a DC offset and low-frequency noise that may limit system performance.

Another limitation of single-channel CW homodyne systems is spectrum folding at DC that results in inability to discriminate approaching and receding targets. In Doppler radar physiological monitoring, due to this effect single-channel receiver performance will be subject to null and optimum position variations. To overcome this issue, either quadrature homodyne, or heterodyne receiver should be used. In this section, three different approaches to overcome DC offset and spectrum folding are discussed.

8.1.1 Single-Channel Homodyne System with Phase Tuning

An RF-based DC cancellation technique can be used to eliminate DC offset. This method significantly simplifies the receiver architecture, and eliminates the demodulation step of processing baseband data. The proposed system has been tested and shown to be capable of producing accurate life sign estimates of the human subject.

In this section, a phase tuning method is introduced to remove the baseband DC by introducing an RF feedback loop and reducing the two-channel *IQ* demodulation to a single-channel receiver that will operate at the optimum point. A similar technique was proposed in Yamada et al. [2008] for clutter cancelation in an AC-coupled system. It is demonstrated that a phase-tuning approach can be used to simultaneously eliminate DC offset and achieve the optimum demodulation point. Thus, this architecture provides a single DC-coupled output with no distortion, and effectively performs analog linear signal demodulation simplifying signal-processing steps. It will be shown that this single-channel receiver estimates the heart rate of a human subject within 1-BPM accuracy of a reference signal.

Figure 8.1 illustrates the proposed single-channel architecture and the experimental setup. DC offset cancellation path consisting of phase shifter 1 and an attenuator is added between the Tx and Rx, and phase shifter 2 is used for tuning to the optimum demodulation point. The signal generator output is assumed to be

$$x_g(t) = \cos(\omega t) \tag{8.1}$$

where ω is the radian frequency of the RF generator's output. Once this signal is reflected from the human subject and received at the receive antenna it becomes

$$x_{rx1}(t) = a\cos\left(\omega\left(t - \frac{d_1 + d_2}{c}\right) + \psi(t)\right) + \sum_i b_i \cos\left(\omega\left(t - \frac{l_i}{c}\right)\right) \tag{8.2}$$

where a is the received signal amplitude, ψ is the phase change caused by physiological motion, c is the speed of light, d_1 is distance between the Tx antenna and the

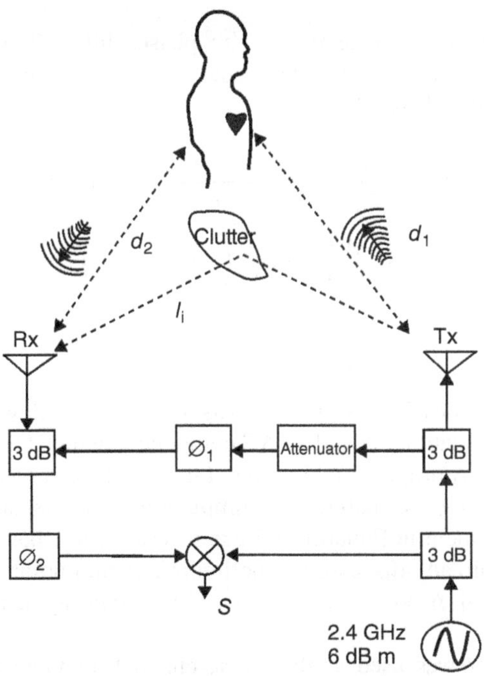

Figure 8.1 Measurement setup showing the single-channel mixer together with the DC cancellation path utilizing RF.

subject, and d_2 is distance between the subject and the Rx antenna. The second term in Equation 8.2 is caused by clutter reflections, direct transmit to receive antenna signal paths (cross talk), and internal local oscillator (LO) leakage [Yamada et al., 2008]. b_i is the clutter signal amplitude and l_i is the corresponding distance the signal travels from Tx to Rx with reflection at each clutter object. Signal coming out of phase shifter 1 is added to the received, resulting in

$$x_{rx2}(t) = aA\cos\left(\omega\left(t - \frac{d_1 + d_2}{c}\right) + \psi(t)\right) + A\sum_i b_i \cos\left(\omega\left(t - \frac{l_i}{c}\right)\right) + A\cos(\omega t + \Phi_1) \tag{8.3}$$

where A is the amplitude change caused by the attenuator. Φ_2 will be used to cancel most of the second term in Equation 8.2. Now signal in Equation 8.3 will go through phase shifter 2 to result in

$$\begin{aligned} x_{rx3}(t) = {} & aA\cos\left(\omega\left(t - \frac{d_1 + d_2}{c}\right) + \psi(t) + \Phi_2\right) \\ & + A\sum_i b_i \cos\left(\omega\left(t - \frac{l_i}{c}\right) + \Phi_2\right) \\ & + A\cos(\omega t + \Phi_1 + \Phi_2) \end{aligned} \tag{8.4}$$

where Φ_2 is the phase change caused by phase shifter 2. Ultimately, the signal (Eq. 8.4) and the local oscillator signal (Eq. 8.1) are mixed to result in the single-channel baseband signal:

$$\begin{aligned} x_b(t) = {} & \frac{aA}{2}\cos\left(-\omega\frac{d_1 + d_2}{c} + \psi(t) + \Phi_2\right) \\ & + \frac{A}{2}\sum_i b_i \cos\left(-\omega\frac{l_i}{c} + \Phi_2\right) \\ & + \frac{A}{2}\cos(\Phi_1 + \Phi_2) \end{aligned} \tag{8.5}$$

It can be seen that the baseband signal consists of a DC level showing up as the second and third terms in Equation 8.5 and an AC component, which has the desired physiological motion data. A proper combination of Φ_1 and Φ_2 can result in removal of the DC component. Since ψ is small (2–5°) summation of the first and last expression in the first cosine argument in Equation 8.5 can be crucial to what ultimately x_b would be. This has been already discussed as optimum and null point effects in Chapter 5 and Park et al. [2006]. A proper choice of Φ_2 can help bring the receiver to optimum point.

ZFM-4212 mixer was used as the single-channel demodulator. Power splitters are ZFSC-2-2500 and the phase shifters are Pulsar ST-21-444A. A Broad Wave 751-002-030 attenuator (0–30 dB with 1-dB adjustment steps) was also used in

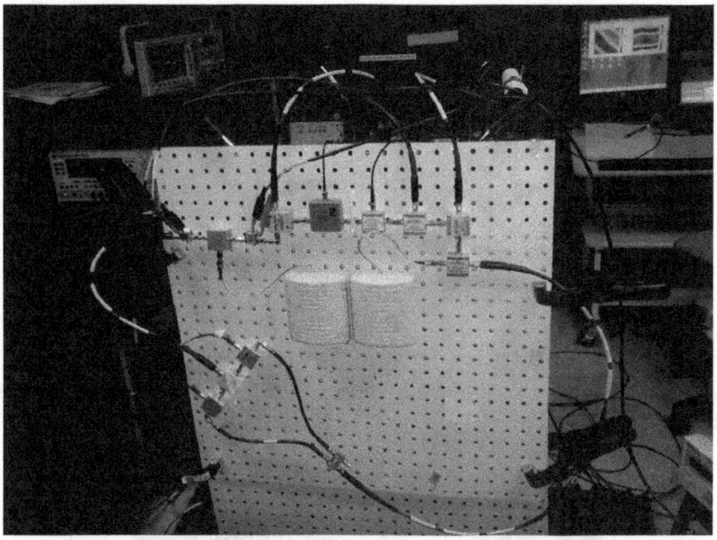

Figure 8.2 Measurement setup showing the antennas, phase shifters, mixers, and splitters.

the setup. Antennas are ASPPT2988 with 8-dBi gain and about 60° beamwidth. Baseband signal (S) is passed through DC-coupled SR560 low noise amplifiers (LNAs) with the low-pass cutoff at 10 Hz, and 46-dB gain. The mixer output can be DC coupled since DC is removed by adjusting the phase shifters; otherwise any significant amplification would result in amplifier saturation. LNA output is sampled at 100 Hz and stored in the computer for further processing using an NI USB6259 16-bit data acquisition (DAQ) device. The photograph of the setup is shown in Fig. 8.2.

The Φ_1 and Φ_2 tuning effects on the output DC and AC level were initially tested using an artificial target consisting of a metallic ball mounted on a handle controlled by a servo motor as shown in Fig. 8.3. The ball was moving back and forth with the amplitude of about 1 cm, at a frequency of 1.3 Hz. Φ_1 and Φ_2 were varied independently between 0° and 180° with a 7.2° step, requiring 25 steps. A MATLAB program was written to set Φ_1 and Φ_2 for each phase pair using the same DAQ as output device, measure the output signals and calculate the DC and AC output of baseband signal for 10 s and then step to the next phase pair. A total of $25 \times 25 = 625$ separate measurements were made taking about 1 h and 42 min. The effect of tuning Φ_1 and Φ_2 on the DC output level are shown in Fig. 8.4, and on the AC output level in Fig. 8.5. As seen, using phase shifter 1, one can tune the DC voltage level to zero in order to cancel the DC and then use phase shifter 2 to bring the demodulation to optimum point and receive the strongest signal in baseband.

As the next step, a measurement with a human subject has been performed. The subject was seated at about 1.5-m distance from the antennas. In order to receive the optimum signal from the single-channel output, phase shifter 1 and the attenuator are tuned until a point of zero DC is obtained from the single-channel output. Once the

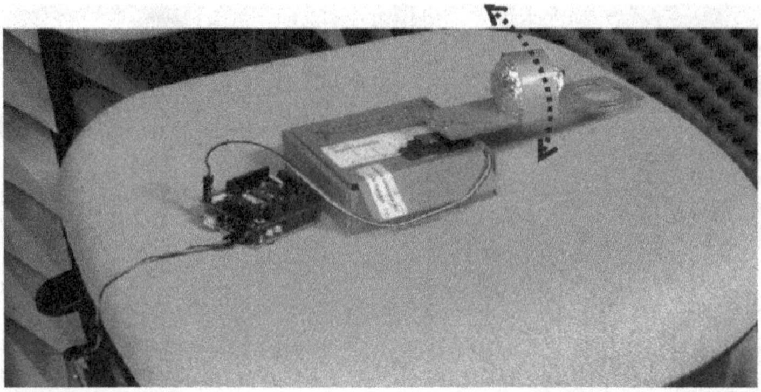

Figure 8.3 Artificial target used in the experiment. The arrow depicts the trajectory of the motion of the target.

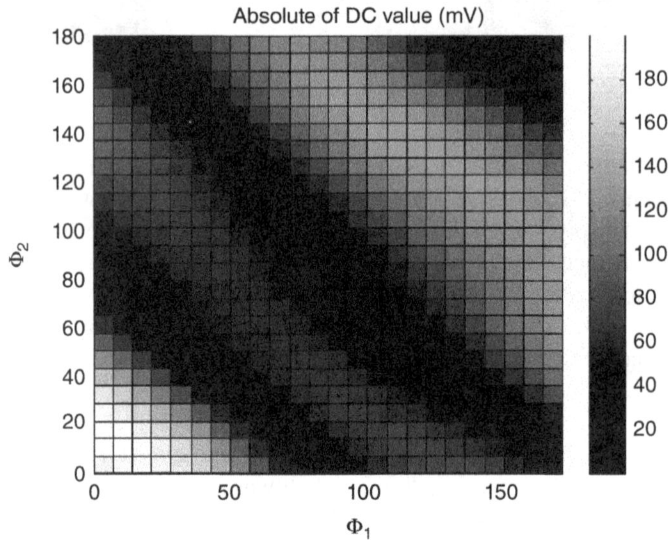

Figure 8.4 DC voltage output of the single-channel mixer, while phase 1 and phase 2 have been swept over a 180° phase shift. © 2011 IEEE. Reprinted, with permission, from Mostafanezhad and Boric-Lubecke [2011].

DC has been cancelled, phase shifter 2 was used to bring the mixer to the optimum point of demodulation [Park et al., 2006]. Figure 8.6 shows the radar output before any processing (top), 0.8 Hz high-pass filtered radar signal to obtain heart signal (middle), and a simultaneously recorded finger pressure pulse for heart signal reference (bottom). The top trace in Fig. 8.6 clearly shows subject's respiratory signal without any distortion and after filtering heart signal is easily extracted from the radar output.

A heart rate estimation tool discussed in Massagram et al. [2009] is used to estimate the heart rate from the radar output and reference signals. Heart rate versus time

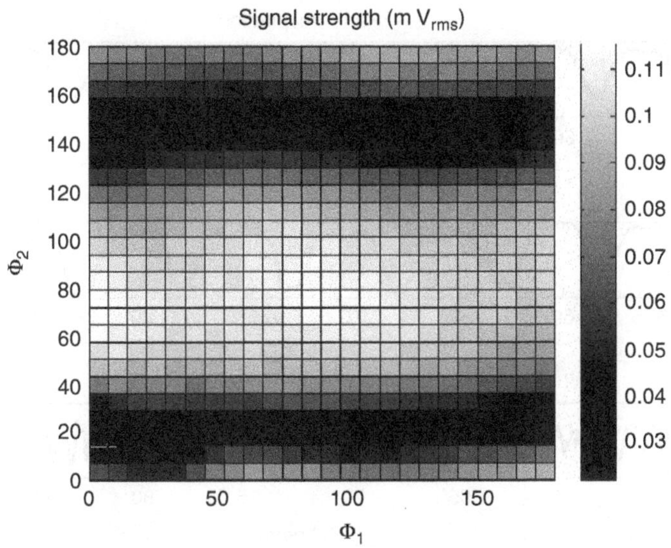

Figure 8.5 Baseband signal strength from the single-channel mixer, while phase 1 and phase 2 have been swept over a 180° phase shift. © 2011 IEEE. Reprinted, with permission, from Mostafanezhad and Boric-Lubecke [2011].

is plotted in Fig. 8.7. It can be seen that heart rate obtained from the radar output closely follows that of reference. Root mean square of error (RMSE) between heart rate obtained from the reference signal and radar signal is less than 1 BPM.

8.1.2 Heterodyne System with Frequency Tuning

Another way to eliminate the position sensitivity of a single-channel Doppler radar system is by having a double-sideband transmission with frequency tuning. A Ka-band heterodyne transceiver using low-power double-sideband transmission to detect human heartbeat and respiration signals is demonstrated. The Ka-band electromagnetic wave offers higher detection sensitivity, and indirect-conversion receiver architecture is chosen to reduce the DC offset and $1/f$ noise that can degrade signal-to-noise ratio (SNR) and detection accuracy. Furthermore, the double-sideband signals at the transmitter output can be in quadrature by choosing a proper frequency separation to relieve the severe null point problem that occurs at high frequency. As a result, the detection accuracy is significantly improved with low transmitted power. This radar sensor system achieves better than 80% detection accuracy at the distance of 2.0 m with a combined transmitted power of only 12.5 μW in both sidebands [Xiao et al., 2006]. Heterodyne architecture also overcomes the issue of quadrature LO generation, and phase and amplitude imbalance effects on signal distortion.

The block diagram of the Ka-band transceiver remote monitoring system is illustrated in Fig. 8.8. The receiver chain includes a receiving antenna (Rx_Antenna),

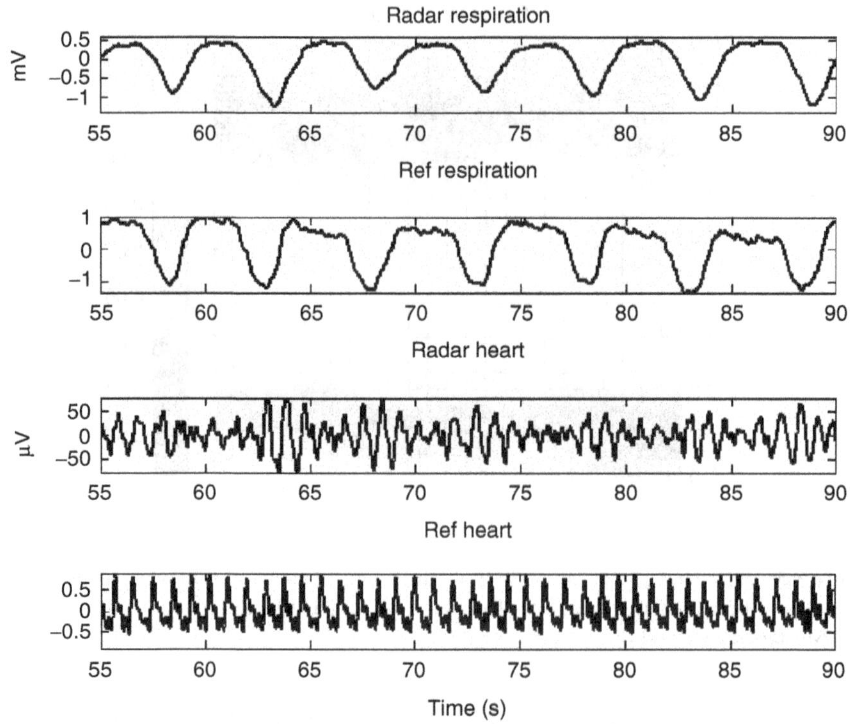

Figure 8.6 Respiration signal from radar (top), respiration reference, heart signal filtered from the radar output (middle), and a finger pressure pulse (bottom) recorded simultaneously as a reference heart signal. © 2011 IEEE. Reprinted, with permission, from Mostafanezhad and Boric-Lubecke [2011].

an LNA, two down-converters (Rx_Mixer1 and Rx_Mixer2), and an IF amplifier (IF_AMP). The transmitter chain contains a transmitting antenna (Tx_Antenna) and an up-converter (Tx_Mixer). Baseband circuits are composed of a preamplifier (PreAMP), a band-pass filter (BPF), and a low-frequency amplifier (LF_AMP).

As shown in Fig. 8.8, the circuits inside the dashed box form a broadband Ka-band radio transceiver, which uses commercial parts as individual building blocks. Their specifications and manufacturers are listed in Table 8.1. The antennas and the baseband circuits are custom-designed for our experiment.

The receiver in the Ka-band radio uses an indirect-conversion architecture that employs two-step conversion. Two local oscillators (LOs) generate signals $S_1(t)$ (with frequency f_1) and $S_2(t)$ (with frequency f_2). Two 3-dB power splitters are used to divide the power of $S_1(t)$ and $S_2(t)$, with half of the power sent to the transmitter chain and the other half sent to the receiver chain.

Since there is no filter following the Tx_Mixer, the output $T(t)$ of the Tx_Mixer has two main frequency components: lower sideband $f_L = f_2 - f_1$ and upper sideband $f_U = f_2 + f_1$. Normally, there is one more frequency component f_2 in the output

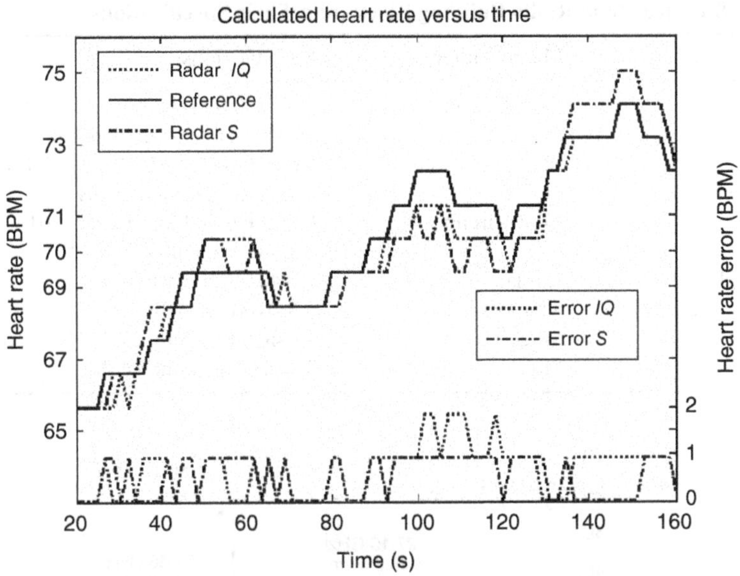

Figure 8.7 Estimated heart rate from the traditional radar *IQ* channels, reference finger pulse outputs, and the proposed single-channel radar, and the absolute error of heart rate. © 2011 IEEE. Reprinted, with permission, from Mostafanezhad and Boric-Lubecke [2011].

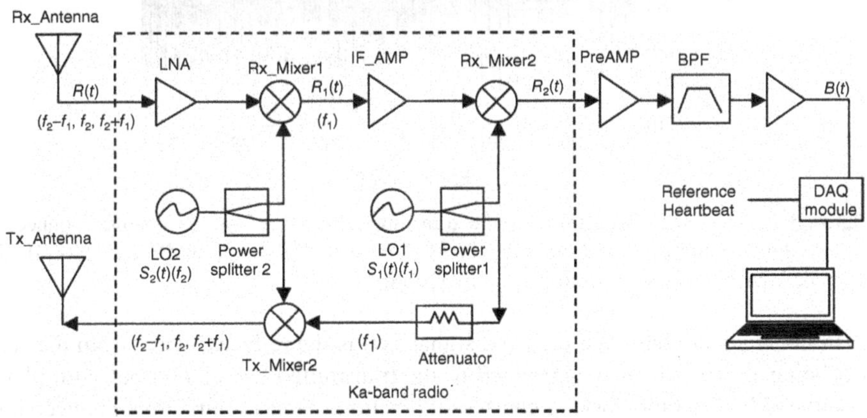

Figure 8.8 Block diagram of the Ka-band remote monitoring system. © 2006 IEEE. Reprinted, with permission, from Xiao et al. [2006].

of Tx_Mixer, which is the leakage from LO2. The output power spectrum of the transmitter measured at antenna connector is shown in Fig. 8.9. The lower sideband and upper sideband frequencies are 26.54 and 27.66 GHz with power levels of −21.1 and −23.3 dB m, respectively. The 27.10 GHz signal in between is the LO2 leakage due to nonideal port-to-port isolation of Tx_Mixer. Although the LO leakage is evident, it does not affect the baseband signal detection, which is discussed later.

TABLE 8.1 Ka-Band Radio Building Blocks and Their Specifications

Blocks	Manufacturer	Specifications
LO1	Mini-Circuit	450–800 MHz; power: 11 dB m
LO2	Avantek	20–40 GHz; power 10 dB m
Tx_Mixer	Miteq	RF/LO: 4–40 GHz; IF: 0.5–20 GHz;
Rx_Mixer1		conversion loss: 10 dB
Rx_Mixer2	Mini-Circuit	RF/LO: 0.3–4.3 GHz; IF: 0.01–2.4 GHz; conversion loss: 6.42 dB
Power splitter1	Narda	10–40 GHz; 3 dB
Power splitter2	Narda	0.5–18 GHz; 3 dB
LNA	Miteq	26–40 GHz; gain: 27 dB; NF: 3 dB
IF_AMP	Miteq	0.1–8 GHz; gain: 33 dB; P_{1dB}: 13 dB

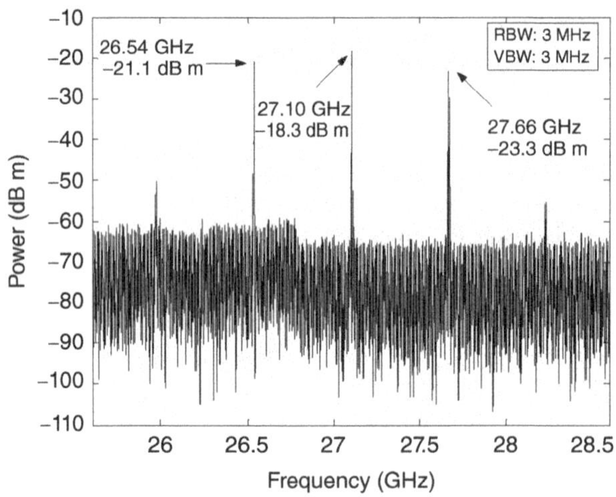

Figure 8.9 The output spectrum of the transmitter, measured at the antenna connector. The resolution bandwidth and the video bandwidth were both set at 3 MHz. © 2006 IEEE. Reprinted, with permission, from Xiao et al. [2006].

In the receiver chain, the received signal $R(t)$ is the reflected wave from the subject being monitored. It is correlated to the transmitted signal $T(t)$ but with phase modulated by the time-varying chest-wall position. After the first down conversion, signal $R_1(t)$ consists of two modulated signals at f_1, down-converted from lower sideband $f_L = f_2 - f_1$ and upper sideband $f_U = f_2 + f_1$, respectively. The chest-wall motion information is modulated on the phases of these two signals at f_1. In addition, it also has a DC offset due to the self-mixing of LO2 leakage transmission, and a baseband signal carrying chest-wall motion information, down-converted from the f_2 component in the received signal $R(t)$. Due to typically higher mixer LO leakage at higher frequencies [Razavi, 1997], DC offset issue in homodyne receivers is more significant at Ka-band as compared with at 2.4 GHz. However, indirect-conversion architecture eliminates the DC offset due to the LO leakage. The large DC offset and the near DC signals are removed by the band-pass frequency response of the IF_AMP before second down-conversion to baseband. Therefore, in the following discussions, the f_2 component in the transmitted wave will be ignored because it does not affect the baseband signal. After the second down conversion, the output $R_2(t)$ consists of baseband signals carrying the subject's chest motion information and other unwanted high frequency spurs, which will be filtered out by the baseband circuits.

Two types of low-profile printed patch antenna were designed and fabricated for use in the measurement. One is a printed single-patch antenna fabricated on a high-frequency substrate material, GML1000, with dielectric constant ε_r of 3.2 and substrate thickness of 0.762-mm. This antenna achieves a maximum antenna gain of 3.9 dB at 30 GHz and an estimated beamwidth of $60° \times 80°$. The other antenna is a 4×4 printed patch antenna array fabricated on the Rogers RO3003 PTFE/Ceramic laminates with ε_r of 3.0 and substrate thickness of 0.508 mm. The total size is $20.9 \times 28.2\ mm^2$. This antenna array achieves a maximum antenna gain of 12.9 dB at 28 GHz and an estimated beamwidth of $10° \times 10°$. Same types of antennas were used in transmitting and receiving. At distances below 2 m, the detection accuracy between these two antenna systems is comparable [Xiao et al., 2006].

The baseband circuits were designed using LM324 low-power operational amplifiers. The BPF has a passband of 0.1–10 Hz. The preamplifier PreAMP and the low-frequency amplifier LF_AMP use the same circuitry, both having a variable gain from 20 to 40 dB. During the measurement, a 22-bit USB DAQ module (IOtech Personal DAQ/54) samples the baseband signal, and a LabVIEW program processes the sampled data and further filters out unwanted spurious responses due to the subject's random motion.

This radar sensor system cannot work properly at higher frequency if it transmits only a single-tone wave since the detection accuracy varies dramatically with even a very small movement of the subject, making it extremely difficult to achieve reliable detection accuracy under this condition. For a 30-GHz wave, the distance between the adjacent null point and optimum point is only 1.25 mm ($\lambda/8$). This distance is so small that a reliable measurement at optimum point is difficult to achieve. This problem is solved by taking advantage of double sideband transmission. There are two Ka-band frequency components f_L and f_U in the transmitted signal $T(t)$, so the received signal $R(t)$ has these two frequency components f_L and f_U as well. Let $B_L(t)$ and $B_U(t)$ represent the baseband signals corresponding to f_L and f_U, respectively. In this case,

$$B(t) = B_L(t) + B_U(t) \tag{8.6}$$

$$B_L(t) = \cos\left[\theta_L + \frac{4\pi x(t)}{\lambda_L} + \Delta\phi_L(t)\right] \tag{8.7}$$

$$B_U(t) = \cos\left[\theta_U + \frac{4\pi x(t)}{\lambda_U} + \Delta\phi_U(t)\right] \tag{8.8}$$

and

$$\theta_L = \frac{4\pi d_0}{\lambda_L} + \theta_{0L}, \quad \theta_U = \frac{4\pi d_0}{\lambda_U} + \theta_{0U} \tag{8.9}$$

where λ_L and λ_U are wavelengths of lower sideband and upper sideband, which equal to c/f_L and c/f_U, respectively. θ_L and θ_U are fixed phase shifts of the lower sideband signal and the upper sideband signal, respectively.

When θ_L and θ_U are separated by an even multiple of π, $B_L(t)$, and $B_U(t)$ are in-phase and synchronized. Therefore, $B(t)$ will give almost the same optimum points and null points at the same places as those given by either $B_L(t)$ or $B_U(t)$ alone, and has the same problem of closely spaced null points that degrade the detection accuracy and reliability. When θ_L and θ_U are separated by an odd multiple of π, $B_L(t)$, and $B_U(t)$ are out of phase. Since $B_L(t)$ and $B_U(t)$ have almost the same amplitudes but with an opposite phase, they cancel each other. Therefore, the amplitude of $B(t)$ is very small and hard to be detected.

As a result, when the phase difference between θ_L and θ_U is the integer of π, a new null-point condition occurs in the measurement. If the null point of the single sideband (SSB) transmission is defined as the *local* null point, then this new null point condition is defined as the *global* null point. At this global null point, the detection accuracy is the lowest.

Let

$$\theta_U - \theta_L = \frac{4\pi d_0}{\lambda_U} - \frac{4\pi d_0}{\lambda_L} + \Delta\theta_0 = k\pi, \quad k = 0, \pm 1, \pm 2, \ldots \tag{8.10}$$

where

$$\Delta\theta_0 = \theta_{0U} - \theta_{0L} \tag{8.11}$$

Substituting $\lambda_L = c/f_L$, $\lambda_U = c/f_U$, then

$$f_U - f_L = \frac{c}{4\pi d_0}(k\pi - \Delta\theta_0), \ k = 0, \ \pm 1, \ \pm 2, \ldots \tag{8.12}$$

Substituting $f_U = f_2 + f_1$ and $f_L = f_2 - f_1$ in Equation 8.12, then

$$f_1 = \frac{k}{d_0} \cdot 37.5\,\text{MHz m} - \frac{c}{8\pi d_0} \cdot \Delta\theta_0, \quad k = 0, \pm 1, \ \pm 2, \ldots \tag{8.13}$$

where d_0 is the distance.

When θ_L and θ_U are separated by an odd multiple of $\pi/2$, $B_L(t)$ and $B_U(t)$ are in quadrature. At least one of $B_L(t)$ and $B_U(t)$ is not at the null point. The one that is not at the null point will determine the final output $B(t)$. Therefore, in this case, the overall detection accuracy will be high. This point is defined as the *global* optimum point. Let

$$\theta_U - \theta_L = \frac{4\pi d_0}{\lambda_U} - \frac{4\pi d_0}{\lambda_L} + \Delta\theta_0 = k\pi + \frac{\pi}{2}, \quad k = 0, \pm 1, \ \pm 2, \ldots \tag{8.14}$$

Repeat the same process as in Equations 8.12 and 8.13,

$$f_1 = \frac{2k+1}{d_0} \cdot 18.75\,\text{MHz m} - \frac{c}{8\pi d_0} \cdot \Delta\theta_0, \quad k = 0, \pm 1, \ \pm 2, \ldots \tag{8.15}$$

where d_0 is the distance.

From the above-mentioned discussions, either $B_L(t)$ or $B_U(t)$ has the severe null point problem and cannot give a reliable detection at high frequency. However, when $B_L(t)$ and $B_U(t)$ simultaneously exist, $B(t)$ is the superposition of $B_L(t)$ and $B_U(t)$. $B_L(t)$ and $B_U(t)$ are similar but with a phase difference between them. If their phase difference is arranged properly, the baseband output $B(t)$ will not have the severe null-point problem as either $B_L(t)$ or $B_U(t)$ alone. If the LO1 frequency f_1 is arranged properly, the null points from lower sideband and optimum points from upper sideband, or vice versa, can overlap each other. Good detection accuracy is, therefore, achieved over a wide distance range. The frequency difference between f_U and f_L is $2f_1$. Therefore, the selection of f_1 will determine whether θ_L and θ_U are separated by $k\pi$ or $k\pi + \pi/2$, and thus if the subject's position is at a null point or an optimum point.

To overcome the null point problem in the measurement, and to obtain high detection accuracy, the best way is to adjust the LO1 frequency f_1. For Ka-band wave, the transmission loss over distance is much higher than for low-frequency wave. For the same power level at receiver input port, the propagating distance for the Ka-band wave will be much shorter. In this measurement, using low-power transmission of 12.5 μW, the detection accuracy starts to drop quickly when the distance is increased to 2.5 m. If a null point occurs at $d_0 = 2.5$ m, in order to switch this null point to an optimum point, the f_1 will need to be changed at least 7.5 MHz according to Equation 8.21. However, if a null point occurs at $d_0 = 0.1$ m, the smallest tuning step will be 187.5 MHz, which is quite a large tuning range for LO1. Therefore, the selection of f_1 and the voltage controlled oscillator (VCO) tuning range need to be considered together when the null point appears at a distance close to radar. Therefore, in this system, a VCO with tuning range from 450 to 800 MHz was selected as f_1 source. At the same time, this VCO frequency provides about 75-mm null point separation, so it also provides a possibility to avoid the null point by adjusting the radar position.

The subject, facing the antenna, was seated at a distance away and breathed normally. A wired fingertip pulse sensor (UFI_1010 pulse transducer) was attached to the index finger during the measurement to provide the reference heartbeat signal.

When performing signal processing, the heartbeat and breath signals were first separated by a fourth-order Butterworth BPF with pass band from 0.1 to 0.7 Hz (for breathing rate of 6–42 breaths/min) and a fourth-order Butterworth BPF with pass band from 0.9 to 3 Hz (for heartbeat rate of 54–180 BPM). These two filtered signals were then windowed and autocorrelated to find the periodic respiration and heartbeat signals. After that, fast Fourier transform (FFT) was applied to the autocorrelated signals to obtain the respiration and heartbeat rate [Lohman et al., 2001; Yang and Rhee, 2000]. Finally, the detected heartbeat signal was evaluated by "heart-rate accuracy." Heart-rate accuracy is calculated as the percentage of time the calculated rate is within 2% of the reference rate.

As discussed earlier, the detection accuracy depends on the subject's position that might be in the null point, the optimum point, or somewhere in between. However, the optimal point can always be achieved by tuning the f_1, thereby high detection accuracy can always be achieved no matter where the subject is. An experiment was set up at a distance of about 1 m. Theoretically, a null point can be switched to an optimum point when tuning f_1 by an odd multiple of 18.75 MHz. One null point was determined experimentally, at $f_1 = 616$ MHz, where a low detection accuracy of 54.5% was observed. Based on the theory, a frequency step $\Delta f = 3 \times 18.75 = 56.25$ MHz was subtracted from 616 MHz to give $f_1 = 559.75$ MHz. The measurement made at $f_1 = 560$ MHz shows that a high accuracy of 94% was achieved. This experiment verified the theory that a null point can be changed to an optimum point by tuning f_1 by an odd multiple of 18.75 MHz. The measurement results are shown in Fig. 8.10(a) and (b), respectively.

8.1.3 Low-IF Architecture

In this section, the use of coherent low-IF architecture is proposed to minimize the effects of low-frequency noise, and eliminate DC coupling and spectrum-folding issues [Mostafanezhad et al., 2010]. Unlike communication low-IF receiver systems that use received signal and local oscillator signals derived from independent sources, CW radar systems require phase coherence for accurate phase detection [Saunders, 1990]. Thus, received signal and local oscillator must originate from the same source, to take advantage of range correlation effect [Droitcour et al., 2004].

A low intermediate frequency (IF) receiver architecture is commonly used in wireless communications [Crols and Steyaert, 1998] to avoid DC offset issues inherent to the direct conversion receiver. It is a type of heterodyne architecture with an IF low enough to be digitized. The main advantage of the low IF receiver is that it avoids the region of highest Flicker noise in the mixer output. As shown in Yamada et al. [2008], mixer Flicker noise is about 50 dB lower at an offset frequency of 100 Hz as compared with near-DC.

Direct conversion receivers need two separate RF receive paths to account for I and Q channels and since baseband outputs of the IQ demodulator have large DC components [Crols and Steyaert, 1998], they require proper DC cancellation methods as simple AC coupling circuits will not be sufficient [Park et al., 2007]. Range correlation of transmit and receive signals plays an important role in a homodyne Doppler radar, by significantly reducing baseband noise [Droitcour et al., 2004]. The coherent low-IF method can be used to down-convert the RF signal to a frequency of about 1 kHz and not directly to DC, thus, it can avoid amplifier and mixer's $1/f$ noise. It will be shown that in addition to a simpler receiver system, particularly suitable for multiple receiver systems, a better noise performance is achieved.

Figure 8.11 shows a simple diagram of the low-IF receiver system and its coherent counterpart. It is assumed that the input signal from the antenna has content around a carrier frequency of ω_{RF}, so the received RF signal is

$$r(t) = A(t)\cos(\omega_{\mathrm{RF}} t + \varphi(t)) \tag{8.16}$$

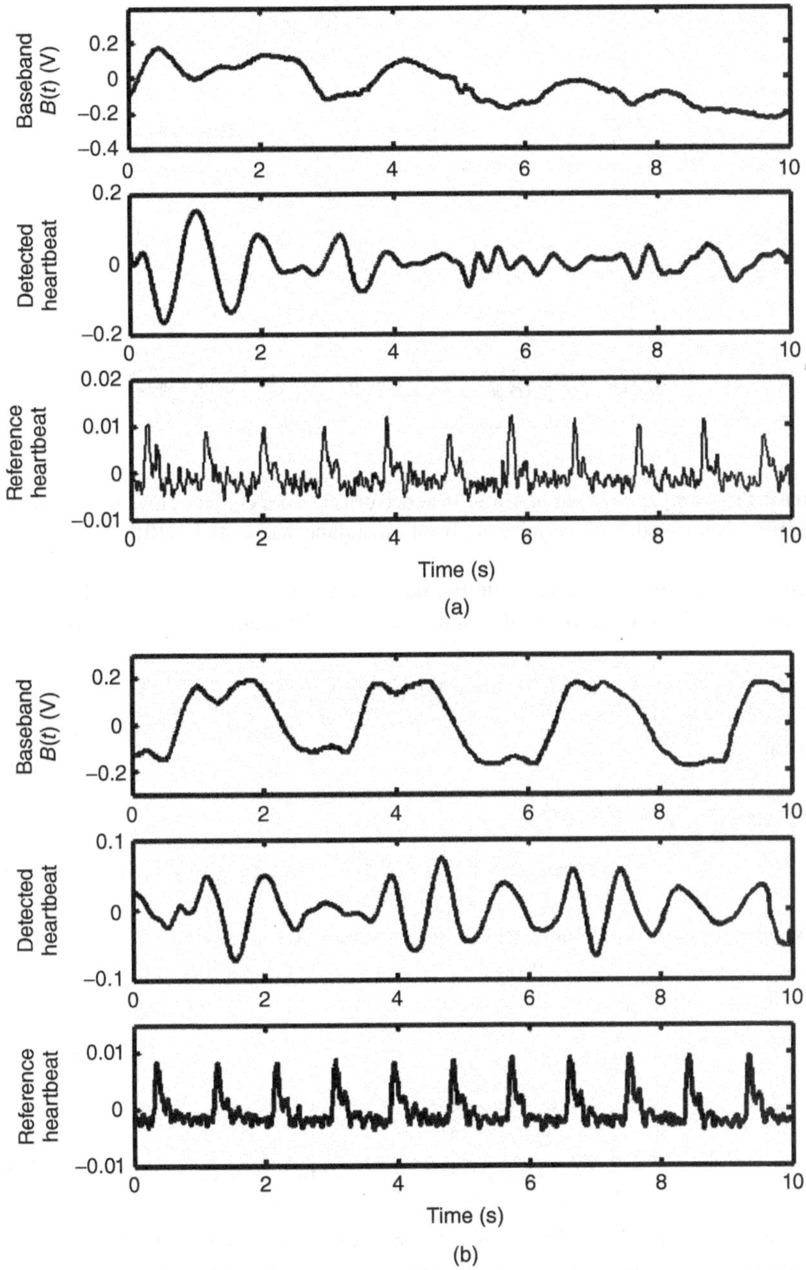

Figure 8.10 Heartbeat detection at null point (a) and optimum point (b). The heart-rate accuracy is 54.5% at the null point while 94% at the optimum point. The frequency difference between them is only 56 MHz. © 2006 IEEE. Reprinted, with permission, from Xiao et al. [2006].

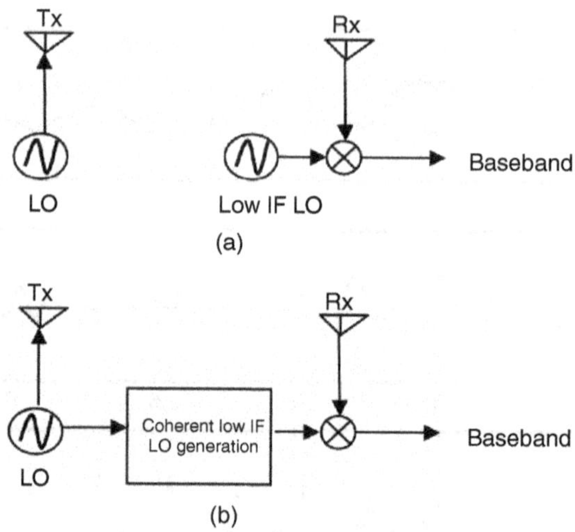

Figure 8.11 Simple diagram of a low-IF receiver (a) and a coherent low-IF receiver (b). © 2010 IEEE, Reprinted, with permission, from Mostafanezhad et al. [2010].

where A and φ generally determine the baseband content of the signal. This signal is then mixed with an local oscillator sine wave at a frequency of ω_{LO}, where

$$\omega_{\mathrm{LO}} = \omega_{\mathrm{RF}} - \omega_{\mathrm{LIF}} \tag{8.17}$$

resulting in

$$\begin{aligned} r_{\mathrm{LIF}}(t) &= r(t)\cos(\omega_{\mathrm{LO}}t) \\ &= \frac{1}{2}A\,(t)\cos((\omega_{\mathrm{LIF}}t + \varphi(t)) + \frac{1}{2}A(t)\cos\left(\left(2\omega_{\mathrm{RF}} - \omega_{\mathrm{LIF}}\right)t + \varphi(t)\right) \end{aligned} \tag{8.18}$$

The second term in Equation 8.18 is a high-frequency out-of-band RF signal in which we are not interested and will be rejected by the low-pass filter at the output of the mixer. From this stage on, since ω_{LIF} is a rather small frequency, the IF signal is digitized and DSP is applied to recover A and φ:

$$\begin{aligned} r_b(t) &= \frac{1}{2}A(t)\cos\left(\omega_{\mathrm{LIF}}t + \varphi\,(t)\right)\exp(-j\omega_{\mathrm{LIF}}t) \\ &= \frac{1}{8}A(t)\exp\left(j\varphi\,(t)\right) + \frac{1}{8}A\,(t)\exp(-2j\omega_{\mathrm{LIF}}t - j\varphi(t)) \end{aligned} \tag{8.19}$$

The first term is the desired complex down-converted signal, while the second term is easily removed using a real filter. The process and the effect on the signal spectrum are depicted in Fig. 8.12.

The measurement setup is shown in Fig. 8.13. A two-antenna bistatic configuration has been used. As it can be seen, both the coherent low IF and direct conversion I and Q paths are implemented concurrently for comparison. Off the shelf coaxial

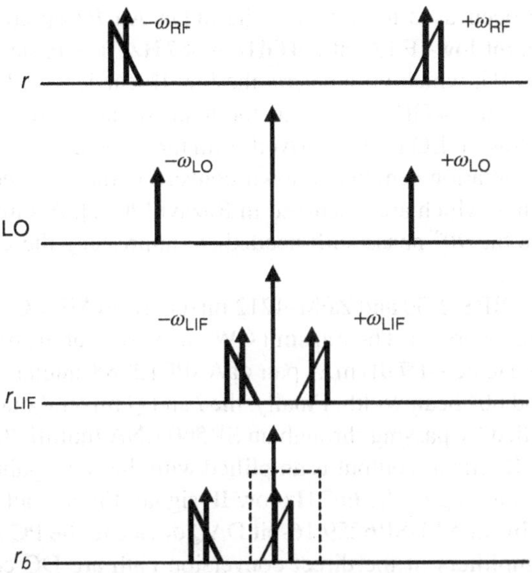

Figure 8.12 Signal spectrum in low IF receiver. © 2010 IEEE, Reprinted, with permission, from Mostafanezhad et al. [2010].

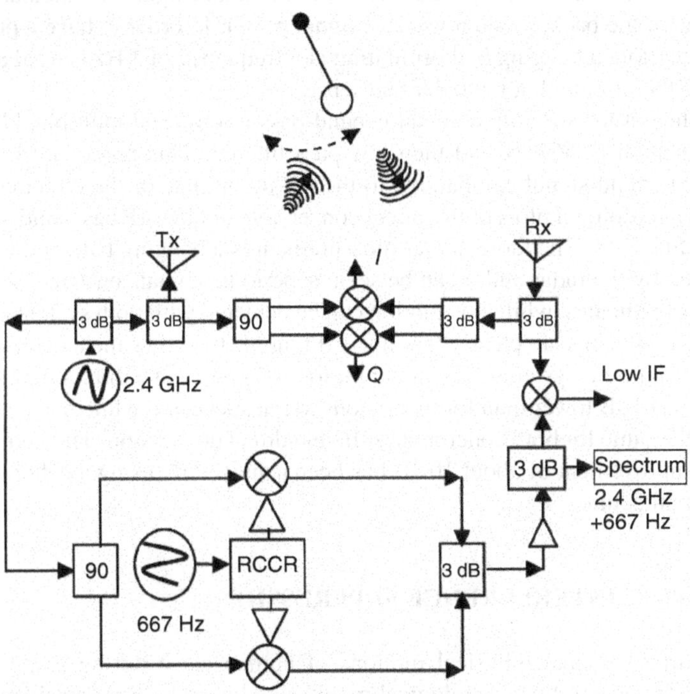

Figure 8.13 Measurement setup. Note the coherent low-IF generation path. © 2010 IEEE, Reprinted, with permission, from Mostafanezhad et al. [2010].

components have been used to construct the radar. An *IQ* up-converter is used to generate the coherent low-IF LO at 2.4 GHz + 667 Hz. It is basically an SSB modulator [Proakis, 2001], which up-converts the low-IF carrier. An HP 83640B signal generator operating at 2.4 GHz is used as the main signal source. It is important to keep in mind that low-IF LO is also derived from the same 2.4-GHz source. The reason is the range correlation benefits of down-converting the received signal using the same transmit signal, which are discussed in Razavi [2001]. A simple RCCR circuit is used to generate the 90° phase shift needed or suppressing the 2.4 GHz to 667 Hz component.

The splitters are RPS-2-30 and ZFM-4212 mixers from Mini-Circuits are used for the down- and up-conversion. The transmit CW microwave at the transmitter antenna input was measured to be −1.7 dB m. A pair of ASPPT2988 antennas was used, which have 8 dBi gain and 60° beamwidth. Finally, the I and Q mixers' outputs are low-pass filtered and amplified by passing through an SR560 LNA (cutoff 30 Hz, 46 dB gain). The coherent low-IF mixer's output is amplified with the same gain but a bandwidth of 1 kHz, to allow passing of the 667 Hz low IF signal. Finally, all baseband signals are then recorded by an NI USB6259 16-bit DAQ device to the PC at a sampling rate of 10 kHz. The amplifiers of the direct conversion path are DC coupled, since the signal has a significant amount of spectral content around DC.

Goal of this measurement is to benchmark the noise performance of the two different configurations. Power spectral density of baseband signals is calculated to give an estimate of the background noise and signal power. In order to have a predictable source of motion, a hanging ball oscillating at a frequency of 1 Hz is used as a target to produce the motion that the radar can sense.

The coherent low-IF output signal is band-pass filtered and multiplied by a complex exponential at 667 Hz and then low-pass filtered. This procedure results in a complex baseband signal comparable to the $I + jQ$ outputs of the direct conversion channel. Time-domain plots of the direct conversion and low-IF baseband signals are plotted in Fig. 8.14. The target under measurement is a hanging ball, and the motion is picked up by the radar and it can be seen as periodic signals on I and Q channels. As Fig. 8.14 indicates, while this motion can be detected with both systems, coherent low-IF I and Q traces are clearly less noisy. Figure 8.15 shows the calculated output spectrum from both systems. The coherent low-IF spectrum exhibits the noise floor that is about 10 dB lower than for direct conversion spectrum, while the signal power at 1 Hz is the same for both coherent low-IF and direct conversion. This indicates that the SNR improvement of about 10 dB has been achieved by using a coherent low-IF receiver architecture.

8.2 MOTION INTERFERENCE SUPPRESSION

Due to extremely small physical motions of human chest during respiration, the Doppler radar system for physiological monitoring has to be very sensitive to phase changes, which are caused by path length variations. For a stationary transmitter and subject, this path length only varies with a subject's heart beat and respiration.

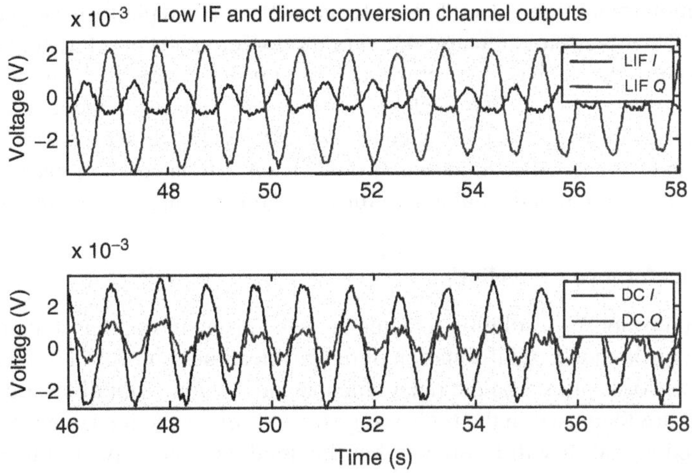

Figure 8.14 Baseband I and Q signals from direct conversion and coherent low-IF receiver paths. © 2010 IEEE, Reprinted, with permission, from Mostafanezhad et al. [2010].

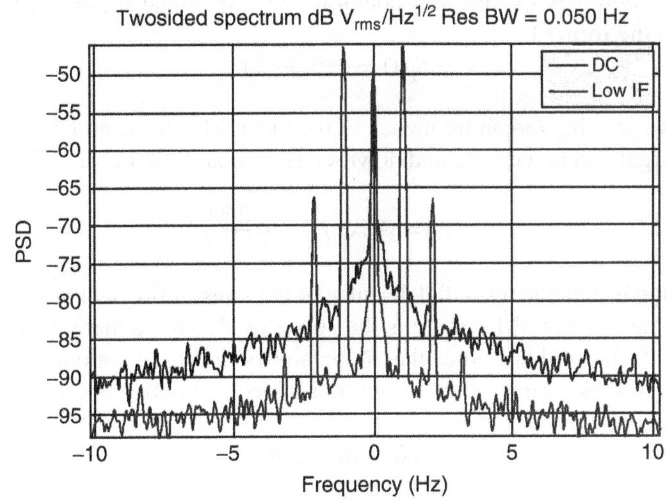

Figure 8.15 Baseband signal spectrum for the two receiver paths. © 2010 IEEE, Reprinted, with permission, from Mostafanezhad et al. [2010].

A nonstationary transceiver antenna or a fidgeting subject will produce additional path variations causing interference in radar signal. There are cases where the transmit/receive antenna cannot be assumed to be physically still. For example, if the transceiver is a handheld unit used in search and rescue operations or sense through the wall military applications, "hand shake" of the operator will introduce path length variations, which phase modulate the received signal in addition to that of target's cardiopulmonary motion. Also, if the antenna is placed on a platform subject to some vibration, the vibration can cause antenna motion creating interference with radar signal. Monitoring a driver in a vehicle is an example of such application. Thus, the final demodulated signal will include components resulted from operator's unwanted handshake to the extent that life signs monitoring will become impossible due to excessive interference. The fidgeting of the subject, which can be caused by coughing, sneezing, itching, and similar activities, is another source of this interference.

8.2.1 Interference Cancellation

This section proposes a solution to antenna shake problem in monostatic radar configuration, through use of information of the physical motion itself. For this purpose, a motion sensor is placed on the radar antenna and its slight vibrations are recorded. These data are then used in the processing step to remove the interference caused on the received signal. It will be shown that this method is effective to the extent that heart rate of the human subject can be accurately extracted from radar data despite the mechanical shake of the antenna.

Figure 8.16 shows the geometry and relative positions of a transceiver antenna and a human subject in two dimensions for simplicity.

Using the complex exponential notation, the CW signal transmitted toward the subject is in the form of

$$S_t(t) = \exp(j\omega_0 t) \tag{8.20}$$

where ω_0 is operating radian frequency of the radar. This transmitted signal bounces off of the target and is received and down-converted at the same antenna as follows:

$$S_r(t) = A \exp j\left(-\omega_0 \frac{2R}{c}\right) \tag{8.21}$$

where R is the instantaneous radial distance of the transmitting antenna from the subject as depicted in Fig. 8.16 and c is the speed of electromagnetic waves (same as speed of light). It should be noted that R contains both subject's motion and transmitter's unwanted shake. Since both antenna and subject are stationary, R can be assumed to have two components:

$$R = R_0 + \Delta R \tag{8.22}$$

where R_0 is the fixed distance and ΔR is its variation, which contains important information. It can be seen that using proper phase demodulation, ΔR is retrievable

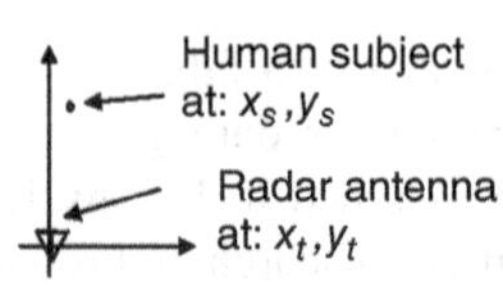

Figure 8.16 Location of antenna and the subject. © 2008 IEEE, Reprinted, with permission, from Mostafanezhad et al. [2008].

[Droitcour, 2006; Park et al., 2007]. Also, assuming the transmit antenna is located in the origin, R can be expanded as

$$
\begin{aligned}
R &= \sqrt{(x_t - x_s)^2 + (y_t - y_s)^2} \\
&= \sqrt{(x_t - X_s - \Delta x_s)^2 + (y_t - Y_s - \Delta y_s)^2}
\end{aligned}
\tag{8.23}
$$

where x_t, y_t are the locations of the antenna, and X_s, Y_s are the locations of the subject, and Δx_s, Δy_s are the slight changes modeling physiological motion of the human subject. R can be further expanded as follows:

$$
R = \sqrt{X_s{}^2 - 2X_s(x_t - \Delta x_s) + (x_t - \Delta x_s)^2 + Y_s{}^2 - 2Y_s(y_t - \Delta y_s) + (y_t - \Delta y_s)^2}
\tag{8.24}
$$

The components $x_t, y_t, \Delta x_s,$ and Δy_s are assumed to be very small compared with the distance of the subject and antenna. It is estimated that Δx_s, Δy_s, resulting from cardiopulmonary activity are in the order of less than a centimeter. Also, x_t and y_t or the unwanted shake motion of the antenna is considered to be in the order of millimeters. Reordering Equation 8.24 and using a Taylor series results in

$$
\begin{aligned}
R &\approx R_0 \sqrt{1 - \frac{2X_s\left(x_t - \Delta x_s\right)}{R_0^2} - \frac{2Y_s\left(y_t - \Delta y_s\right)}{R_0^2} + \cdots} \\
&= R_0 - a\left(x_t - \Delta x_s\right) - b\left(y_t - \Delta y_s\right)
\end{aligned}
\tag{8.25}
$$

The constant part of Equation 8.25 is R_0 and thus:

$$
m = \Delta R = z - a x_t - b y_t \tag{8.26}
$$

where $z = a\Delta x_s + b\Delta x_s$ is a measure of the subject's physiological motion in which we are interested. The phase demodulation of the output of the radar yields m, thus in order to remove the effect of transmitter antenna's shake, components x_t and y_t must be known in advance to be removed from m. Once these signals are known, a and b can be calculated and z can be computed. Assuming signals are sampled and represented in column vectors, an orthonormal basis can be developed as follows:

$$
\left[x_T \ y_T\right]^T \left[x_T \ y_T\right] = \begin{bmatrix} 1 & 0 \\ 0 & 1 \end{bmatrix} \tag{8.27}
$$

in which superscript T is matrix transpose, and x_T and y_T are

$$
\left[x_T \ y_T\right] = \left[x_t \ y_t\right] \mathbf{A} \tag{8.28}
$$

where **A** is a 2×2 matrix. Matrix **A** can be easily calculated to yield Equation 8.35, details of which can be found in Meyer [2001]. Including Equation 8.28 in Equation 8.26 results in

$$\begin{aligned} m &= z - x_T\left(aB_{11} + bB_{21}\right) - y_T\left(aB_{12} + bB_{22}\right) \\ &= z - ux_T - vy_T \end{aligned} \tag{8.29}$$

where $\mathbf{B} = \mathbf{A}^{-1}$. Antenna's shake motion and the subject's physiological motion are two signals originating from two separate sources, which makes them independent. That is, $x_T^T z = y_T^T z = 0$. Thus,

$$x_T^T m = x_T^T z - x_T^T x_T u - x_T^T y_T v \tag{8.30}$$

yielding u as

$$u = -\frac{x_T^T m}{x_T^T x_T} \tag{8.31}$$

Similarly, v can be calculated as

$$v = -\frac{y_T^T m}{y_T^T y_T} \tag{8.32}$$

Once u and v are computed, z can be easily recovered from Equation 8.29.

The radar used for these measurements has a single antenna monostatic configuration with I and Q phase-modulated outputs resulting from direct conversion of received RF signal. Radar's block diagram is shown in Fig. 8.17. Off-the-shelf components have been used to construct the radar. A JTOS-2700 VCO operating at 2.4 GHz has been used as the signal source and RPS-2-30 splitters, SKY-42 mixers from Mini-Circuits, and a Narda 4923 circulator are included in the design. The transmit CW microwave signal at the antenna input was measured to be 0 dB m. An ASPPT2988 antenna was used, which has 8-dBi gain and 60° beamwidth. Finally, the I and Q mixers' outputs are low-pass filtered and amplified by passing through an SR560 LNA (cutoff 100 Hz, 40 dB gain) and are then recorded by an NI USB6259 16-bit DAQ device to the PC at a sampling rate of 1 kHz.

A chest band (Pneumotrace 1132 piezoelectric respiration transducer) was used on the subject as a reference for respiratory motions of the subject. Also a UFI 1010 finger pulse sensor was used as a heart signal reference. The radar's output will be compared with these reference signals. In order to be able to perform shake cancellation, antenna's shake or vibration must also be recorded. An ADXL203 dual-axis accelerometer chip was attached to the transmit antenna. Outputs of ADXL203 correspond to physical acceleration in x and y (parallel to the ground) directions. These outputs are passed through the SR560 LNA and then recorded by the acquisition device, then postprocessed to provide instantaneous displacement values. The human subject was seated 1.5 m away from the transceiver antenna. The antenna

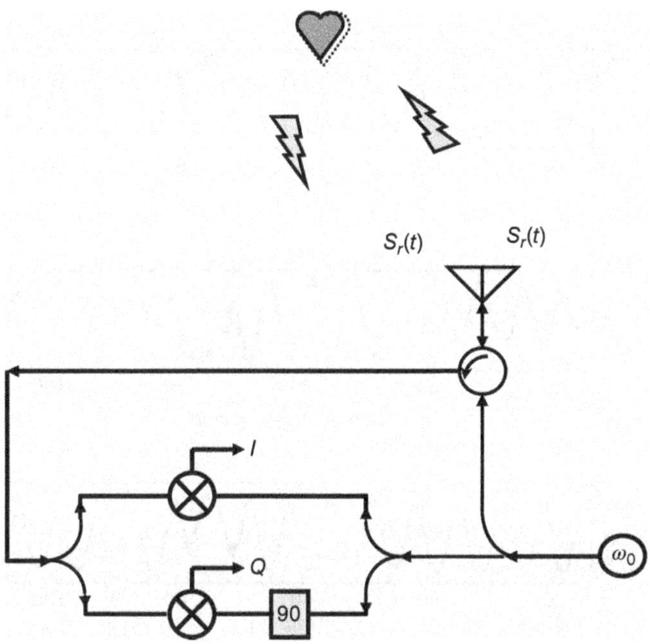

Figure 8.17 RF configuration of the radar.

was placed on a fixture and the fixture was vibrated by hand during parts of the measurement.

I and *Q* outputs of the radar are linearly demodulated using the method in Massagram [2008]. Figure 8.18 depicts *I*, *Q*, and the demodulated phase while the subject was breathing normally. Demodulated motion and antenna's shake in *x* and *y* directions are plotted in Fig. 8.19. As shown in Figs. 8.18 and 8.19, antenna's mechanical motion occurs between 65 and 82 s. In measurements shown in Fig. 8.18, linear demodulation (or principal component analysis) is used to calculate *m*, denoted in Equation 8.26, from *I* and *Q* signals. *I* and *Q* signals are added such that the desired signal content is maximized. The received signal is defined as

$$S = \left[S_I \; S_Q\right] \tag{8.33}$$

where S_I and S_Q are the *I* and *Q* components with their mean values removed. *M* is a linear mapping of *S* and is calculated using

$$M = \left[M_1 \; M_2\right] = S_R \mathbf{V} \tag{8.34}$$

where $\mathbf{V}$ is the 2×2 matrix containing eigenvectors of $S^T S$. Variance of M_1 and M_2 are compared and the signal with the largest variance is selected as *m*.

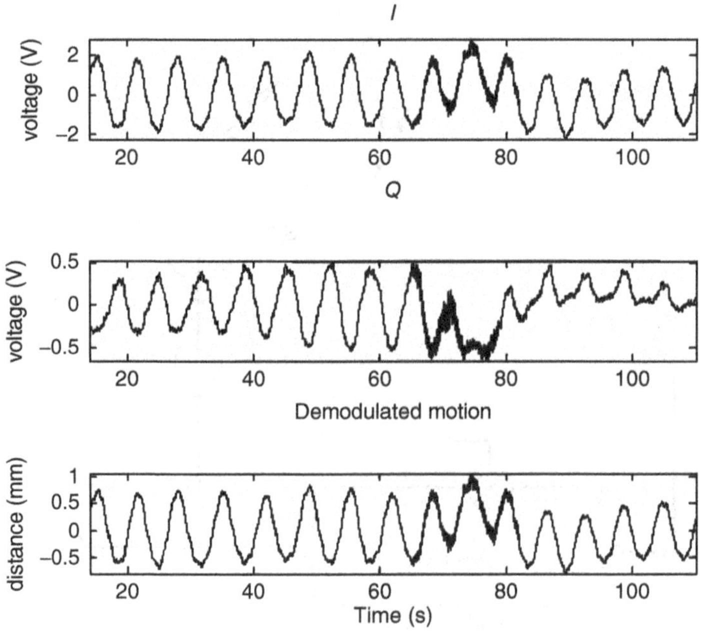

Figure 8.18 Radar channel outputs (top) and the demodulated signal (bottom). © 2008 IEEE, Reprinted, with permission, from Mostafanezhad et al. [2008].

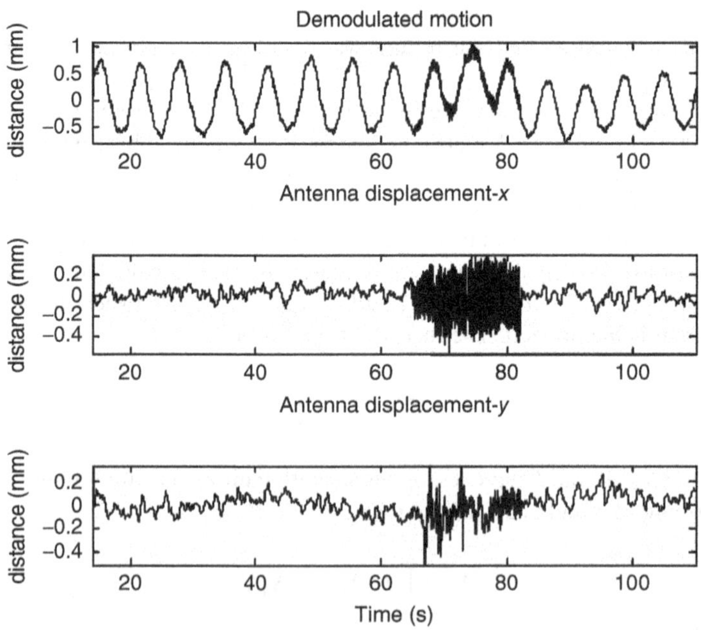

Figure 8.19 Demodulated motion (top) and antenna's recorded mechanical motion in x and y direction (bottom). © 2008 IEEE, Reprinted, with permission, from Mostafanezhad et al. [2008].

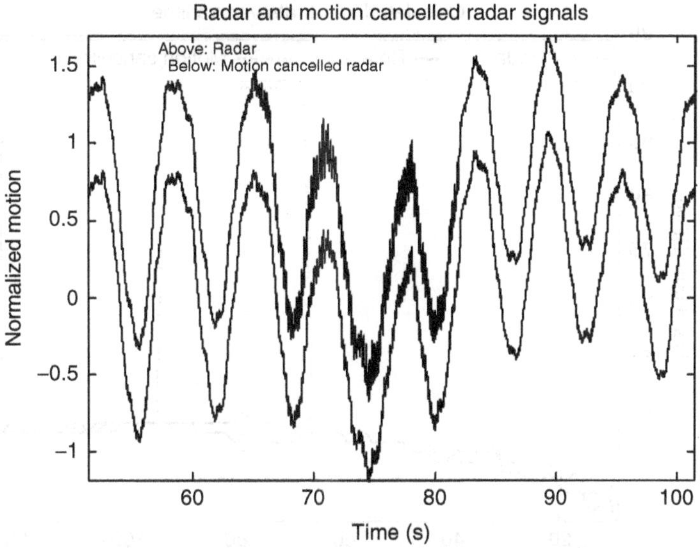

Figure 8.20 Radar signal and motion-cancelled signal, zoomed in the area of antenna motion. © 2008 IEEE, Reprinted, with permission, from Mostafanezhad et al. [2008].

The motion-cancelled data are plotted in Fig. 8.20. Top trace shows the demodulated radar output, and bottom trace compensated radar output. It is evident that noise is significantly lower in the compensated radar output trace.

Success of the motion cancellation was evaluated by comparing the heart rates calculated from finger pulse reference, demodulated radar, and motion-cancelled radar signals. Heart rates were calculated from a combination of high-pass filtering and window-based FFT computations [Lohman et al., 2001].

Figure 8.21 shows the extracted heart rates from finger pulse reference, demodulated radar, and motion-compensated radar signals, using accelerometer data. It is evident that, during the time period when antenna is shaking, it becomes impossible to calculate the heart rate from radar's signal, while the signal from motion cancelled radar tracks the finger pulse reference with an RMSE of 0.97 BPM. This method can be further expanded to compensation of three-dimensional motion of larger amplitude, and can be used for handheld and platform-mounted and vehicle-mounted devices.

8.2.2 Bistatic Radar: Sensor Nodes

Phase stability of the Doppler radar measurement system plays an important role in successful life signs detection. As shown in the previous section, small unwanted mechanical motions of the transmit antenna may result in unrecoverable phase errors in the received signal. Another proposed technique for overcoming this issue is a bistatic radar with a sensor node receiver placed in the vicinity of the human subject. Theoretical and experimental results confirm the benefits of using sensor nodes.

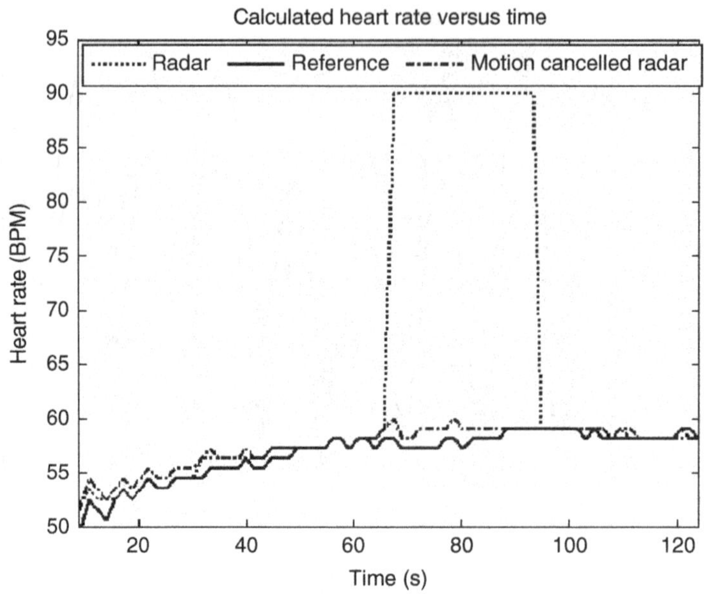

Figure 8.21 Time-variant heart rate calculated from the radar, reference, and motion-cancelled signals. The motion-cancelled radar signal is in good agreement with the reference finger pulse signal, while the radar signal before motion cancellation cannot be used to retrieve the heart rate. © 2008 IEEE, Reprinted, with permission, from Mostafanezhad et al. [2008].

A simple phase-demodulation technique of mixing the received signal with the portion of the transmitted signal (Fig. 8.22(a)), results in a baseband signal that can be processed to yield the heart and respiration rates of the human subject. Phase stability of the measurement system affects the accuracy of the phase demodulation. It has been shown that if the transmitted signal and the LO are derived from the same source, the range correlation effect greatly reduces detrimental effects of electrical phase noise of the signal source [Skolnik, 1962]. This reduction in output signal's noise is inversely proportional to the phase delay between the local oscillator and the received phase-modulated signal. If the transceiver is a handheld device, which could be for example used for search-and-rescue operations or sense through the wall military applications, "hand shake" of the user will introduce path length change that will appear as phase noise in the demodulated baseband signal. In case of the monostatic radar (Fig. 8.22), this noise does not appear in the LO path and thus there is no benefit of range correlation. Therefore, such "shaking" typically results in signal degradation that obstructs the detection of cardiopulmonary signals.

We propose to use a bistatic radar with a receiver (sensor node) placed in the vicinity of the human subject (Fig. 8.22) to overcome this issue. Sensor node consists of an antenna and a mixer, similar to the tag used in Lubecke et al. [2002]. It receives both the direct signal from the transmitter (LO), and the signal reflected from a human subject, and both of these signals are subject to the same "mechanical" phase noise.

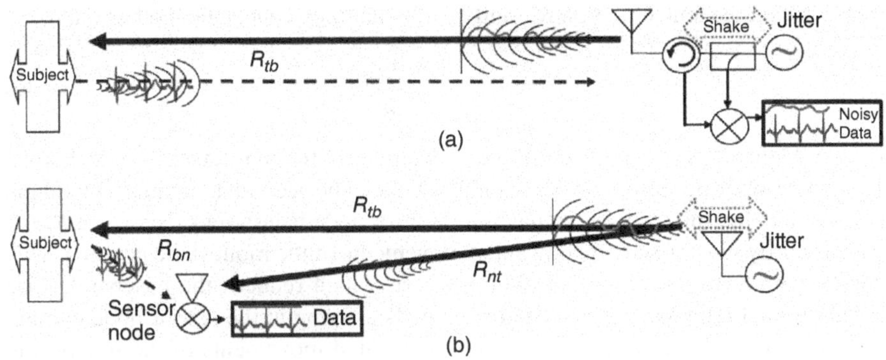

Figure 8.22 A monostatic radar (a), and a sensor node configuration (b). © 2007 IEEE, Reprinted, with permission, from Mostafanezhad et al. [2007].

If these path lengths are similar, there will be a significant phase noise reduction due to the range correlation effect, thus enabling accurate detection of life signs.

Transmitted signal from an ideal CW radar has the form

$$S_t(t) = \cos(\omega_0 t) \tag{8.35}$$

where ω_0 is the radian oscillation frequency. This signal once reflected from the human body will be demodulated at the monostatic end as

$$S_r(t) = A\cos\left(-\frac{4\pi R_{tb}}{\lambda}\right) \tag{8.36}$$

where λ is the wavelength and R_{tb} is the time-varying distance of the subject's chest from the transmitting antenna, as indicated in Fig. 8.22. However, the total RF signal received at the sensor node antenna is

$$\begin{aligned} S_{n\mathrm{RF}}(t) = {} & B\cos\left(\omega_0 t - \frac{\omega_0}{c}R_{nt}\right) \\ & + C\cos\left(\omega_0 t - \frac{\omega_0}{c}R_{tb} - \frac{\omega_0}{c}R_{bn}\right) \end{aligned} \tag{8.37}$$

where R_{tb} is the time-varying distance of transmitter to the subject and R_{bn} is the time-varying distance of the subject to the node. If we neglect amplitude variation due to propagation loss, mixing $S_{n\mathrm{RF}}(t)$ by itself by passing it through a nonlinear device results in the following baseband component:

$$S_n(t) = BC\cos\left(\frac{2\pi}{\lambda}\left(R_{tb} + R_{bn} - R_{nt}\right)\right) \tag{8.38}$$

If the monostatic antenna is located at a large distance from both the human subject and the node, such that $R_{tb} \approx R_{nt}$, slight physical movements of the monostatic antenna have the same effect on R_{tb} and R_{nt}, so that they cancel each other out:

$$S_n(t) \approx BC \cos\left(\frac{2\pi}{\lambda} R_{bn}\right) \tag{8.39}$$

Figure 8.22 displays the geometry of the node, monostatic transceiver, and the human subject. Considering Equations 8.36 and 8.39, it can be seen that, compared with the monostatic radar, the received signal at the sensor node is less sensitive to the $R_{tb}(t)$, which is partly given rise to by unwanted movements of the monostatic antenna. This effect is similar to the range correlation effect, which reduces the baseband noise caused by the LOs phase noise. The two signals arriving at the sensor node contain nearly the same phase variation caused by unwanted movements of the monostatic antenna. The closer the node and the subject are, the better these two phase variations cancel out, resulting in a less noisy baseband signal providing more accurate life signs detection.

Figure 8.23 shows the simulated outputs of the radar node and monostatic radar, for subject displacement of 5 mm. Node is placed at 20 cm from the subject. The

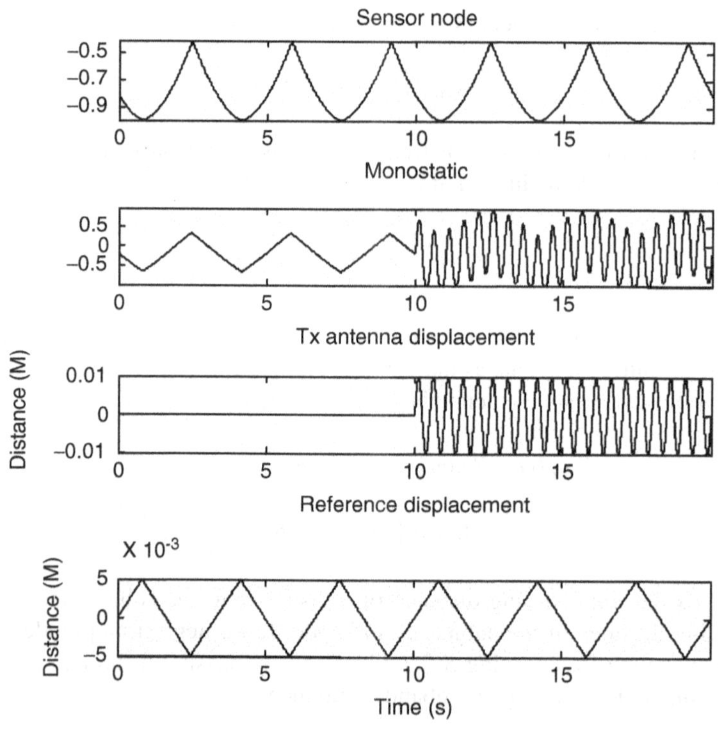

Figure 8.23 Simulated outputs of the sensor node and monostatic radar for subject displacement of 5 mm. The monostatic antenna begins shaking with the amplitude of 10 mm after 10 s. While sensor node output (top) remains unchanged, antenna's shake greatly alters monostatic radar output. The bottom trace is the reference displacement. © 2007 IEEE, Reprinted, with permission, from Mostafanezhad et al. [2007].

monostatic transceiver is located 2 m away from the subject and begins to shake with a 10-mm displacement after 10 s. During the first 10 s of the simulation interval, both the monostatic radar and the sensor node track the subject motion. Once the transmitter antenna begins to shake, this shaking is clearly visible at the monostatic radar output, while sensor node output is not affected and it continues to accurately track subject motion.

This concept may be used to implement a network of sensor nodes covering an area of interest to, for example, facilitate search and rescue operations using a handheld device. In order to have a better vision of how the sensor node reduces mechanical shake of the antenna and how effective this method is, another simulation has been performed, in addition to the results of Fig. 8.23.

In the next simulation, location of the transmit antenna has been changed while the human subject and the sensor node are fixed. Tx antenna is moved to six different locations on a circle around the subject while it was vibrating in the x direction. These locations are depicted in Fig. 8.24 using circles. This helps understanding whether angle of arrival of the transmitted wave makes major difference in the signal level on the sensor node. Figure 8.25 contains the received signal pattern for each of the six transmit positions of Fig. 8.24. The transmitter stays at each location for a total of 16 s. It is stationary for the first 8 s, then shakes for another 8 s and then moves on to the next position. This helps compare the effect of shake/no shake for each position. As shown in Fig. 8.25, signal from the node stays very robust to the mechanical shake of the transmit antenna, while simulated I and Q outputs from the monostatic radar suffers

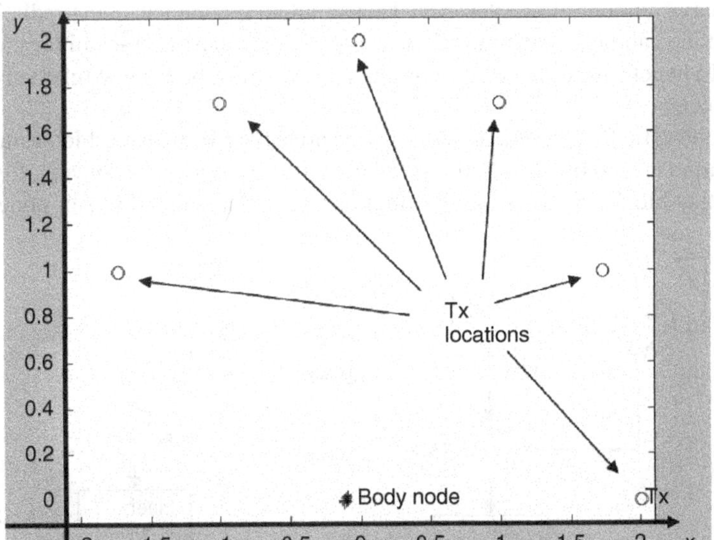

Figure 8.24 Simulation: human subject fixed in a location next to sensor node, while Tx antenna (circle) is moved to various positions to simulate how effective sensor node will be. The transmit antenna is shaking in the x direction.

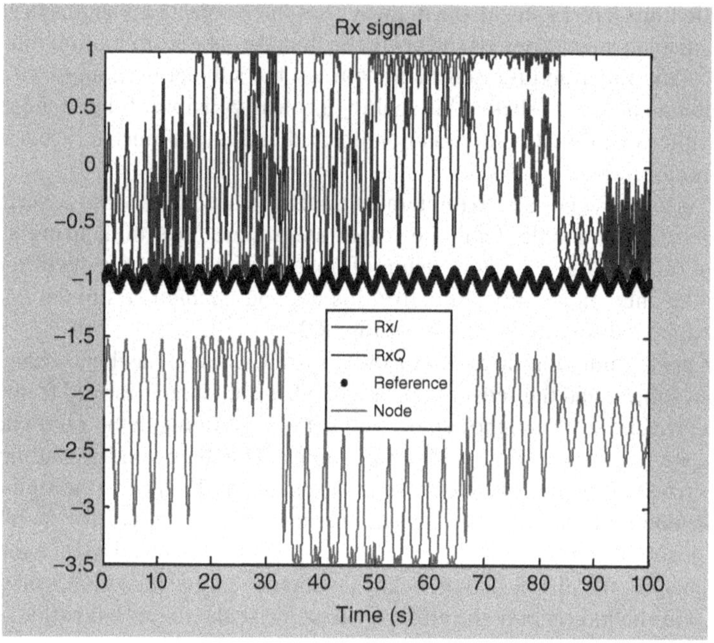

Figure 8.25 Received signal pattern for various locations of the Tx antenna. *I*, *Q*, reference, and node signals are depicted in this plot.

from a large amount of interference. This simulation completes the prediction that a sensor node can help make the whole system very robust to transmitter mechanical shake. To benefit from range correlation, the node must be located relatively close to the subject.

A monostatic Doppler radar and a sensor node have been assembled using laboratory equipment and off-the-shelf coaxial components to test this concept. Figure 8.26 shows the block diagram of the monostatic radar (a) and sensor node (b). An

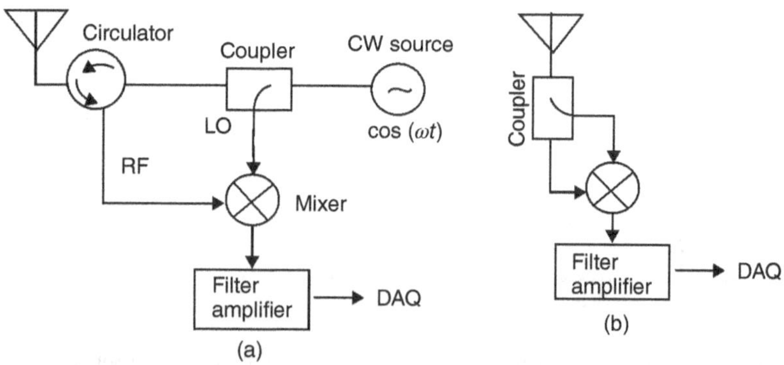

Figure 8.26 Block diagrams of the monostatic transceiver (a) and sensor node (b). © 2007 IEEE, Reprinted, with permission, from Mostafanezhad et al. [2007].

HP-E4433B signal generator was used as signal source for the monostatic radar providing a CW signal at 2.4 GHz with a power of 0 dB m. Mini-Circuits ZFSC-2-2500 coupler was used to split the signal source output into the transmit and LO paths.

The transmit power at the monostatic antenna connector was measured to be −6 dB m. The antenna specialist ASPPT2988 antenna was used with 8-dBi gain and 60° *E*-plane beamwidth. Narda 4923 circulator was used for isolating the transmit and receive signals. The received signal was fed into a Mini-Circuits ZFM-4212 mixer. Finally, the mixer's output is low-pass filtered and amplified by passing through an SR560 LNA (cutoff 10 Hz, 30 dB gain) and is then recorded by a TI DAQ9801 acquisition device to the PC with the sampling rate of 1 kHz.

The sensor node consists of an antenna, a power splitter, and a mixer. The received signal at the sensor node antenna goes through a splitter and is mixed with itself as depicted in Fig. 8.26(b). The printed circuit board (PCB) in Fig. 8.34(a) is designed to produce the fabricated sensor node can be seen in Fig. 8.27 to perform this function. All traces are 50 Ω for an FR-4 substrate. The antenna is a compact vertically polarized 2.45-GHz antenna designed for wireless LAN applications with a 0 dBi typical gain and a 50 Ω termination.

Sensor node uses the same type of coupler, mixer, baseband amplifier, and DAQ card. The signal received at the sensor node antenna, consisting of direct path signal from the transmitter, and the signal reflected from the subject, is equally divided to feed the mixer LO and RF ports. Output of the mixer device is processed like its monostatic counterpart.

Goal of the experiment is to investigate the effect of mechanical motion of the monostatic antenna on the output signals. In order to quantify hand-shaking motion of the monostatic antenna, an ADXL203 dual-axis accelerometer chip was attached to it. Outputs of ADXL203 correspond to physical acceleration in x and y directions, which are passed through the SR560 LNA and then recorded by the acquisition device, then postprocessed to provide instantaneous displacement values. A chest band (Pneumotrace 1132 piezoelectric respiration transducer) was used on the subject as a reference to log physiological respiratory motions of the subject.

Figure 8.27 The sensor node used in measurements.

The subject was seated at a distance of 2 m away from the monostatic transceiver, whereas the sensor node antenna was placed at about 30 cm of the subject's chest, nearly at the same height as the monostatic antenna. During a part of the measurement interval, the transceiver antenna was being shaken in the *x*–*y* plane, which was recorded by the accelerometer. A separate PCB has been used to mount the accelerometer and its connectors. The accelerometer board is then attached to the transmit antenna to measure its relative motion.

Figure 8.28 shows the output signals of the sensor node, monostatic radar, chest band, and *x* and *y* antenna displacements recorded using the accelerometer. The monostatic antenna begins shaking after 31 s. The sensor node output appears to be more noisy for stationary antenna, due to very low local oscillator level at the node mixer. Once the antenna starts shaking, the benefits of sensor node are clearly demonstrated. As shown in Fig. 8.28, even at very low antenna displacements with a

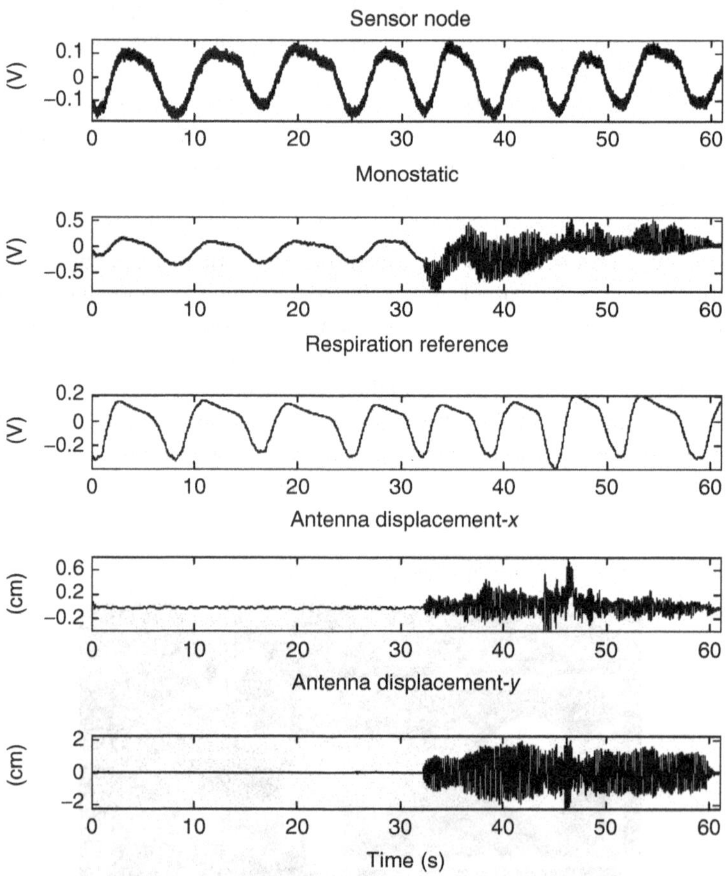

Figure 8.28 Measured signals from the radar node, monostatic radar, chest band reference, and *x*–*y* displacement. The monostatic antenna begins shaking after 31 s. © 2007 IEEE, Reprinted, with permission, from Mostafanezhad et al. [2007].

peak-to-peak amplitude of 2 cm, the received signal at the monostatic end contains a large amount of unwanted perturbations. At the same time instance, signal received from the sensor node shows very little effect of the monostatic antenna movements. The effect of shaking on the monostatic received signal is so large that it completely obscures useful information in the baseband signal.

Output signals from monostatic radar, sensor node, and chest band were post-processed using MATLAB to obtain respiration rates. The signals were low-pass filtered (at a 0.3 Hz cutoff frequency) and then a moving FFT window of 12 s was applied to calculate respiration rate during the measurement interval. Figure 8.29 shows the detected respiration rate from the monostatic radar (dotted line), sensor node (dashed line), and the chest band reference (solid line). Before shaking begins, both monostatic radar and sensor node outputs are in close agreement with the chest band reference. During the shaking period, sensor node output continues to accurately track reference output, to within 1 breath/min, while the monostatic output exhibits dropout regions that can no longer provide respiration rates.

Theoretical and experimental results confirm that this sensor node configuration is much less sensitive to the mechanical motion of the transmit antenna. It has been demonstrated that the sensor node output maintains 1 breath/min accuracy in respiration rate measurement when the transmitting antenna is shaking, while the monostatic output can no longer track the respiration signal. A network of such sensor nodes

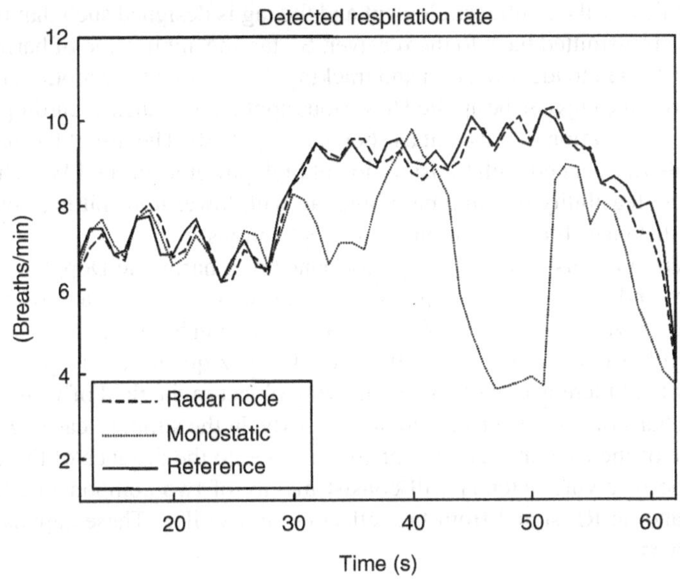

Figure 8.29 Detected respiration rate from the monostatic radar (dotted line), sensor node (dashed line), and reference (solid line) outputs. It can be seen that the monostatic output, once the antenna is physically shaking, loses track of the reference signal while the sensor node output remains in good agreement with the reference. © 2007 IEEE, Reprinted, with permission, from Mostafanezhad et al. [2007].

could be used to facilitate life detection in search-and-rescue operations. For this scenario, a handheld illuminator can transmit (illuminate) RF signal in the operation scene and then several sensor nodes can be placed in various locations in the area. These nodes can connect to a main processing unit, which retrieves and compares signals from all nodes. Deployment of such a system will still depend on radio wave propagation within the environment.

8.2.3 Passive RF Tags

Harmonic tags can be used to isolate desired targets from clutter in Doppler radar systems. A system that can isolate small motion for a tagged target in the presence of motion of untagged objects would provide means for improved SNR for isolating cardiopulmonary motion from other body motion. With a harmonic tag being placed on the subject's chest, only the tag motion is being detected and any other motion in proximity will be subdued by the receiver to a reasonable extent and would cause very minor interference, if not completely suppressed. In addition to giving us a positive identity on the person of interest, it also gives us the information on motion of the tag from which we can calculate the subject's respiration rate.

Harmonic tags have been successfully used earlier for uniquely separating a target from the environment [Colpitts et al., 1999; Colpitts and Boiteau, 2004]. A harmonic tag consists of a tag antenna with a strongly nonlinear element at its port. In most cases, the nonlinear element is a Schottky diode. The incoming signal is converted into harmonics by the nonlinear element and the tag is designed such that the second harmonic is transmitted back to the receiver. So far, the application of harmonic tags has been restricted to identification and tracking. By sensing their motion, the advantages of harmonic tags can be applied to various applications such as cardiopulmonary monitoring and motion assessment as shown in Fig. 8.30. The use of harmonic tag is generally also associated with transmission of high power levels (~4 W). This section discusses the feasibility of using harmonic tag with lower transmitting power levels and closer distances for health-monitoring applications.

This section discusses the range performance of a harmonic Doppler radar at 30 and 100 cm and also presents a qualitative analysis of the accuracy of respiratory measurements with respect to a reference. A comparison has been made between the results from harmonic radar to that from a 2.45-GHz quadrature Doppler radar for scenarios where there is more than one moving object in the field of view of radar.

When a harmonic tag is placed on human body, in the simplest case, ignoring the phase noise of the oscillator and the phase shift due to the distance of the target, the signal at the receiving antenna will consist mainly of two components: leakage at 2.45 GHz and an RF signal from tag reflection at 4.9 GHz. These signals could be represented as

$$A_{rf}\cos\left[\omega t-\frac{2\omega d}{c}-\frac{2\omega x(t)}{c}\right]+A_{rh}\cos\left[2\omega t-\frac{4\omega d}{c}-\frac{4\omega x(t)}{c}\right] \tag{8.40}$$

where the term ωt represents the fundamental frequency of 2.45 GHz, d represents the nominal distance between the transmitting antenna, and $x(t)$ is the periodic motion

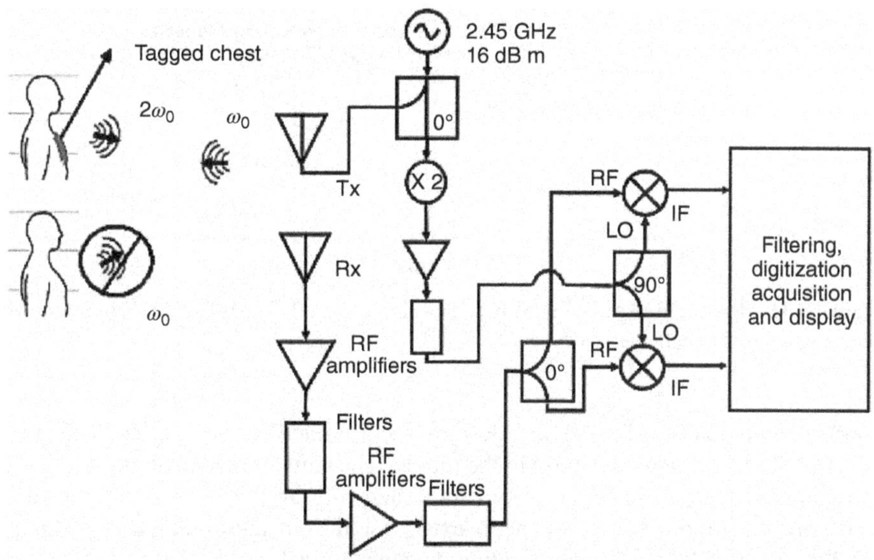

Figure 8.30 Proposed RF tagging for Doppler radar respiratory monitoring using harmonic tags. © 2012 IEEE. Reprinted, with permission, from Singh and Lubecke [2012].

of the target. The terms A_{rf} and A_{rh} represent the amplitude variations corresponding to the received fundamental and harmonic components of signal, respectively. The LO signal without the phase noise can be represented as

$$A_L \cos(2\omega t) \tag{8.41}$$

After mixing, the required baseband signal

$$A_L A_{rh} \cos\left[\frac{4\omega d}{c} + \frac{4\omega x(t)}{c}\right] \tag{8.42}$$

is filtered out and decoded to yield the respiration rate.

A quadrature receiver is used to alleviate measurement issues with null points [Park et al., 2006].

For efficient operation, the tag antenna in radio-frequency identification (RFID) circuit must be designed to present an appropriate impedance match to the RFID chip. In a wearable tag, the human body presents a large conducting mass in close proximity, and is thus an integral part of the antenna design. The effect is detrimental, in that the body blocks and absorbs RF energy, and complicates impedance matching in a variable manner that is difficult to quantify [Occhiuzzi et al., 2010; Rajagopalan and Rahmat-Samii, 2010; Sanad, 1994].

A novel harmonic tag was designed and fabricated to facilitate the sensing of human respiration through Doppler shift in a 4.9 GHz harmonic backscatter signal. RFID tag design theory was used to design the harmonic tag [Seshagiri Rao et al.,

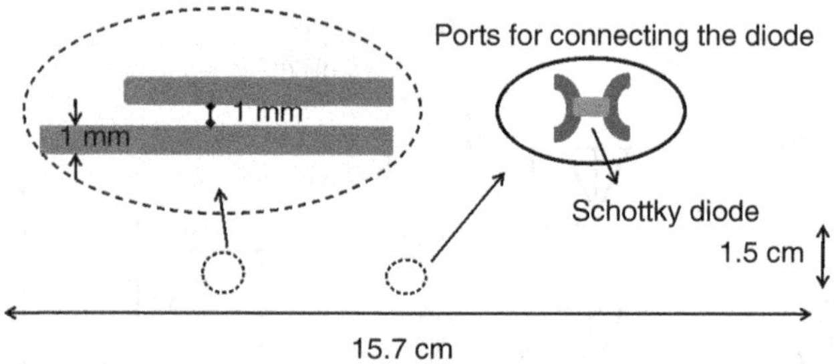

Figure 8.31 RF tag used as an on-body sensor (not to scale). © 2011 IEEE. Reprinted, with permission, from Singh and Lubecke [2011].

2005; Dobkin, 2008]. Earlier tag designs incorporated a wire dipole with diode attached across an inductive loop in the dipole. The simulations involving tag design were limited to the use of an ideal dipole and the tag design was optimized by trimming the dipole wire. The need exists for design and simulation technique for harmonic radar tags incorporating unconventional tag antenna designs. For the application intended, a planar tag design was needed. In order to enable the operation of the tag at lower powers, a match between the diode and tag element was needed. The tag was designed to keep its impedance somewhat close to being the conjugate of the diode reactance at both 2.45 and 4.9 GHz. Agilent ADS 2006 was used to design the antenna element and simulate the tag. The network parameters of the antenna were exported to the schematic and simulated with diode. Thus, both the harmonic performance and the overall scattering parameter of the tag could be evaluated. The tag was constructed with a copper tape. The performance of the tag has been evaluated in Singh and Lubecke [2011]. The gain of the designed tag antenna at 2.45 GHz is 5 dB and at 4.9 GHz is 5.2 dB.

The tag was placed over a 0.5 cm styrofoam substrate in order to minimize the effect of the human body on the EM field of the antenna. The tag along with its dimensions is shown in Fig. 8.31. For receiving the tag signal at 4.9 GHz, an array of microstrip antenna having a gain of 5.82 dB was also designed using ADS. The microstrip antenna was fabricated in-house on Rogers Duroid 6002 substrate.

A simple power budget was performed using the tag. The received signal from the tag was observed on a spectrum analyzer for various transmitted power levels. For a transmitted power of 10 dB m at 2.45 GHz, the received power at 4.9 GHz as seen using a spectrum analyzer was −87 dB m with tag distance being ~0.7 m. The source power was selected to maintain acceptable power levels for close range monitoring applications without affecting the tag activation range. Filtering and amplification requirements for motion detection were then calculated.

8.2.3.1 Experimental Validation Doppler radar was set up using connectorized components mostly from Mini-Circuits. Two targets were moved at two different frequencies to distinguish the data from each other. The two targets consisted of the tag and a styrofoam ball having a diameter of an inch. The styrofoam ball was wrapped in aluminum foil to increase the scattering. The motion of the tag was very close to 0.3 Hz while the motion of the ball was at 1.3 Hz. The distance between the targets and the antennas was approximately 30 cm. The transmitted power to the antenna was 10 dB m. Five high-pass filters (Mini-Circuits VHF-3300 and VHF-3500) and two RF amplifiers (Mini-Circuits ZX60-6013E-S) were used in the receiving circuit to condition the signals. The LO for the mixers was generated by Mini-Circuits ZFSC-2-2500 splitter splitting the transmitted signal from the signal generator (HP E3344B) and running it through a commercial frequency doubler (Mini-Circuits ZX-90-2-36). Two high-pass filters and an RF amplifier (ZX60-6013) were used to generate the LO input. Measurements were taken for three scenarios.

Experiment I:

a. Tag in motion/ball stationary
b. Tag stationary/ball in motion
c. Tag in motion/ball in motion

The three scenarios have been considered to evaluate three aspects of performance of the radar. First, how well can the radar detect the motion of the tag. Secondly, how well can the radar reject the motion due to fundamental backscatter signal and thirdly, how well can the radar isolate the motion of the tag from the motion of other non-harmonic scattering objects. The styrofoam ball was controlled by a rotational servo with a plexiglass arm attached to it [Hafner and Lubecke, 2009]. The plexiglass arm had slots to hold the styrofoam ball in place. The servo was controlled by a controller board (Arduino Duemilanove) that could be programmed using a PC. The movement of the styrofoam ball was set to be around 1 cm. The tag was placed in between the line of sight of transmitting and receiving antenna, while the ball was in front of the receiving antenna [Singh and Lubecke, 2011].

The data obtained from I and Q channel were combined using linear demodulation technique and then Fourier transform was performed on the data using MATLAB. Some DC offset present in the data can be attributed to electronics used. However, it does not affect rate extraction. Figure 8.32 is a combined plot of data gathered from moving tag and moving target in a separate instance (Experiment I(a) and I(b)). The FFT data shows that the received signal from the tag is ~25 dB more than that of the nontagged target. The IQ plot complements the FFT data. This test is measure of the effectiveness of the radar in rejecting the reflected fundamental frequency. This test could also be used as a calibration procedure before conducting measurements on human subjects. The interaction between two moving objects and radar is a little more complicated than when only a single moving object is present in front of the radar. Figure 8.33 presents the results obtained from Experiment I(c) where the tag and target are moving simultaneously at frequencies of 0.2 and 1.3 Hz, respectively.

An interesting FFT data shows an increase in detected target motion while maintaining similar levels for the tag. In spite of the increase in 1.3 Hz frequency content (target motion), the tag signal is ~13 dB greater. The I–Q plot shown in Fig. 8.33 also shows both the detected frequencies.

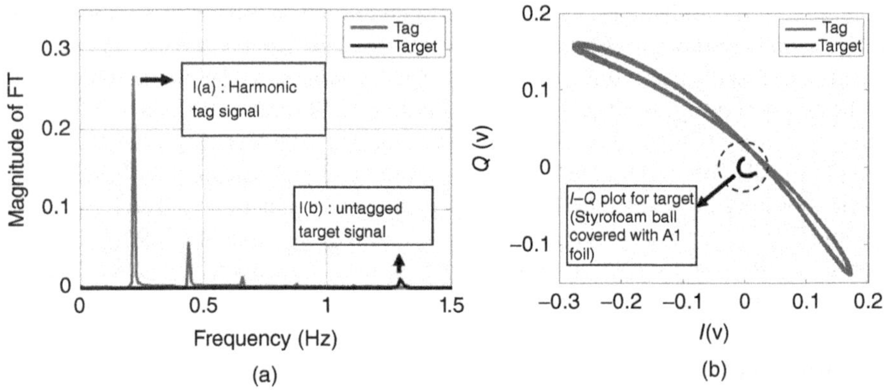

Figure 8.32 (a) Fourier transform of data from Experiment I(a) and I(b) (tag and mechanical target motion, respectively) and (b) their *I–Q* plots showing the relative amplitudes of the motion. © 2012 IEEE. Reprinted, with permission, from Singh and Lubecke [2012].

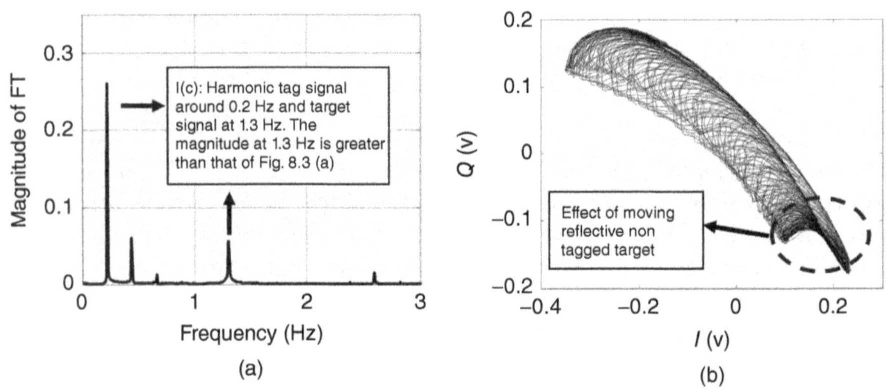

Figure 8.33 (a) FFT data for Experiment I(c) (tag and mechanical target moving together) showing an increase in detected target motion compared with case I and (b) the *I–Q* plot showing the presence of two frequency components and the phase relation between the two. © 2012 IEEE. Reprinted, with permission, from Singh and Lubecke [2012].

After the initial testing and calibration with mechanical targets, a system was set up to measure the respiration of a human subject in the presence of a large moving object that would scatter back 2.45 GHz. The object was again a styrofoam hemisphere with a radius of 10 cm covered with aluminum foil. The target motion was set to 0.2 Hz with a linear displacement of about a centimeter in order to observe radar response to interference near respiratory frequencies.

The distance between the subject and the antenna was approximately 1 m. The received antenna was connected to two chains of two high-pass filters (VHF-3500, VHF-3100) and an RF amplifier (ZX60-542LN-S+). The amplified and filtered signal was split and then fed to mixers (Mini-Circuits ZX05-14-S+). The signal from

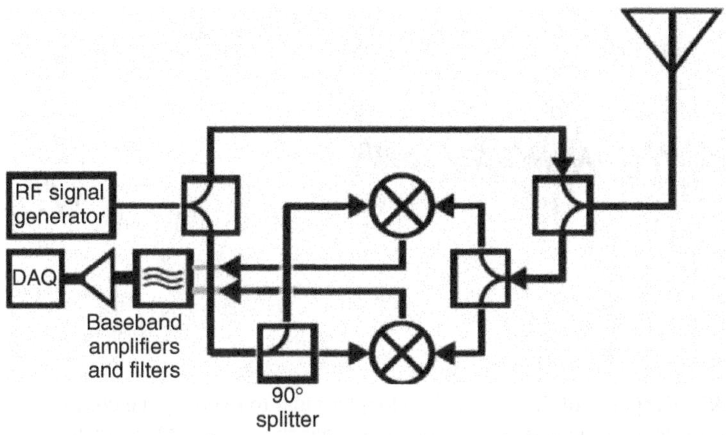

Figure 8.34 Schematic of the dual-channel 2.45-GHz Doppler radar. © 2012 IEEE. Reprinted, with permission, from Singh and Lubecke [2012].

signal generator was filtered through a band-pass filter and split and fed to the commercial frequency doubler (Mini-Circuits ZX-90-2-36-S+) through a variable attenuator (0–30 dB). The output from the doubler was amplified using an RF amplifier (ZX60-6013) and passed through three high-pass filters. This signal was then split using commercial hybrid (Pasternack PE 2058) and fed as the LO to the two mixers. The IF signals were fed to LNAs using AC coupling, and gain setting of 200 was used. NI-DAQ 6289 was used to acquire the data at a sampling rate of 100 Hz. The same experimental setup was then used to evaluate the response of a 2.45 GHz quadrature Doppler radar system. The 2.45 GHz radar system consisted of single antenna with a gain of 8 dBi connected to a splitter that was used as a circulator (Fig. 8.34). The signal from the signal generator was split using a two-way splitter (Mini-Circuits ZFSC-2-2500). The experiments are labeled as follows:

Experiment II:

a. 2.45 GHz Doppler radar human testing with noise source
b. Harmonic Doppler radar human testing with noise source

Doppler radar is very sensitive to motion, which enables us to detect even heart rate. Hence, it is logical to assume that any other motion in the vicinity of the subject should affect the measurements in an adverse way. Figure 8.35 displays the response of 2.45-GHz radar when two objects are simultaneously moving in front of it (Experiment II(a)). One is a human subject while the other is linear target with 0.2 Hz of motion. The raw data and the beat rate of the radar data obtained from 2.45-GHz quadrature Doppler radar shown in Fig. 8.35 show that radar is not able to track the motion of any one of the objects. The radar data in Fig. 8.35 show the same trend as the reference but it is centered at 12 BPM, which is the rate at which the styrofoam target is moving. The results obtained from harmonic Doppler radar for the same

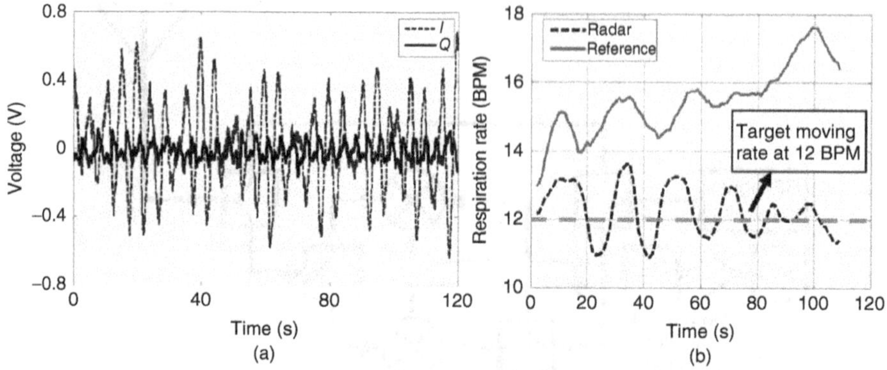

Figure 8.35 Response of 2.45-GHz CW Doppler radar to two simultaneous moving objects in its view. (a) Raw data showing the amplitude changes due to EM interaction between the two targets and (b) the rate indicating the inability of the radar to clearly isolate the motion of any of the two moving objects. © 2012 IEEE. Reprinted, with permission, from Singh and Lubecke [2012].

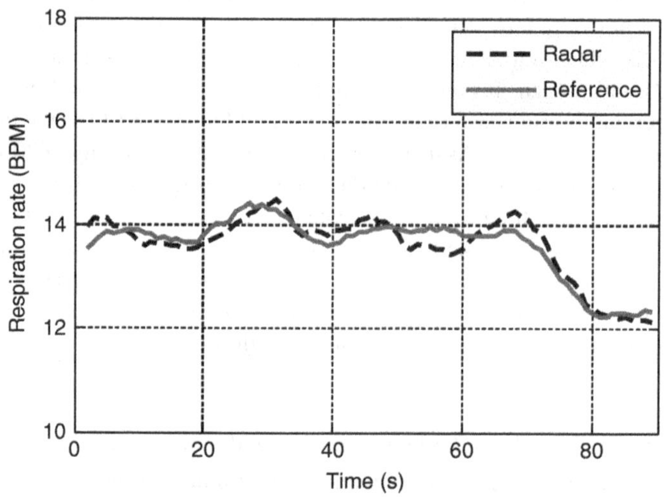

Figure 8.36 Respiration rate of a human subject with an untagged moving scattering object in front of harmonic radar. © 2012 IEEE. Reprinted, with permission, from Singh and Lubecke [2012].

experiment scenario (two simultaneous moving objects with one being tagged) are shown in Fig. 8.36 (Experiment II(b)). The harmonic radar clearly tracks the respiration rate as accurately as the reference.

After evaluating the response of two radar systems separately, experiment was performed with both the radar systems connected simultaneously to see the time response of the radars together. The measurements were first made for a tagged human subject breathing normally in front of radar with the untagged target present but motionless. Then, the target was moved at 0.2 Hz and data were recorded. The experiments are labeled as follows:

Experiment III:

a. Both radars, human testing without noise source
b. Both radars, human testing with noise source moving 1 cm (amplitude)
c. Both radars, human testing with noise source moving 2 cm (amplitude).

Figure 8.37 shows the time synchronous response of both the radars to a tagged person and a stationary/nonstationary object in front of it (Experiment III(a)). From Fig. 8.37(a), we can see that the radars track the respiration rate of the human subject very well when the target is stationary. When the target starts moving at 0.2 Hz (Experiment III(b)), the 2.45 GHz radar is unable to track the respiratory motion while 4.9 GHz radar tracks it with sufficient accuracy (Fig. 8.37(b)). The error rates for both radars have been shown in Fig. 8.37(c). The maximum error rate for 4.9 GHz radar is around 0.5 BPM. It is interesting to note the relatively low error rate for 2.45 GHz radar between 40 and 50 s. This might result due to EM interaction between the two targets and also depends on their relative motion signature. However, this result serves to show that even a small scattering target can affect the performance of a Doppler radar. With multiple moving objects in front, it would be difficult to distinguish the source of motion using a conventional medical Doppler radar. A moving person would present an even bigger RCS. Another factor to consider is that error resulting from the above-mentioned sources is not a constant and hence cannot be treated as a priori information.

Figure 8.38 shows the worst-case scenario for a 2.45 GHz Doppler radar where it tracks the mechanical target instead of tracking the respiration rate (Experiment III(c)). The measurement was performed for 5 min. The results indicate the possibility of erroneous detection and triggering of false alarms when two moving objects are present in front of 2.45-GHz radar. The radar might be detecting motion but it is difficult to interpret the source of the detected motion. However, harmonic radar still tracks the tag and thus respiration of the human subject.

All this data suggest that in an environment where respiratory motion has to be tracked in presence of other moving people or objects, use of a harmonic tag with a Doppler radar would prove to be more robust.

8.2.3.2 Discussion The harmonic radar was tested at different ranges for the same amount of transmitted power (10 mW) [Singh and Lubecke, 2011]. The range of such radar primarily depends upon the ability of the tag to generate and reflect back harmonics, which is influenced by tag–diode matching at lower incident powers and the incident power. The current system has been tested at 1 m for 10 dB m (10 mW) antenna transmitting power. The system should function at several meters with transmitted power (antenna) still below 100 mW. The novelty of the tag lies in its design, planar structure and the fact that it can be used on the body. The tag can successfully

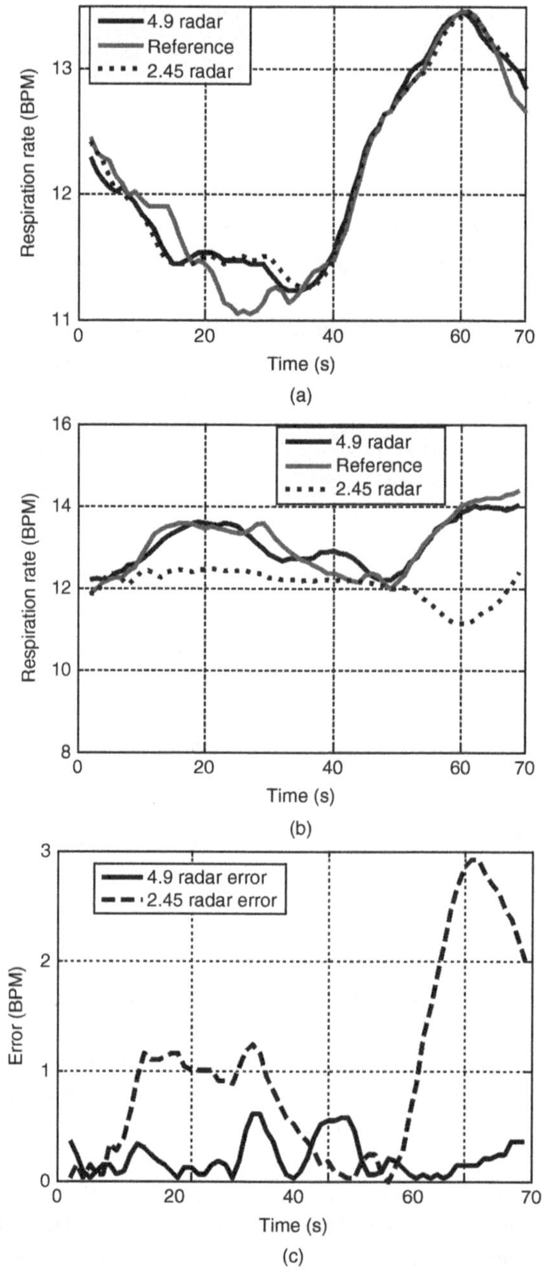

Figure 8.37 The response of 2.45 and 4.9 GHz radar to (a) a tagged human subject in front of radar when the target is not moving (Experiment III(a)), (b) tagged human with the target moving at 0.2 Hz (Experiment III(b)), and (c) the error rate in the detected respiration rate for the two radars. As expected, both the radar can track respiration accurately when mechanical target is stationary (53 (a)) but 2.45 GHz radar cannot track the respiration accurately in (53 (b)) when the mechanical target starts moving. © 2012 IEEE. Reprinted, with permission, from Singh and Lubecke [2012].

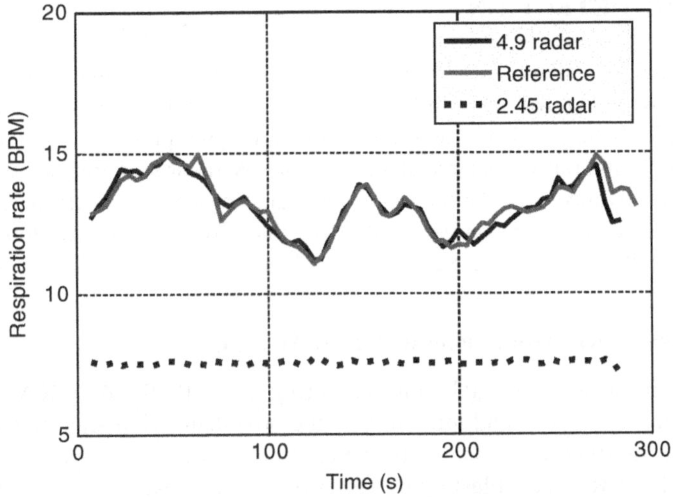

Figure 8.38 Response of 2.45 and 4.9 GHz radar to a tagged human and a mechanical untagged object. The mechanical target is moving 2 cm at a frequency of approximately 0.15 Hz. This condition represents the worst-case scenario where 2.45-GHz radar would completely detect the undesired motion. © 2012 IEEE. Reprinted, with permission, from Singh and Lubecke [2012].

reject any clutter motion in the vicinity and also any other body motion that does not affect the respiratory motion. For respiration activity estimation, the tag could be placed anywhere on the chest as long as it is facing the receiving antenna or in the field of view of receiving antenna. The tag could effectively be placed anywhere on the body where motion is caused due to respiration such as on abdomen or shoulders. The radar response and SNR depends on the magnitude of the motion of the tag and its orientation with the receiving antenna.

The rate extraction was based on a moving average FFT where a window of 12–18 s (4–6 respiration cycles) was used. A lesser window length could be used for real-time applications. A Kaiser window was used to reduce ripples. The peak frequency in the window and its amplitude was stored. The method was tested on a sample sine waveform and for a constant moving target before it was applied for respiration estimation.

In addition to effectively rejecting clutter motion in the environment that could cause inaccuracies in detected respiration rate in a conventional Doppler radar system, the use of tags also leads to the idea of possibly sensing other physiological information such as temperature. The harmonic Doppler radar system is able to detect small signals from harmonic tags in the presence of large clutter motion and the system works well at a range of 1 m using a low transmit power of 10 mW, which is well suited for monitoring a subject in a home or hospital environment.

8.3 RANGE DETECTION

A fundamental limitation of CW radar is the inability to discriminate range to target. Either FMCW or UWB radar can be used to detect both range to target, and target micro-Doppler behavior. Both radar hardware and signal processing are more complex in FMCW and pulsed radar systems than in CW radar systems; however, recent technological advances in radar hardware and signal-processing platforms have made those types of radar more compact and affordable. To detect micro-Doppler, these systems typically add a separate coherent receiver.

8.3.1 Physiological Monitoring with FMCW Radar

FMCW has been used for vital sign monitoring since 1980s. A FMCW was used to improve the sensitivity and reliability of the broadband vital sign monitor [Seals et al., 1986]. In 1997, the Georgia Tech Research Institute developed a microwave radar, called the RADAR Flashlight. This system was designed to detect the respiration of a motionless human behind a wall. The laboratory unit is a homodyne FMCW radar, which operates at a frequency near 10.525 GHz. The RADAR Flashlight can detect the respiration of a human standing up to 5 m away behind 20 cm hollow-core concrete block wall [Bestak et al., 2007]. In 2001, GTI developed a portable RADAR Flashlight which can measure motion and respiration activity from 3 m distance behind 20-cm thick wall [Greneker and Geisheimer, 2000].

In Postolache et al. [2010], 24-GHz FMCW Doppler radar has been used for both cardiopulmonary signals and range measurements. Minimum measurable distance of 0.75 m is a distance between the radar and the 24 GHz antenna. Digital filtering and detrending algorithms, based on digital wavelet transform, were used to extract the heart rate and respiratory rate. Measurements are done with stationary subjects sitting on a wheelchair and with low amplitude motion.

In Anitori et al. [2009], they investigate the use of X-band FMWC radars for detecting human life-sign. Recumbent subjects were tested using a radar at a distance of 2 m. While respiration rates were extracted successfully, the heart rate measurements had errors in the presence of respiration signal.

UWB FMCW radar is presented in Maaref et al. [2009a, 2009b]. Propagation modeling through different types of walls and RCS measurements of human beings are given. The used frequency bandwidth is 1–5 GHz. Human RCS has been measured in different positions. The results are given for 20-cm hollow brick, and target range of 20 m with 1-m resolution. It has been assumed the target is not fixed and moves, and only range of target has been measured. In Loschonsky et al. [2009] simulated respiration motion of buried people with externally driven pendulum was tested. It operates with a transmitting power of less than 2 mW and possesses a penetration depth of approximately 1 m (reinforced concrete) or up to 30 m (free space). FFT and continuous wavelet transform in real time, was developed for a Doppler radar system. In Ivashov et al. [2004], the experiments with using of continuous-wave subsurface radar are described for measuring heart rate and respiration behind the wall; 1.6 GHz frequency is used; wall thickness is 10 cm; and the subject was at a distance of 1 m from the antenna.

8.3.2 Physiological Monitoring with UWB Radar

Motion sensing and imaging with UWB radar systems have strong potential for use in the medical field. A UWB system transmits narrow impulse-like signals that span several gigahertz of frequency range and the pulse width is typically within a range of about 100 ps to several nanoseconds with rise times as fast as 50 ps. Since the energy of the pulse spreads across broad frequency ranges, the power spectral density is much lower than narrowband systems [Paulson et al., 2005]. Therefore, UWB signals appear below the noise floor to the conventional narrow-band systems (Fig. 8.39).

The FCC has provided several guidelines on the use of UWB signals for different applications. The UWB bandwidth is the frequency band bounded by the points that are 10 dB below the highest radiated emission, as based on the complete transmission system including the antenna. The upper boundary is designated f_h and the lower boundary is designated f_l. The frequency at which the highest radiated emission occurs is designated fm. The center frequency f_c is the average of f_l and f_h, that is,

$$f_c = \frac{f_l + f_h}{2} \tag{8.43}$$

The fractional bandwidth is defined as

$$\text{FB} = 2\frac{f_h - f_l}{f_h + f_l} \tag{8.44}$$

A UWB transmitter has been defined as an intentional radiator that, at any point in time, has a fractional bandwidth equal to or greater than 0.20 or has a UWB bandwidth equal to or greater than 500 MHz, regardless of the fractional bandwidth.

The advantages of a UWB system include excellent range resolution, immunity to multipath fading, better security, low-power consumption, and faster data rate.

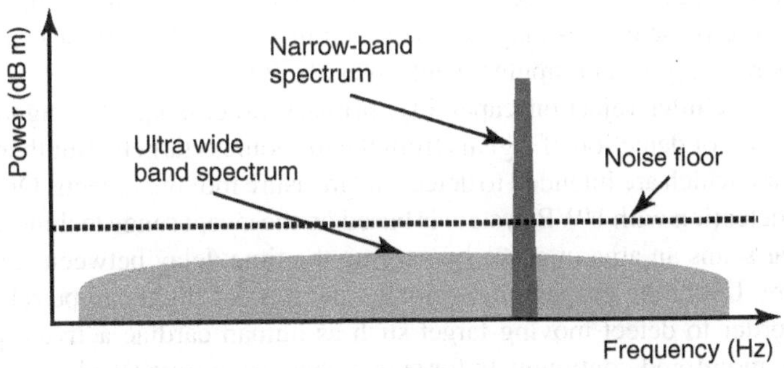

Figure 8.39 Spectrums of UWB and narrow-band systems.

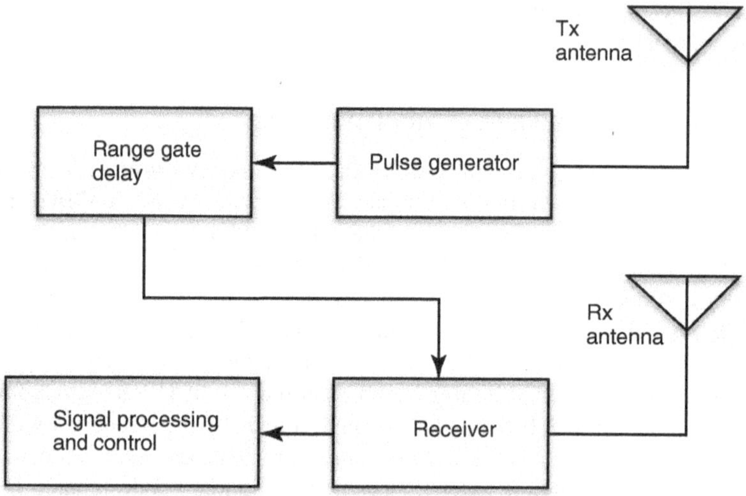

Figure 8.40 UWB radar block diagram.

The combination of low power consumption with excellent range resolution makes UWB radar a very good candidate for medical applications such as imaging and physiological measurements. UWB radars transmit extremely narrow pulses and analyze received reflection signals for characteristic indicators of material boundaries and movements. Since the transmitting power of a UWB system is very low, the radars are safe for medical use. The basic UWB radar system consists of a transmitter including pulse generator, a receiver, and a signal processor (Fig. 8.40) [Ossberger et al., 2004]. The transmitter generates a series of short pulses. At dielectric interfaces, portions of the transmitted pulse reflect back toward the receiving antenna. The receiver uses a range gate to sample the echo signals during a specific time interval corresponding to the round trip time. A sensor with a fixed range gate can only detect echo signals from a single radial distance. By sweeping the range gate across a time span, or equivalent time sampling, targets can be detected within a specified distance range. Multiple pulses are integrated to achieve a sufficient SNR. Signal processing of the received pulse echoes may be performed in analog circuitry or using software algorithms running on a computer [Paulson et al., 2005].

Due to its clutter rejection capability, primary processing of a signal in UWB radars is done for detection of signals from the motionless targets. But there are some UWB radars, which are intended to detect and measure moving objects. One of human vital sign detection with UWB system is based on the swept range technique. A swept range radar scans an area in space by varying the time delay between signal launch and capture. Using the propagation velocity, the time of flight can be related to distance. In order to detect moving target such as human cardiac activity, pulse echo signals are monitored continuously for signal changes in time [Paulson et al., 2005].

Another technique to detect vital signs is Doppler radar using UWB system. The simplified block diagram of UWB Doppler radar system is shown in Fig. 8.41. The

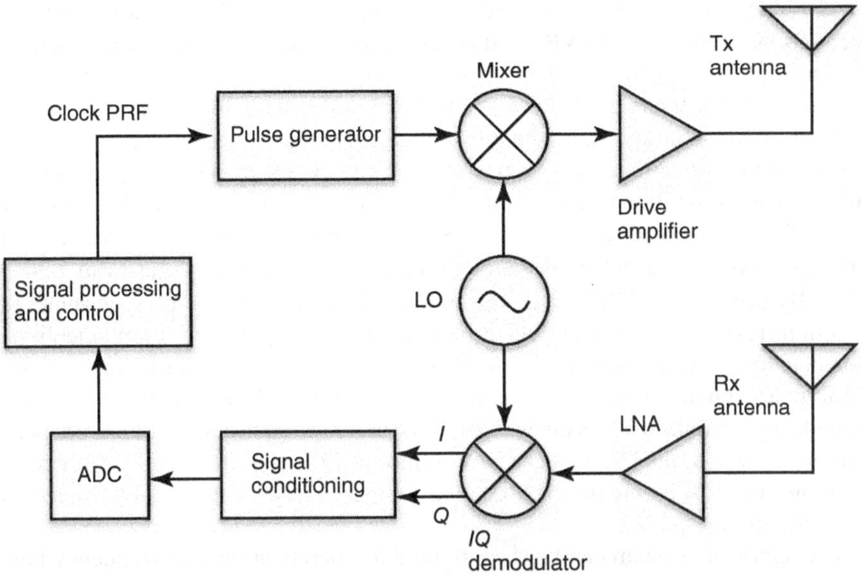

Figure 8.41 UWB quadrature Doppler radar block diagram.

radar is made by scheme with separate receiving and transmitting antennas. The receiver has rigid synchronization with the transmitter to maintain the coherency of reflected signal by using same local oscillation signal.

Staderini [2002] has provided a good overview of possible applications of UWB radars in medicine. Ossberger et al. [2004] have shown respiratory movement detection at 5 m without a wall and have also detected respiration of a person behind a wall at 85 cm using a UWB radar. Paulson et al. [2005] have demonstrated the feasibility of using UWB radar systems for many medical applications including monitoring of respiratory and cardiac functions. In Venkatesh et al. [2005], a very nice analytical framework for the development of signal-processing algorithms for respiration and heart-rate estimation has been presented. Respiratory data and cardiac signal obtained with UWB radar has been verified with that of one dimensional in vivo MRI by Thiel et al. [2008]. They have also described the use of principal component analysis to reduce the redundancy and obtain the data of greater interest from all the UWB signatures. More recently, Leib et al. [2010] have compared the heart rate obtained from a UWB radar with ECG and have also discussed a method to distinguish between targets of different RCSs.

In the following sections, we describe a UWB impulse radar that can provide high-resolution range profiles, as well as micro-Doppler detection capability [Wang and Fathy, 2011].

8.3.2.1 UWB Radar Prototype A reconfigurable radar imaging system has been developed to operate at either the lower UWB band (2–4 GHz) when seeing through a lossy media such as brick/concrete walls to minimize the through-wall associated attenuation, or the upper UWB band (8–10 GHz) for seeing through low loss media such as drywall to achieve a better image resolution [Yang and Fathy, 2009; Wang and Fathy, 2010; Wang et al., 2009]. The developed system is a carrier-based UWB transceiver architecture, where the transmitted pulse signal is up-converted through a mixer, and then down-converted at the receiver side. A 700-ps Gaussian pulse signal is utilized with a repetition frequency of 10 MHz. The system occupies an RF bandwidth of 2 GHz to allow double-sideband transmission of the 700-ps Gaussian pulse. The radar prototype has a detection range of 12 m and a scanning angle from -30° to $+30^{\circ}$. By using a USB2.0 for data transfer, the developed radar prototype achieves a system refresh rate of 280 Hz, which is adequate for Doppler vibration detection of human activities (e.g., human breathing, arm swing, walking, and running) as well.

A typical chain of such a reconfigurable system is shown in Fig. 8.42, where redundancy is minimized, as the wideband Vivaldi array, the synthetic aperture beamformers, as well as the DAQ blocks are common; while the LOs and mixers are reconfigurable. The UWB antenna is based on a compact design with a smooth impedance transformation.

The aggregate system can be reconfigured to operate at the two frequency bands by using four single-pole single-throw (SPDT) switches with maximum utilization of the common blocks. A 10 MHz clock (pulse repetition frequency, PRF) generated by the field-programmable gate array (FPGA) is used to drive a Gaussian pulse generator. The pulse is then modulated by a carrier signal of either a 3 GHz (LO1) or a 9 GHz (LO2). The modulated signal chosen by the SPDT Switch2 passes through two stages of amplification and is then transmitted via a wideband Vivaldi subarray.

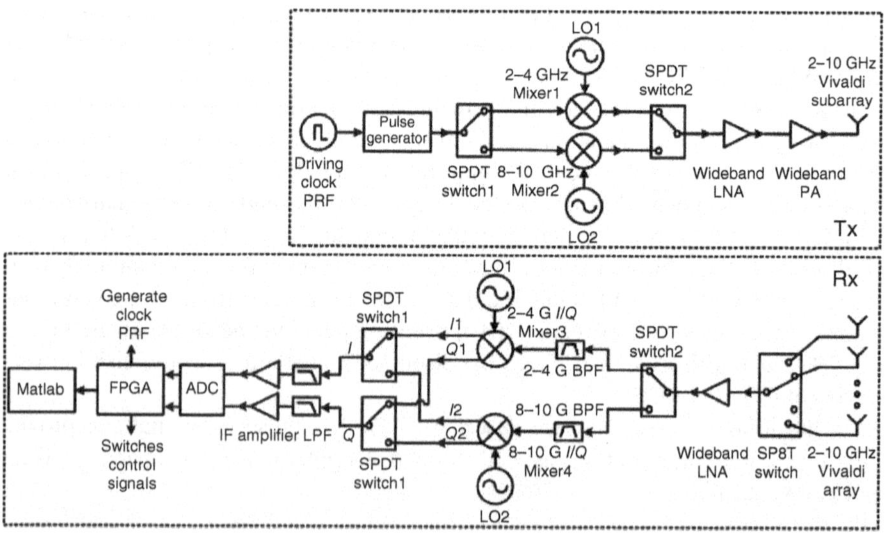

Figure 8.42 Detailed block diagram of the reconfigurable UWB pulse radar system. © 2010 IEEE, Reprinted, with permission, from Wang et al. [2010].

At the Rx link, the signal received by the wideband Vivaldi array passes through the SP8T switch and is then amplified by a wideband LNA. Next, the signal is selected and band-pass filtered before being down-converted into I and Q channels by mixing the same carrier signal with the received signal. This system is coherent, that is, system transmitter and receiver use the same carrier signal. Then, the recovered I and Q data are opted, filtered, and amplified before being sent to the analog-to-digital converter for sampling using an equivalent time sampling scheme. Next, all the sampling data are sent to the common FPGA circuitry for data uploading using USB2.0 communication link. Last, the image/Doppler signature is recovered by either microwave imaging algorithm or time-frequency analysis implemented using MATLAB.

The developed radar prototype utilizes a 1×8 linear Vivaldi antenna array shown in Fig. 8.43 as the receiving antenna. The 8-element antenna array is moved mechanically in a vertical plane to collect the signals at different positions for 3D microwave imaging recovery. Each Vivaldi antenna element covers the entire UWB band for UWB imaging applications from 1.99 to 10.6 GHz and has a good input match and directional radiation patterns over the entire UWB band. The antenna element spacing is reconfigurable and can be adjusted to be $0.8\lambda_0$ at the highest the high frequency end, to balance the radar aperture and scanning angle.

At the baseband, a DAQ and transfer module is used and shown in Fig. 8.44. Off-the-shelf Xilinx Virtex-4 FPGA board with a Texas Instrument CDC5801 low-jitter clock multiplier/divider, and a two 8-bit MAX104 ADC evaluation boards with a 2.2 GHz analog input bandwidth are utilized for signal digitization. According to Nyquist's sampling theorem, a minimum sampling rate of 2 GS/s is required to digitize the 700 ps pulse signal, which occupies an SSB baseband of approximately 1 GHz. However, the MAX104 ADC only provides a maximum conversion rate of 1 GS/s while faster ADCs are very expensive.

Subsequently, a low-cost equivalent-time sampling method [Yang and Fathy, 2009] is applied. The signal is digitized using the 100 MHz clock generated by an FPGA board and 10 samples are collected at the first signal cycle. Then, a 13.02 ps (i.e., 10 ns/768) time delay generated by a CDC5801 chip is applied to the sampling trigger clock before sampling the next signal cycle. After 768 signal cycles, a complete pulse signal is sampled and acquired. This equivalent-time sampling method leads to a 76.8 GS/s equivalent conversion rate. Finally, the collected data are transferred to a laptop through USB2.0 communication, which provides an uploading data rate of 32 MB/s.

8.3.2.2 Through Wall Imaging Experiments Different experiments have been performed, and the first is a real-time experiment, shown in Fig. 8.45, to investigate the performance of the radar imaging system when detecting static targets. The experiment is performed in the hall, which occupied an area of approximately 10 m by 3 m. A 2-cm-thick cement block is applied in the experiment – a larger dihedral, a smaller dihedral, and a cylindrical bucket, all metallic, stand on the floor as the static targets. The images are obtained and displayed on the monitor. The reconstructed image, depicted in Fig. 8.46, indicates the positions of all three targets accurately.

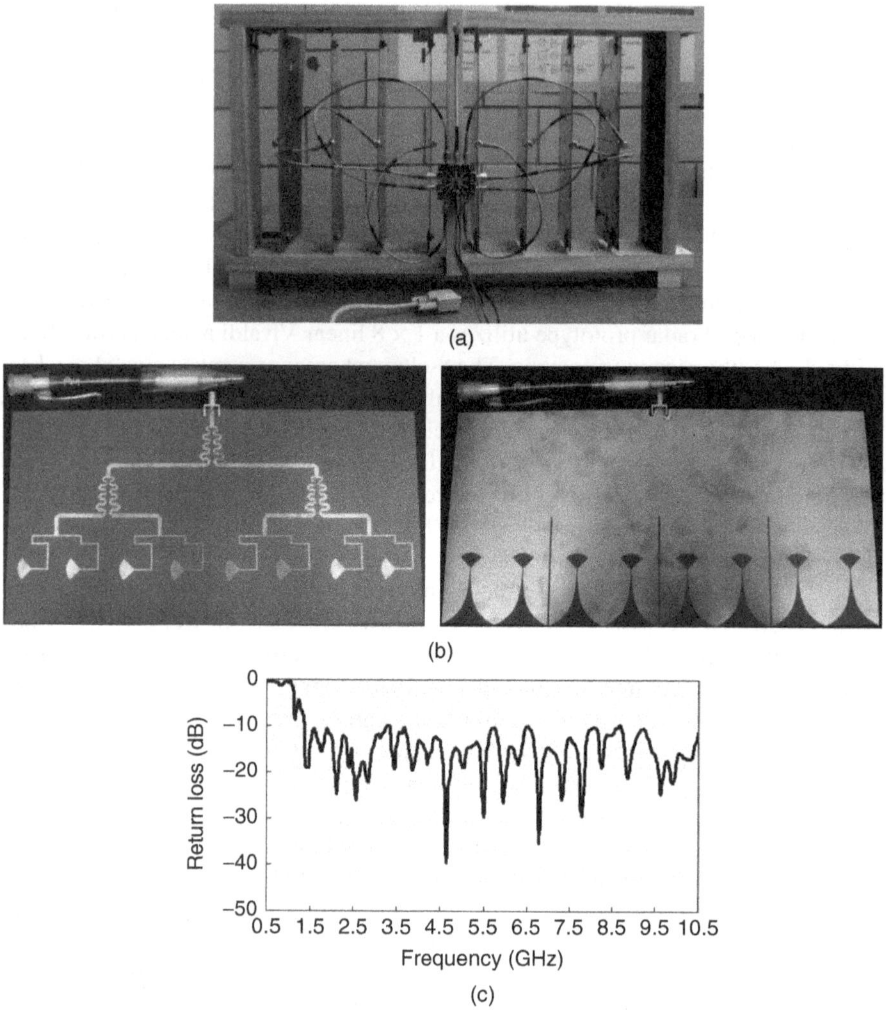

Figure 8.43 Fabricated 2–10 GHz Vivaldi Array. © 2010 IEEE, Reprinted, with permission, from Wang et al. [2010]. (a) 1×8 linear Vivaldi full array; (b) prototype of single Vivaldi subarray; (c) measured return loss of the Vivaldi subarray.

The second experiment is also a real-time experiment but performed to recover the image of a dynamic target, as presented in Fig. 8.47. A person walks behind a 2-cm-thick cement block back and forth in the surveillance space, with an area of 10 m by 3 m. Figure 8.48 presents snapshots of the radar imaging result in detection of the moving person. The real-time image on the monitor indicates the positions of the person precisely.

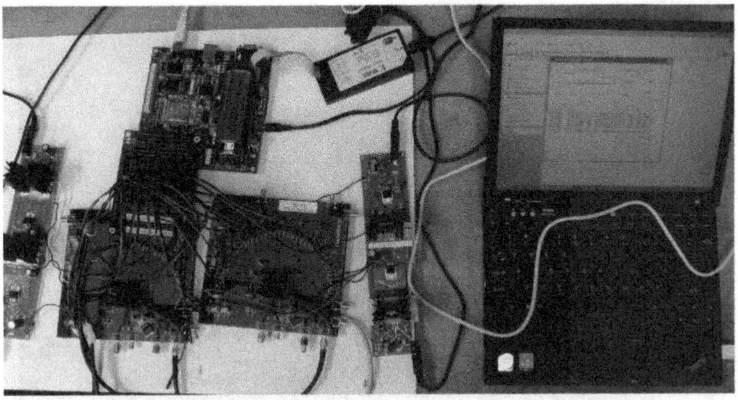

Figure 8.44 Data acquisition and transfer module including two MAX104 ADCs, a Xilinx Virtex-4 FPGA evaluation board, a USB cable, and a laptop. © 2011 IEEE, Reprinted, with permission, from Wang and Fathy [2011].

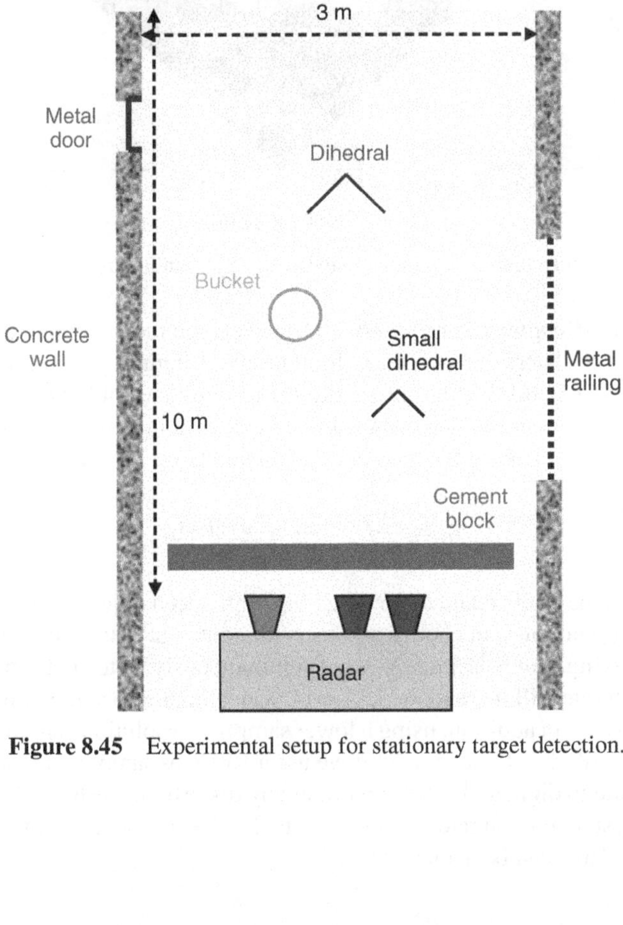

Figure 8.45 Experimental setup for stationary target detection.

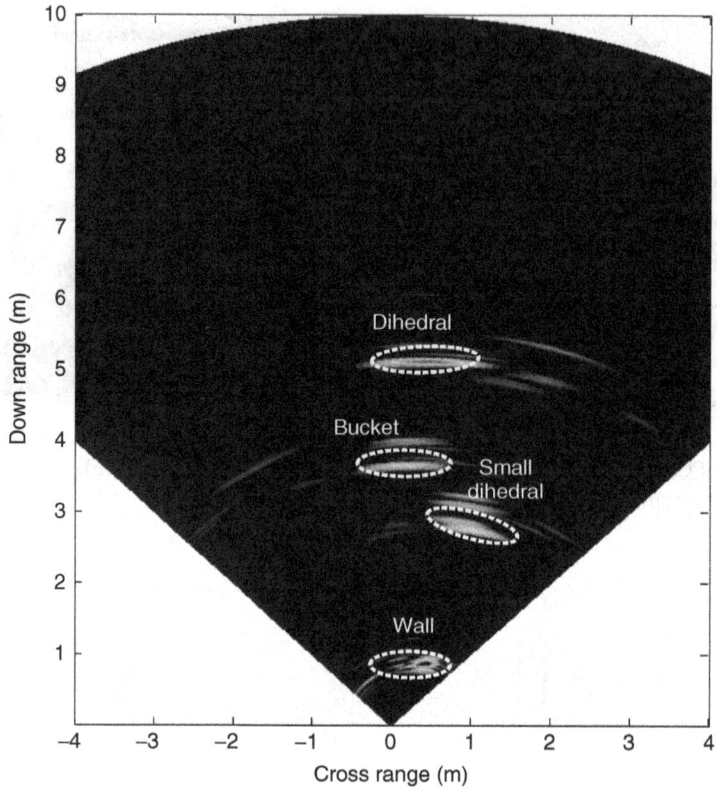

Figure 8.46 Real-time image of multiple stationary targets.

8.3.2.3 Micro-Doppler Experiments The developed radar Doppler ambiguity has been analyzed to determine operation limitations of human gait analysis. The system refresh rate of 280 Hz determines the maximum detectable Doppler frequency to be ±140 Hz for both moving forward and backward targets. Next, the maximum detectable radial velocity of the targets is calculated to be 7 m/s using Equation 8.45:

$$f_D = \frac{2v}{c} f_c = \frac{2v}{\lambda} = \pm 140\,\mathrm{Hz} \tag{8.45}$$

The system can be easily adapted for detecting high-speed objects, for example, vehicle, helicopter, and missile. If only one receiving antenna is used in the system or if the eight receiving channels are processed simultaneously instead of sequentially, the system refresh rate will increase to 2.24 kHz, which relates to a maximum detectable velocity of 56 m/s. In addition, using a lower sampling resolution can further improve the system refresh rate. For example, if we use a 130.2-ps sampling resolution, which is also adequate to digitize the pulse without any distortions, instead of 13.02-ps resolution, the system refresh rate will increase to 22.40 kHz and a maximum detectable velocity of 560 m/s can be achieved.

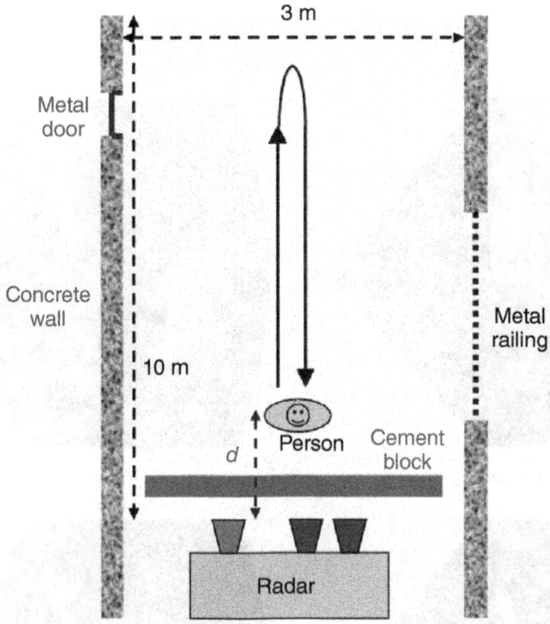

Figure 8.47 Experimental detection of a moving person.

The advantage of above UWB Doppler radar compared with most of other Doppler radar systems is that it achieves a high-range resolution by using a wideband signal and a fine sampling, in addition to the Doppler capability. Several experiments have been performed to acquire the m-D signatures of human arms swing using the developed radar. During the experiment, a human object is either marching or walking with arms swing. To make the movements of arms more conspicuous initially to the radar, the human also carries corner reflectors in the hands. The motions of arms have been successfully observed from both range–time plot and the spectrogram of radar returned signal. Subsequently, some experiments have also been performed without carrying any reflectors in hands. Very promising results have been acquired based on the Doppler spectrogram. In the following sections, we discuss some of the experiments and the acquired results.

A. Human marching with one-arm swing

In Experiment A, the human is marching on the spot facing the UWB Doppler radar, with one-arm swing. When arm moves forward/backward, the distance between the arm and the radar (i.e., range information) is changing. By achieving a high-range resolution, the developed radar can clearly present the change of range as a function of time, as shown in Fig. 8.49(a). Doppler spectrogram of the radar returned signal is presented in Fig. 8.49(b). Comparing the two plots, it is observed that positive Doppler is always acquired when arm moves forward (e.g., time range 2.4–3.1 s), while negative Doppler is acquired when

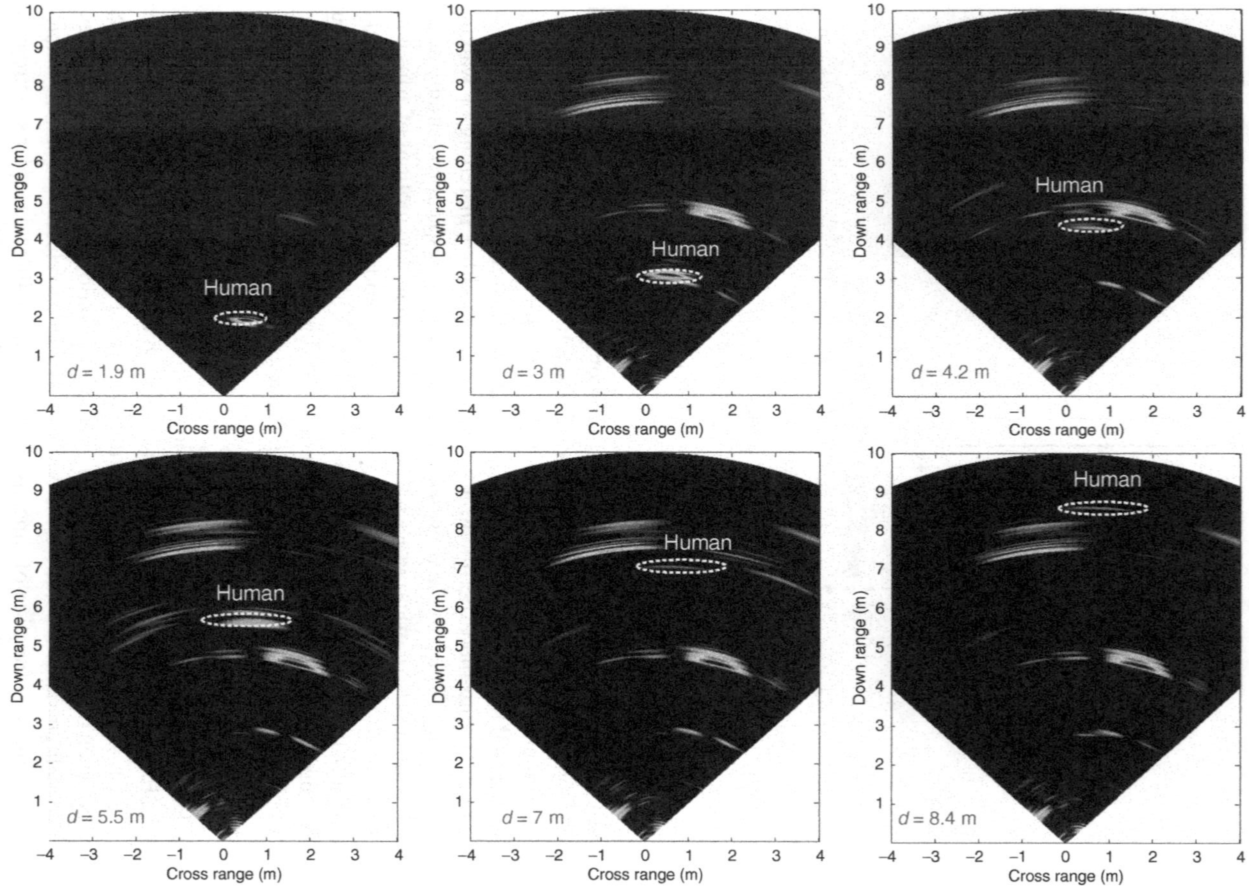

Figure 8.48 Real-time images of a moving person.

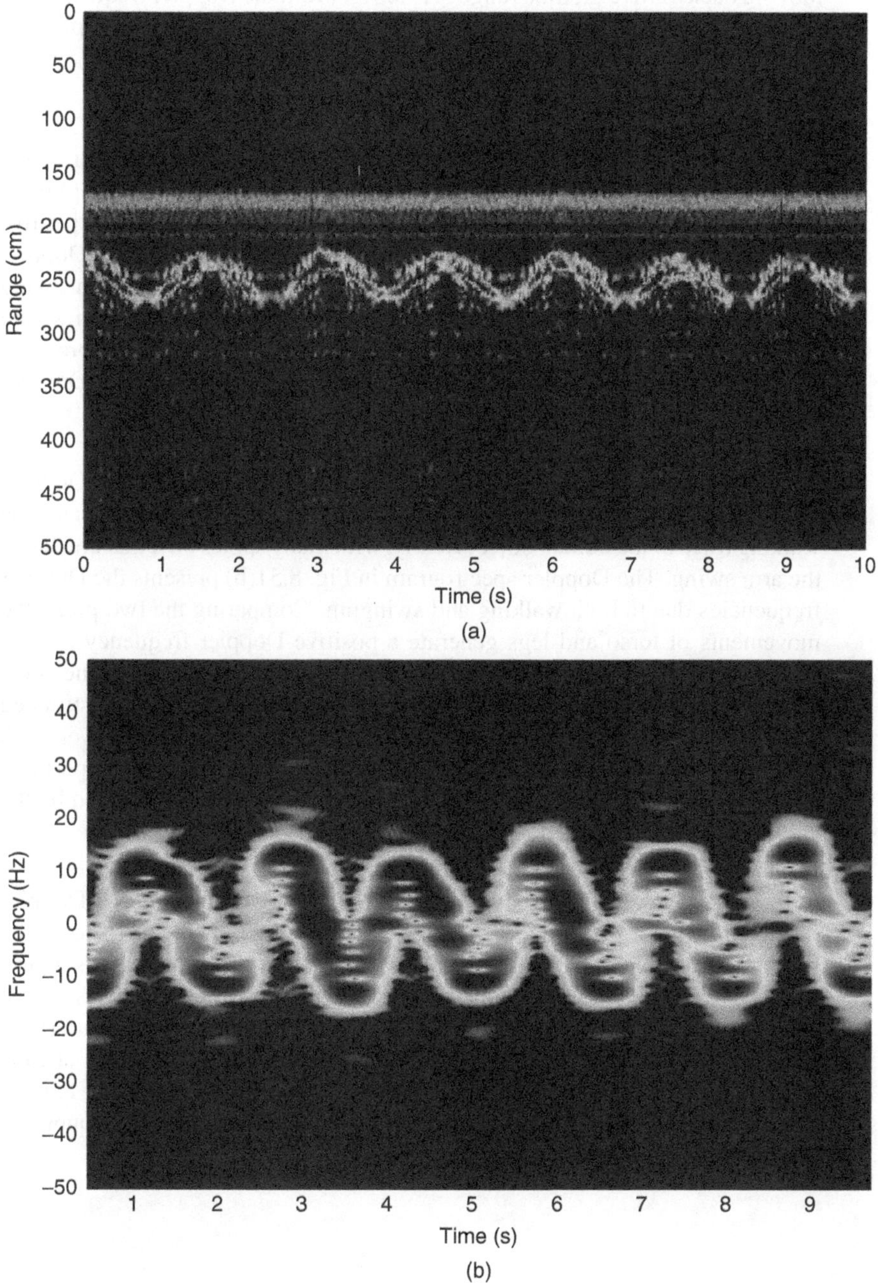

Figure 8.49 Human marching on the spot with one-arm swing. © 2011 IEEE, Reprinted, with permission, from Wang and Fathy [2011]. (a) Range of the object versus time and (b) Doppler spectrogram (TF signature).

moving backward (e.g., time range 3.1–3.8 s). From the Doppler frequency, we can estimate the maximum arm swing velocity of 0.5 m/s. From either plot, we can also figure out the arm swing period of approximately 1.5 s.

B. Human marching with two-arm swing
In Experiment B, the human is marching on the spot facing the UWB Doppler radar, with two-arm swing. When swinging two arms, the one closer to the radar always dominates the radar-returned signal in terms of amplitude. This also explains the range–time plot in Fig. 8.50(a), where a repetitive positive half-sinusoid wave is acquired. Figure 8.50(b) presents the Doppler spectrogram of the radar returned signal. Compared with one-arm swing during which the arm moves either forward or backward, two-arm swing includes both forward and backward arm motions at a time. Therefore, it is expected that two-arm swing generates both positive and negative Doppler frequencies at the same time, as shown in Fig. 8.50(b).

C. Human walking with one-arm swing
In Experiment C, the human is walking toward/away from the UWB Doppler radar, with one-arm swing. The range–time plot in Fig. 8.51(a) indicates the walking trace of the human object clearly, with many spikes on it that are due to the arm swing. The Doppler spectrogram in Fig. 8.51(b) presents the Doppler frequencies due to both walking and swinging. Comparing the two plots, the movements of torso and legs generate a positive Doppler frequency at time range 0–10 s when human target is walking toward the radar and a negative Doppler at 10–20 s when walking backward. The walking velocity is estimated to be 0.4 m/s using the range–time characteristics, which also agrees well with the calculation using the Doppler due to torso and legs motions. The periodic positive/negative Doppler frequencies in the spectrogram are generated by the arm movement, as marked in Fig. 8.51(b).

D. Human walking with two-arm swing
In Experiment D, the human is walking toward/away from the UWB Doppler radar, with two-arm swing. The range–time plot in Fig. 8.52(a) shows the movements of torso and legs, as well as the vibrations of arms swings. Compared with Fig. 8.51(a), the number of vibrations from two-arm swing is almost doubled than that from one-arm swing. If we compare Fig. 8.52(a) and (b), a positive Doppler is induced from the approaching of torso and legs at time range 0–15.5 s, while negative from the ascending at 15.5–20 s. As we expect, the two-arm swing generates both positive and negative Doppler frequencies at the same time, as shown in Fig. 8.52(b).

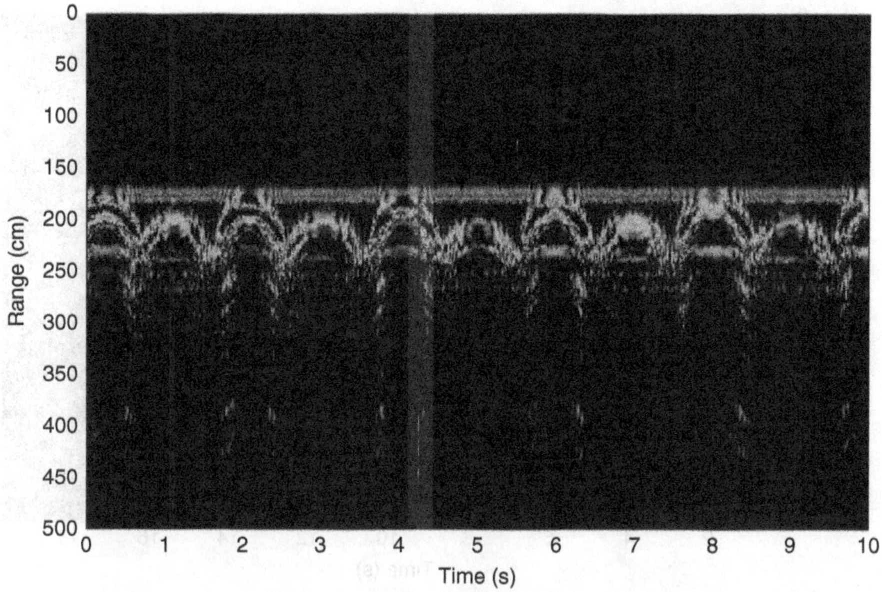

(a)

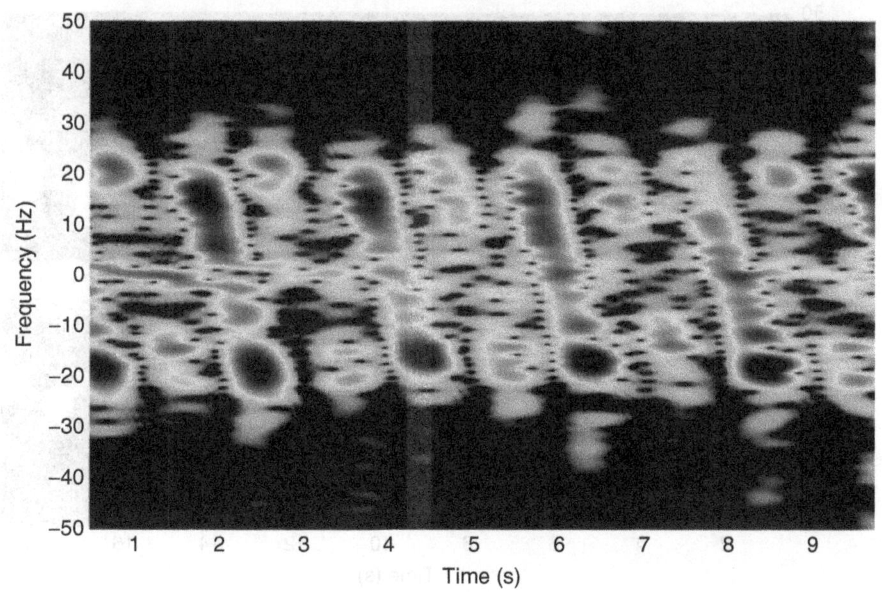

(b)

Figure 8.50 Human marching on the spot with two-arm swing. © 2011 IEEE, Reprinted, with permission, from Wang and Fathy [2011]. (a) Range of the object versus time and (b) Doppler spectrogram (TF signature).

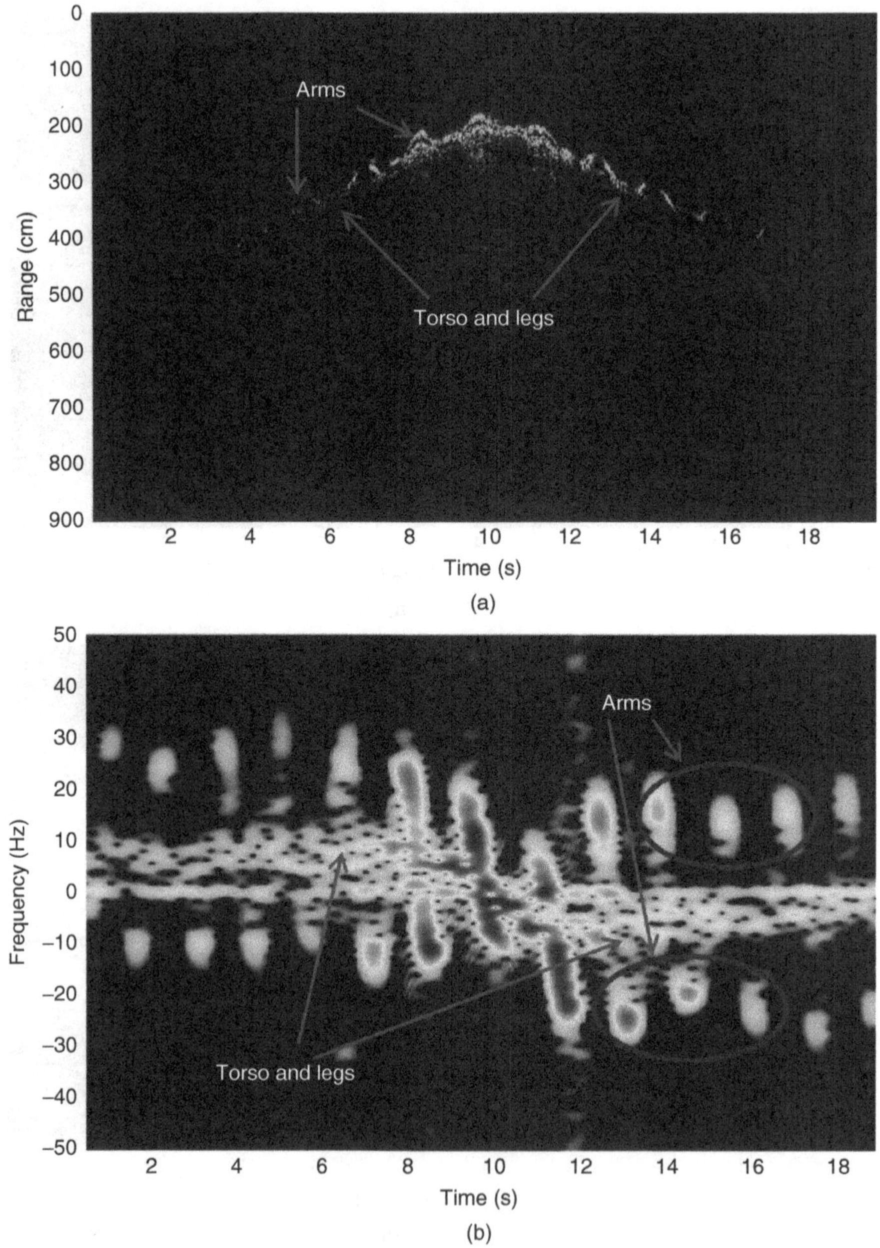

Figure 8.51 Human walking with one-arm swing. © 2011 IEEE, Reprinted, with permission, from Wang and Fathy [2011]. (a) Range of the object versus time and (b) Doppler spectrogram (TF signature).

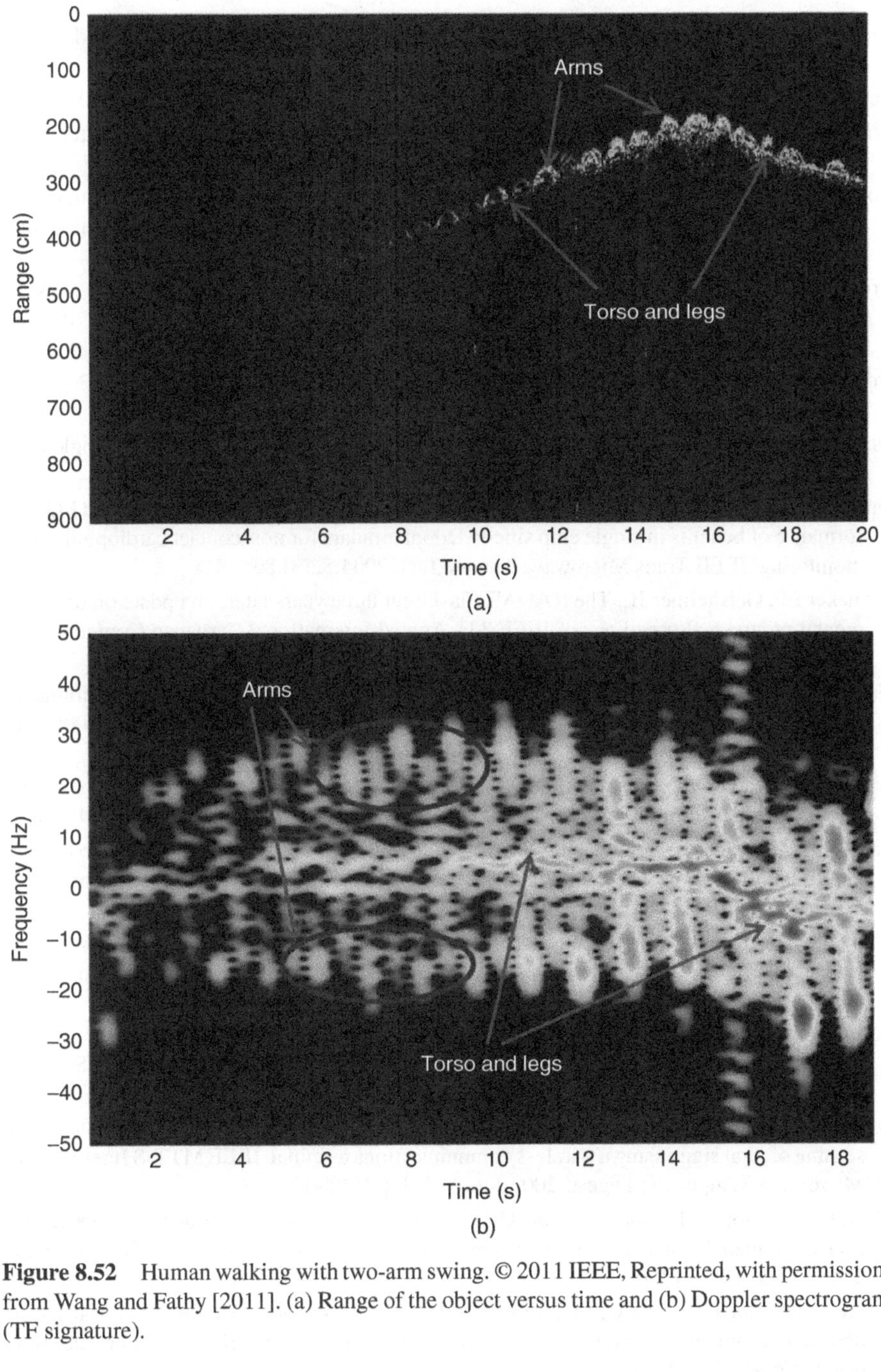

Figure 8.52 Human walking with two-arm swing. © 2011 IEEE, Reprinted, with permission, from Wang and Fathy [2011]. (a) Range of the object versus time and (b) Doppler spectrogram (TF signature).

REFERENCES

Anitori L, de Jong A, Nennie F. FMCW radar for life-sign detection. 2009 IEEE Radar Conference; 2009 May 4–8. p 1–6.

Bestak R, Simak B, Kozlowska E. Personal Wireless Communications. Springer; 2007.

Colpitts BG, Boiteau G. Harmonic radar transceiver design: miniature tags for insect tracking. IEEE Trans Antennas Propag 2004;52(11):2825–2832.

Colpitts B, Luke D, Boiteau G, Doyle G. Harmonic radar identification tag for insect tracking. Proceedings of 1999 IEEE Canadian Conference on Electrical and Computer Engineering, Alberta, Canada; 1999 May 9–12.

Crols J, Steyaert MSJ. Low-IF topologies for high-performance analog front ends of fully integrated receivers. IEEE Trans Circuits Syst II: Analog Digital Signal Process 1998;45(3):269–282.

Dobkin DM. The RF in RFID: Passive UHF RFID in Practice. Oxford, UK: Publisher Newnes; 2008.

Droitcour AD. Non-contact measurement of heart and respiration rates with a single-chip microwave doppler radar [Ph.D. dissertation]. Stanford University; 2006.

Droitcour A, Boric-Lubecke O, Lubecke VM, Lin J, Kovacs GT. Range correlation and I/Q performance of benefits in single chip silicon Doppler radars for non-contact cardiopulmonary monitoring. IEEE Trans Microwave Theory Tech 2004;52(3):838–848.

Greneker EF, Geisheimer JL. The RADAR Flashlight three years later: an update on developmental progress, Proceedings of IEEE 34th Annual International Carnahan Conference on Security Technology, 2000. p 257–259.

Hafner N, Lubecke V. Performance assessment techniques for Doppler radar physiological sensors, 31st Annual International Conference of the IEEE EMBS, Minneapolis; 2009 Sep. 2–6. p 4848–4851.

Ivashov SI, Razevig VV, Sheyko AP, Vasilyev IA. Detection of human breathing and heartbeat by remote radar. Progress in Electromagnetic Research Symposium 2004; 2004 Mar.

Leib MM, Wolfgang S, Schumacher B. Vital signs monitoring with a UWB radar based on a correlation receiver. Proceedings of the Fourth European Conference on Antennas and Propagation; 2010. p 1.

Lohman B, Boric-Lubecke O, Lubecke VM, Ong PW Sondhi MM. A digital signal processor for Doppler radar sensing of vital signs. Proceedings of the 23rd IEEE Annual Engineering in Medicine and Biology Society Conference; 2001. Vol. 4, p 3359–3362.

Loschonsky M, Feige C, Rogall O, Fisun S, Reindl LM. Detection technology for trapped and buried people. IEEE MTT-S International Microwave Workshop on Wireless Sensing, Local Positioning, and RFID, 2009, IMWS 2009; 2009 Sept 24–25. p 1–6.

Lubecke V, Boric-Lubecke O, Beck E. A compact low-cost add-on module for Doppler radar sensing of vital signs using a wireless communications terminal. IEEE MTT-S International Microwave Symposium Digest; 2002 Jun. Vol. 3, p 1767–1770.

Maaref N, Millot P, Pichot C, Picon O. A study of UWB FM-CW radar for the detection of human beings in motion inside a building. IEEE Trans Geosci Remote Sens 2009a;47(5):1297–1300.

Maaref N, Millot P, Pichot C, Picon O. FMCW ultra-wideband radar for through-the-wall detection of human beings. Proceedings of the 2009 International Radar Conference; Bordeaux, France; 2009b Oct.

Massagram WJ. A study of feasibility in long-term cardiopulmonary monitoring via Doppler radar [Ph.D. dissertation]. University of Hawaii at Manoa; 2008.

Massagram W, Lubecke VM, Host-Madsten A, Boric-Lubecke O. Assessment of heart rate variability and respiratory sinus arrhythmia via Doppler radar. IEEE Trans Microwave Theory Tech 2009;57:10.

Meyer C. Matrix Analysis and Applied Linear Algebra. Philadelphia: SIAM; 2001.

Mostafanezhad I, Boric-Lubecke O. An RF based analog linear demodulator. IEEE Microwave Compon Lett 2011;21(7):392–394.

Mostafanezhad I, Boric-Lubecke O, Lubecke V. A low IF receiver architecture for Doppler radar motion detector. IEEE RWS-2010; New Orleans, LA; 2010.

Mostafanezhad I, Boric-Lubecke O, Lubecke V, Host-Madsen A. Cancellation of unwanted motion in a handheld Doppler radar used for non-contact life sign monitoring, IEEE MTT-S International Microwave Symposium Digest 2007; 2007 Jun. p 1241–1244.

Mostafanezhad I, Boric-Lubecke O, Lubecke V, Host-Madsen A. Cancellation of unwanted motion in a handheld Doppler radar used for non-contact life sign monitoring. IEEE MTT-S International Microwave Symposium Digest 2008; 2008 Sep. p 1171–1174.

Occhiuzzi C, Cippitelli S, Marrocco G. Modeling, design and experimentation of wearable RFID sensor tag. IEEE Trans Antennas Propag 2010;58(8):2490–2498.

Ossberger G, et al. Non-invasive respiratory movement detection and monitoring of hidden humans using ultra wideband pulse radar. Ultra Wideband Systems 2004, UWBST; 2004. p 395–399.

Park B-K, Boric-Lubecke O, Lubecke VM. Arctangent demodulation with DC offset compensation in quadrature Doppler radar receiver systems. IEEE Trans Microwave Theory Tech 2007;55(5):1073–1079.

Park B-K, Yamada S, Boric-Lubecke O, Lubecke V. Single-channel receiver limitations in Doppler radar measurements of periodic motion. IEEE Radio and Wireless Symposium; 2006 Jan. Vol. 1, p 17–19.

Paulson CN et al. Ultra-wideband radar methods and techniques of medical sensing and imaging. SPIE International Symposium on Optics East; 2005.

Postolache O, Madeira RN, Girão PS, Postolache G. Microwave FMCW Doppler radar implementation for in-house pervasive health care system. 2010 IEEE International Workshop on Medical Measurements and Applications Proceedings (MeMeA); 2010 April 30–2010 May 1. p 47–52.

Proakis JG. Communication Systems Engineering. 2nd ed. Prentice Hall; 2001.

Rajagopalan H, Rahmat-Samii Y. On-body RFID tag design for human monitoring applications. IEEE Antennas and Propagation Society International Symposium; 2010 July. p 1.

Razavi B. Design considerations for direct-conversion receivers. IEEE Trans Circuits Syst-II: Analog Digital Signal Process 1997;44(6):428–435.

Razavi B. Design of Analog CMOS Integrated Circuits. McGraw-Hill; 2001.

Sanad M. Effect of human body on microstrip antennas. Antennas and Propagation Society International Symposium; AP-S Digest; 1994. Vol. 1, p 298-301.

Saunders K. CW and FM Radar. In: Skolnik MI, editor. Radar Handbook. 2nd ed. San Francisco: McGraw-Hill, Inc.; 1990. p 14.1–14.45.

Seals J, Crowgey SR, Sharpe SM. Electromagnetic vital signs monitor. Georgia Tech. Res. Inst.; Atlanta, GA, USA; 1986. Final Rep. Project A-3529-060.

Seshagiri Rao KV, Nikitin PV, Lam SF. Antenna design for UHF RFID tags: a review and a practical application. IEEE Trans Antennas Propag 2005;53(12):3870–3876.

Singh A, Lubecke V. Body-worn passive planar harmonic tag design for use with Doppler radar. 2011 IEEE Topical Conference on Biomedical Wireless Technologies, Networks, and Sensing Systems (BioWireleSS); Phoenix, AZ; 2011 Jan. p 51–54.

Singh A, Lubecke V. Respiratory monitoring and clutter rejection using a CW Doppler radar with passive RF tags. IEEE Sensors 2012;12(3):558–565. DOI: 10.1109/JSEN.2011.2134083.

Skolnik MI. Introduction to Radar Systems. New York: McGraw-Hill; 1962.

Staderini EM. UWB radars in medicine. IEEE Aerosp Electron Syst Mag 2002;17(1):13–18.

Thiel FH, Sachs M, Schwarz J, Seifert U. Physiological signatures monitored by ultra-wideband-radar validated by magnetic resonance imaging. IEEE International Conference on Ultra-Wideband; 2008. Vol. 1, p 105.

Venkatesh S, Anderson CR, Rivera NV, Buehrer RM. Implementation and analysis of respiration-rate estimation using impulse-based UWB. IEEE Military Communications Conference, MILCOM; 2005 Oct. Vol. 5, p 3314–3320.

Wang Y, Fathy AE. Three-dimensional through wall imaging using an UWB SAR. IEEE AP-S International Symposium on Antennas and Propagation; Toronto, Canada; 2010 Jul.

Wang Y, Fathy AE. Micro-Doppler signatures for intelligent human gait recognition using a UWB impulse radar. IEEE AP-S International Symposium on Antennas and Propagation; Spokane, WA; 2011 Jul.

Wang Y, Yang Y, Fathy AE. Reconfigurable ultra-wide band see-through-wall imaging radar system. IEEE AP-S International Symposium Antennas and Propagation; Charleston, SC; 2009 Jun.

Wang Y, Yang Y, Fathy AE. A reconfigurable UWB system for real-time through wall imaging applications. IEEE Radio and Wireless Symposium (RWS); New Orleans, LA; 2010 Jan, p 633–636.

Xiao Y, Lin J, Boric-Lubecke O, Lubecke VM. Frequency tuning technique for remote detection of heartbeat and respiration using low-power double-sideband transmission in Ka-band. IEEE Trans Microwave Theory Tech 2006;54(5):2023–2032.

Yamada S, Lubecke V, Boric-Lubecke O. Cancellation techniques for LO leakage and DC offset in direct conversion systems. IEEE IMS 2008; 2008 Jun.

Yang Y, Fathy AE. Development and implementation of a real-time see-through-wall radar system based on FPGA. IEEE Trans Geosci Remote Sens 2009;47(5):1270–1280.

Yang BH, Rhee S. Development of the ring sensor for healthcare automation. Robot Autonom Syst 2000;30:273–281.

9

APPLICATIONS AND FUTURE RESEARCH

ADITYA SINGH[1] AND VICTOR M. LUBECKE[2]

[1]*University of Hawaii Neuro-science and MRI research Program, John A. Burns School of Medicine, Honolulu, Hawaii, United States*
[2]*Department of Electrical Engineering, University of Hawaii at Manoa, Honolulu, Hawaii, United States*

Radar technology for sensing of physiological motion has reached the point of adoption for basic commercial applications in medicine and security. There remains, however, ample room for further development to allow more advanced applications. This chapter gives a brief review on some existing US Food and Drug Administration (FDA)-approved and other commercial devices followed by ongoing research efforts.

9.1 COMMERCIAL DEVELOPMENT

9.1.1 Healthcare

Radio-based sensor technologies are being applied to health care on many levels. In remote health care or telemedicine applications, radio transmission has been used to establish a wireless link between the patient and health-care provider, using conventional communications standards such as Wi-Fi or cellular networks. In other applications, the advantages of a tether-free data link have been further exploited to provide a convenient connection between sensors on a subject, and a stationary central terminal or nursing station. Ultimately, wireless transmissions alone can be applied to sense physiological-based motion for a subject.

Doppler Radar Physiological Sensing, First Edition.
Edited by Olga Boric-Lubecke, Victor M. Lubecke, Amy D. Droitcour, Byung-Kwon Park, and Aditya Singh.

Medical devices range from bedpans to surgical lasers. This includes electronic radiating devices such as vital signs radar systems. In the United States, Section 510(k) of the Food, Drug and Cosmetic Act requires medical device manufacturers to notify the FDA in advance of marketing devices to determine whether proper consideration has been made for efficacy and safety. Subsequent changes in design or manufacture require additional review. Thus, such regulatory compliance efforts are regarded as a late-stage move, taken when the technology has a suitable maturity level. To date, there have been only a few approved filings of this type for vital signs radar systems, though with the existence of these precedents, future filings for such approvals can be expected. These early products mainly involve close-range monitoring of sedentary subjects. Similarly, other countries and nation-unions have parallel requirements for the marketing of medical devices within their jurisdictions.

9.1.1.1 Devices Without FDA Compliance Some health-monitoring products can be introduced to market if they do not claim to be a medical diagnostic tool/medical device. Instead, the devices are intended to provide some other pertinent information such as abnormal activity and fall detection through long-term activity monitoring. In the following list, we mention some devices that provide noncontact wireless monitoring but do not provide medical diagnostic information:

1. *GE/Intel QuietCare* (http://www.careinnovations.com/solutions/smart-sensor-analytics/) – This system is a product of Intel-GE Care Innovations LLC and consists of multiple wireless sensors that are placed at certain locations inside the unit and detect the motion of subjects as they move about care facilities. The information from the sensors is relayed to a server, and measurement data are analyzed to detect out-of-the-ordinary events that may indicate trouble. The system descriptions boast of learning algorithms that take into account a subject's personal behavioral patterns for better results than rules-based systems. The website provides some white papers on case studies with their system.
2. *WellAWARE Systems* (http://wellawaresystems.com/index.php) – A company founded in 2000 at the Medical Automation Research Center at the University of Virginia, WellAWARE® Systems offers a monitoring system using various unobtrusive sensors to measure daily activity levels, sleep quality, and other physiological information. Their sensors, which communicate with a secure data manager through a wireless connection, include contact sensors, motion/humidity sensors, and sleep quality sensors. None of their sensors use a camera or a microphone. The data from all the sensors are processed by analytical software to generate reports that could be used to find changes in daily activity levels. These changes can be seen by caregivers who can then address a potential health concern if any.
3. *e-Neighbor* (http://healthsense.com/index.php/products/eneighbor-remote-monitoring) – A product of Healthsense, e-Neighbor remote monitoring system uses sensors (no cameras and microphones are used) that work together to monitor a resident's daily routine. Sensors include contact sensors for usage of bed, doors, windows, and wireless motion and incontinence sensors. Sensors

are placed strategically throughout the residence to detect general activity, or "Activities of Daily Living." The system looks for basic activities, such as movement in the living room or bedroom or the opening and closing of the refrigerator or front door, and establishes a normal range for these activities. The purpose of using the routine as a benchmark is to get a resident assistance in case of an event without having to rely on pushing a button or pulling a cord.

4. *LifeWave Biomed* (http://www.lifewavebiomed.com/) – LifeWave is developing noninvasive medical imaging devices based on ultra-wideband (UWB) radar. UWB employs extremely low-power electromagnetic energy capable of single-organ assessment and trending anatomically correct structural imaging. Their current products include intrapartum and antepartum maternal fetal monitors. They are also developing small wearable and wireless, noninvasive devices to measure respiratory pattern, blood pressure, relative stroke volume and other physiological parameters. Most of their products are geared toward providing accessible health-monitoring solutions in developing world.

9.1.1.2 Devices with FDA Compliance

1. Kai Medical RSpot respiratory rate spot sensor (http://www.kaimedical.com/en2/kaispot.php) – This FDA-approved device can measure respiration through bedding and clothing, using low-power microwave Doppler radar. Clinical validation of this device was performed on 24 patients with respiratory rate accuracy benchmarked against the respiratory rates obtained using Welch Allyn Propaq Encore model 242, the Embla Embletta system with Universal XactTrace respiratory effort sensor and Somnologica for Embletta software, and by counting chest excursions. The difference between simultaneous respiratory rate measurements made with the Doppler radar and with the reference methods were assessed on hospitalized patients. The 95% limits of agreement between the Kai RSpot and reference measurements fall within ±5 BPM. This level of agreement has been shown to be within the repeatability for the reference methods in this study and within the interobserver and intraobserver variability of visual assessment of respiratory rate, which is commonly used to obtain the respiratory rate in vital signs assessments. Therefore, the Doppler radar respiratory rate agrees sufficiently well with the respiratory rates provided by the Welch Allyn Propaq Encore model 242 and the Embla Embletta system with Universal XactTrace respiratory effort sensor and Somnologica for Embletta software that it can be used interchangeably for hospitalized patients.
2. SleepMinder™ – SleepMinder™ is an innovative sensor technology for contactless and convenient measurement of sleep and breathing in the home developed by BiancaMed, which has since been acquired by ResMed. Their noncontact sensor senses the movement and respiration of a subject using an ultra low-power radio frequency (RF) transceiver. Signal analysis is performed by a proprietary software to give information on respiration, sleep quality, and sleep apnea [Zaffaroni et al., 2009]. The performance of SleepMinder as a device for

the monitoring of sleep-disordered breathing (SDB) and the provision of an estimate of the apnea–hypopnea index (AHI) has been discussed by Zaffaroni et al. [2009]. The results from the study that was performed on 129 subjects with suspected SDB reported a correlation of 91% for AHI estimation. They also reported a sensitivity of 89% and specificity of 92% for the detection of clinically significant SDB (AHI > 15).

9.1.2 Defense

Time domain – Time domain has developed their sensors using UWB for applications in defense, security, and robotics. Some of the benefits of UWB radar along with their medical applications have been discussed earlier in Chapter 8. The motivation for their use in military applications comes from the need for a technology that could provide robust ranging, tracking, and communication in all types of terrain and weather. Terrains such as forests and cities are high clutter environments where UWB radar performs well. It could also be used for covert communication owing to its low-power spectral density. Time domain enhances the performance of their UWB radar sensors by pseudorandom encoding of data blocks and synchronizing the clocks of multiple modules through wireless transmission allowing coherent processing.

In addition to personnel tracking, UWB radars could also be used for asset tracking and as unattended ground sensors for area surveillance.

9.2 RECENT RESEARCH AREAS

Ongoing research in the field is directed toward broader, more robust applications of remote sensing of physiological motion. Most published work on cardiopulmonary sensing has been restricted to the measurement of cardiopulmonary rate or activity for isolated sedentary subjects within a few meters of the sensor. These approaches have mostly consisted of the application of monostatic radar systems employing signals with frequencies and power levels similar to those found in commonly encountered radio equipment. Many potential applications could benefit from the development of systems capable of greater stand-off potential, multiple target separation, and passive exploitation of existing signals in the environment. Furthermore, tracking calibration methods can lead the way to quantitative displacement-based volume and pressure measurements.

In addition to technical challenges, a significant amount of research and development effort is being directed at meeting government regulatory compliance standards, for the application of radar to human health applications.

9.2.1 Sleep Study

Sleep is widely understood to play a key role in physical and mental health. Research indicates that 40 million Americans suffer from insomnia and chronic sleep disorders, with over 12 million Americans suffering from obstructive sleep apnea (OSA)

[National Commission on Sleep Disorder Research, 1993]. The quality and quantity of sleep that an individual gets can have a significant impact on learning and memory, metabolism and weight, safety, mood, cardiovascular health, disease, and immune system function. While deprivation is sometimes voluntary, most people who have trouble sleeping do not bring the affliction to the attention of their physicians. Serious consequences including increased mortality can result from untreated sleep disorders. Effective strategies for sleep disorder intervention include methods to increase public awareness of the seriousness and treatability of these problems, combined with technologies that facilitate quick and comfortable diagnosis and treatment. Sleep disorders are generally diagnosed through an overnight polysomnography (PSG) study carried out in a sleep laboratory. PSG involves the measurement of a number of physiological parameters including brain activity (electroencephalography, EEG), eye movements (electrooculography, EOG), muscle activity (electromyography, EMG), heart patterns (electrocardiography, ECG), blood oxygenation, and respiratory effort. These measurements are typically made through a variety of sensors that come in contact with the patient's skin. Typically, these sensors and associated wiring can impose significant discomfort and movement restrictions on the patient, which can adversely affect the results of the study. Wireless technologies have been recently introduced to alleviate some of these complications and make the consideration of home-based sleep studies possible, and emerging technologies promise to revolutionize PSG and other sleep-monitoring procedures to the point where a much broader segment of the population can be reached with convenient and transparent monitoring systems.

A PSG study requires various physiological parameters to be monitored simultaneously, typically with sensors mounted on various parts of the body. Sensors for EEG, ECG, and EMG measurements are typically conductive electrodes attached to the patient's skin. Thermal or pressure sensors attached at the nose or mouth can be used to measure respiratory airflow, while piezoelectric or impedance change straps can be used to measure chest and abdomen movement to assess respiratory effort. Blood oxygenation and limb motion is also often recorded with a pulse oximeter attached to a finger and accelerometers attached to the wrist or ankle, respectively. A simplified example is shown in Fig. 9.1. While comprehensive PSG is performed in a sleep laboratory, categorized as Type 1, there are also other more easily implemented and transported monitoring systems in use, as described in Table 9.1. Wireless technologies have been applied to provide a more versatile approach for monitoring systems, varying with the category of interest.

A straightforward approach to applying wireless solutions to PSG monitoring systems (including Type 1) can be applied by directing the wiring from body-attached sensors to a body-worn wireless communications device. Although sensor-related problems and some wiring issues remain, this approach significantly reduces measurement interference and sensor disconnects by allowing the patient greater degree of unrestricted movement including rolling over and even leaving the bed. While comprehensive diagnostics for sleep disorders generally require comprehensive PSG study, many disorders can be recognized and studied with less comprehensive, yet less intrusive systems.

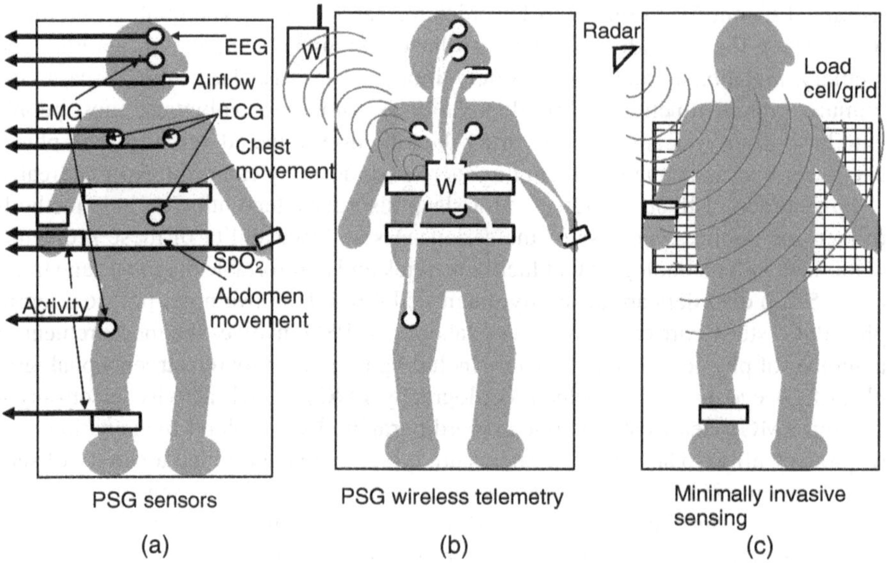

Figure 9.1 Sleep monitoring approaches. Polysomnography (PSG) involves several physiological sensors attached to the body, which can be connected to a body-worn wireless transponder (b). (c) Less comprehensive sensing can be performed without any bodily attachments. © 2009 IEEE, Reprinted, with permission, from Lubecke and Boric-Lubecke [2009].

TABLE 9.1 Types of Sleep Monitoring Systems

Type 1	Monitors perform *full PSG* in-laboratory, *technician-attended*, overnight
Type 2	Monitoring devices can perform full PSG outside of the laboratory. The major difference from Type 1 devices is that a *technologist is not present*. These devices are called comprehensive portable devices
Type 3	Monitoring devices do not record the signals needed to determine sleep stages or sleep disruption. Typically, *four physiologic variables* are measured including two respiratory variables (e.g., respiratory movement and airflow), a cardiac variable (e.g., heart rate or an electrocardiogram), and arterial oxygen saturation. Some devices may have other signals including a monitor to record snoring, detect light, or a means to determine the body position
Type 4	Monitoring devices record *one or two variables* (e.g., arterial oxygen saturation and airflow) and can be used without a technician. These devices are called continuous single or dual bioparameter devices

Various solutions have been proposed for noninvasive sleep monitoring. Mack et al. [2009] have proposed ballistocardiography to measure heart rate and apnea using contact pressure sensors. Hofsoy et al. [2009] have used a headband with 3-axis MEMs accelerometer and to detect snoring and breathing disorder and apply positional therapy using a feedback signal. Nobuyuki et al. [2009] have combined the results from an SpO_2 monitor and a bone conduction microphone to assess the quality of sleep. A wireless solution that has been proposed involves placement of on-body sensors on the patient but no sleep monitoring data have been provided. Wireless sleep monitoring and OSA detection without the use of on-body sensors still remains an attractive solution with minimal interference with the subject/patient and flexibility in use.

Remote sensing of heart rate, respiratory rate, and gross bodily motion has been demonstrated using radar technology in monitoring systems, which makes no contact with the patient. Doppler radar can be used to detect motion ranging from arbitrary limb movement to periodic chest displacement associated with cardiopulmonary activity. Unique advantages to radar-based sleep-monitoring systems include insight provided by combined actigraphy and cardiopulmonary monitoring, and elimination of measurement interference with sleep.

Technology that can capture minute changes in physiological parameters has proven highly effective at assessment of sleep in elaborate and expensive sleep laboratory studies. Wireless technology is emerging as the key to taking sleep studies out of the laboratory and into the home, where more people can receive the benefits of diagnosis and treatment, while providing for more reliable data through minimized interference with a patient's normal sleep activity. Wireless solutions demonstrated range from full PSG studies with significantly reduced interference to cardiopulmonary and activity monitoring with no interference.

9.2.2 Range

Limitations on the range or minimum power requirements for cardiopulmonary radar sensors result from a variety of interrelated factors including transmit power, antenna gain, frequency of operation, range-correlation, receiver sensitivity, signal-to-noise ratio (SNR), and demodulation and signal-processing approaches. Thus, the maximum range at which a subject's heartbeat or respiratory activity can be sensed is effectively determined by a radar system's available power, allowable size, acceptable cost, and the degree to which subject and environment can be controlled.

For example, a 24 GHz radar system using tens of milliwatts and a parabolic dish reflector antenna was used to demonstrate heartbeat and respiration monitoring for potential archery and rifle competition competitors. The system was reported to be capable of measuring a subject's heartbeat at 10 m, and respiration at 20 m. More recently, the rate measurement of a mechanical heartbeat simulator has been demonstrated at a range of 30 m, using a 2.4 GHz Doppler radar system with a 10 dBm source power and a patch antenna array. When monitoring a subject from a long distance, an additional challenge arises as a result of the detection of extraneous motion within the measurement system field of view.

The use of a large antenna with high gain and narrow beamwidth is very common for systems made for long-range measurements. Antenna size is a very important factor in deciding the portability and the performance of a radar system. Parametric analysis performed to assess the trade-offs between antenna size, pattern, transmit power and range for long-range Doppler radar heart rate detection suggests that better system performance can be obtained with the use of a smaller antenna with a lower

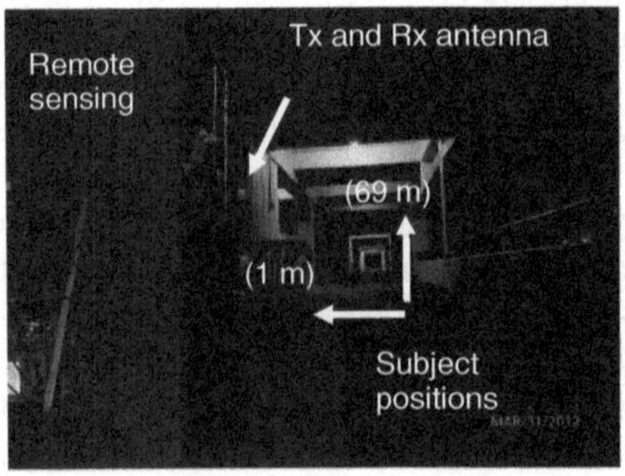

(a)

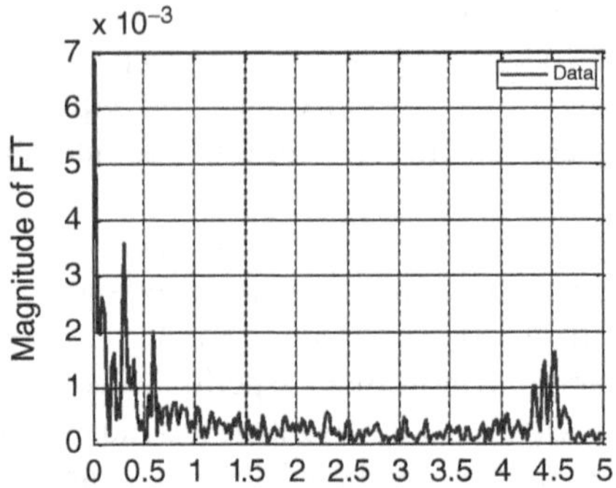

(b)

Figure 9.2 Experiment setup at night in a hallway (a). Note that the building is not completely isolated from variations in the weather outside. FFT of linear demodulated data showing detected respiration rate at approximately 0.27 Hz for a 69-m distance (b). © 2012 IEEE. Reprinted, with permission, from Baboli et al. [2012].

gain as compared with a larger antenna with a high gain. The same study reported the successful detection of human respiration at a distance of 69 m with the setup shown in Fig. 9.2.

9.2.3 Multiple Subject Detection

Separating multiple sources of motion using CW Doppler radar has always been a difficult problem. Multiple sources of motion can arise from one lone subject in the form of respiration, heart and other body motion, and from the presence of multiple subjects. Although it is conveniently possible to separate sources of motion from one subject due to the prior knowledge about such signals [Boric-Lubecke et al., 2005], it is a challenging problem to separate similar forms of motion coming from two different subjects. Various proposed solutions include using MIMO radar systems with blind source separation techniques, UWB radar, and direction of arrival (DOA) techniques. Although most of these systems can successfully separate motion signatures, they cannot identify the particular source of motion for each signature.

In addition to uniquely identifying a subject from clutter, Doppler radars with RF tags could be used in situations where the tagged motion is known and it would be of interest to find other motion that is not tagged. An example of such an application could be a rescue operation where the rescuer is wearing a tag (Fig. 9.3). In such a case, in addition to keeping track of the rescuer, the objective is to find any other person in need of help (casualty). In this case, one can use CW or continuous wave harmonic radar (2.45–4.9 GHz) and CW Doppler radar (2.45 GHz) to keep track of both the tagged and nontagged activity (Fig. 9.4). The harmonic radar receives only the $2f$ (4.9 GHz) tag reflected signal, and thus can readily isolate the rescuer motion. The conventional radar receives the $1f$ (2.45 GHz) signal, which is reflected by both the rescuer and the casualty, and thus does not by itself allow isolation of the casualty.

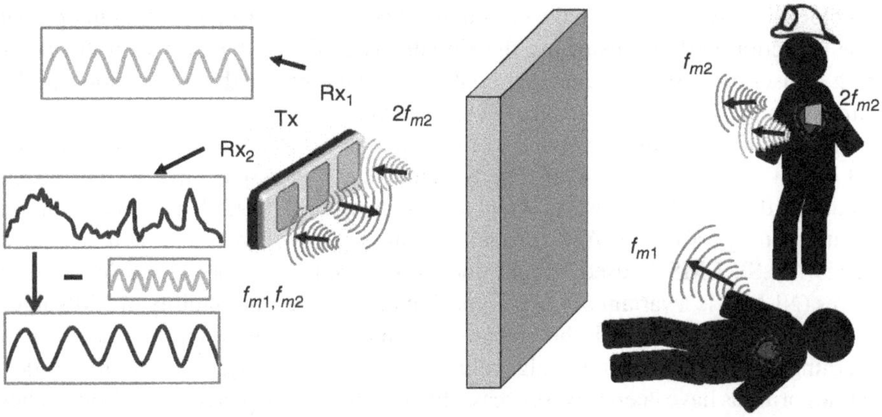

Figure 9.3 Figure depicting a rescue situation where in addition to tracking the rescuer, it is critical to find any victim (conscious or unconscious). f_{m1} and f_{m2} refer to the physiological signal coming from persons $m1$ and $m2$, respectively. Since $m2$ is wearing a tag, he/she is also sending a signal $2f_{m2}$.

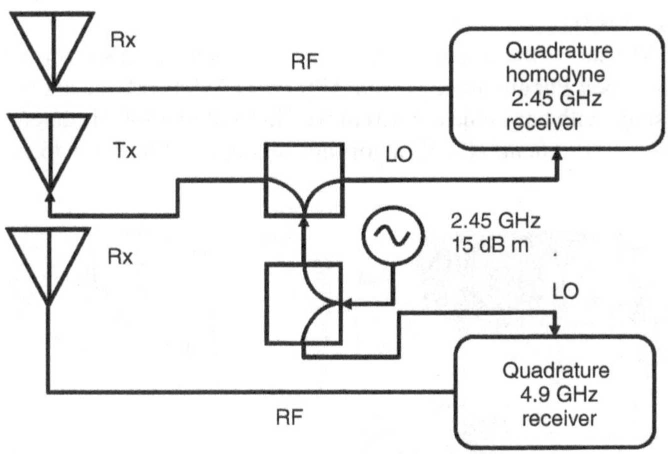

Figure 9.4 A block diagram showing two-frequency radar setup where tag subtraction algorithms could be used to separate sources of motion. © 2013 IEEE. Reprinted, with permission, from Singh and Lubecke [2013].

The $1f$ motion caused by the rescuer can be considered noise. However, the signal from the tag that has been isolated by the harmonic $2f$ receiver is correlated with this noise in the $1f$ system. Subtraction of the nonstationary tag signal from the baseband signal of 2.45 GHz radar could facilitate reliable monitoring of the untagged source of motion.

Fixed filters are not an efficient way of reducing tag signal that is dynamic in nature. Hence, subtraction using adaptive noise cancellation (ANC) techniques would be preferred. ANC is based on the principles of adaptive filtering resulting in optimal noise reduction without distorting the signals, as could be the case with direct filtering. It uses a reference signal that contains signal correlated with the noise in the desired signal. This reference signal is used to generate a varying impulse response by the adjustment of filter weights to minimize an error signal by minimizing the total output power of the system. The principles and techniques of ANC have been adequately described in Haykin [1991] and Widrow et al. [1975]. The two common algorithms for performing ANC are least mean squares (LMS) and recursive least squares (RLS). LMS is used widely due to its simplicity. Normalized least mean squares (NLMS) is a variant of LMS algorithm that ensures the stability of LMS algorithm by normalizing it with the power of the input. The two important parameters governing the behavior of LMS algorithms are the step-size (μ) and the filter order. ANC algorithms have been used successfully to cancel out acoustic noise. Some other applications include cancellation of 60 Hz from biomedical signals and cancellation of radar clutter.

One of the major challenges in the application of ANC to radar signals is the relative strength of the signal to be cancelled (tag). In some cases, the tag signal content might be greater than the untagged motion. A detailed explanation and analysis using simulations and measurement using mechanical movers has been provided by Singh and Lubecke [2013].

Figure 9.5(b) shows the results from the measurement setup shown in Fig. 9.5(a). A piezoelectric belt Pneumotrace II from UFI was used as reference chest belt. The Galil linear stage with harmonic tag was moved in front of 4.9-GHz antenna at 0.4 Hz with 1 cm of displacement. NLMS algorithm was applied to the 2.45 receiver data

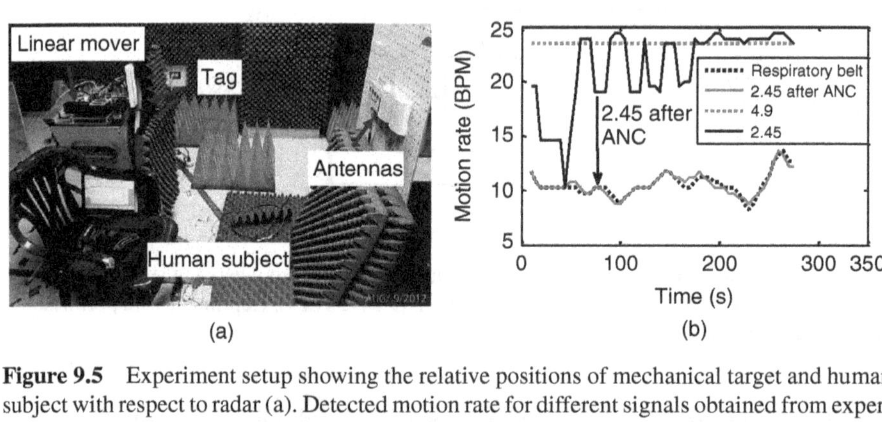

Figure 9.5 Experiment setup showing the relative positions of mechanical target and human subject with respect to radar (a). Detected motion rate for different signals obtained from experiment III (b). The 4.9 GHz trace shows the successful detection of the tag motion at 0.4 Hz (24 BPM) exclusively. The trace from 2.45 GHz radar initially fails to track the respiration rate of human subject. After the application of ANC technique, the data from 2.45 GHz radar tracks the respiration of the human subject exclusively as verified by the reference chest belt. © 2013 IEEE. Reprinted, with permission, from Singh and Lubecke [2013].

and Fourier transform was used to analyze the data with a window length of 18 s (4–6 respiration cycles) and an overlap of 5 s. It is clear from Fig. 9.5(b) that without the application of ANC algorithm, it would be difficult to make any conclusions from the analysis of 2.45 GHz receiver data. However, after cancellation of tag signal, the rate corresponds to respiration rate of human subject as verified by the respiratory belt data.

9.2.4 Animal Monitoring

Activity monitoring of animals in their natural environment can yield important information about energy expenditure, thermoregulation, behavioral patterns, and even population health. As energetics plays a significant role in ecology, behavior, and physiology, accurate methods for activity monitoring are critical for a wide range of animal studies. The standard technique for measuring field metabolic rate is the doubly labeled water technique, which involved injecting animals with radiolabeled water and observing the rate of CO_2 production over several weeks. Because the technique relies on the biological half-life of ^{18}O, which is long relative to the duration of specific behaviors, it is not possible to measure the cost of specific activities such as foraging, mating, or locomotion. Recent advances in the miniaturization of electrical circuits have allowed measurements of activity using continuous heart-rate monitoring, but as this technique uses implantable data-loggers, it is limited to animals 1 kg or larger. For smaller animals, the only available techniques are visual inspection or video recording. Both are extremely time-consuming, labor-intensive, and require extensive postexperiment effort in recording, transcribing, or analyzing the raw data.

Doppler radar motion sensing can provide a better tool for the automated activity monitoring in animals, as well as the detection of multiple behavioral events in real time.

Target motion can be classified based on the changes in the amplitude of baseband signals. A simple way to detect large motion is by the use of eigen demodulation. A sudden change in eigen vectors will indicate nonsedentary motion. Motions can

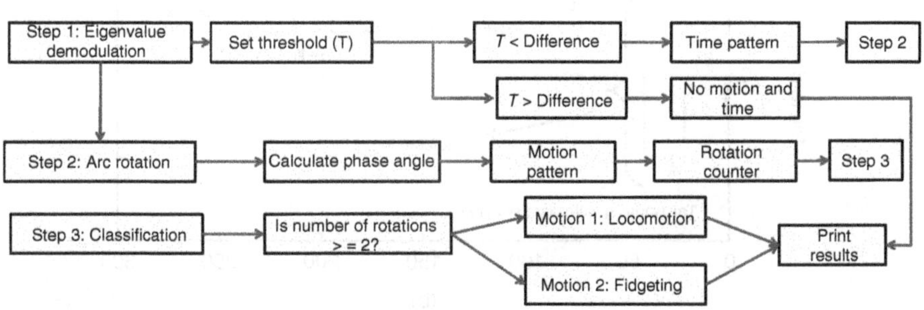

Figure 9.6 Classification algorithm used for each radar to characterize motion. © 2012 IEEE. Reprinted, with permission, from Singh et al. [2012].

Figure 9.7 Photograph showing the setup for monitoring chameleon activity. © 2012 IEEE. Reprinted, with permission, from Singh et al. [2012].

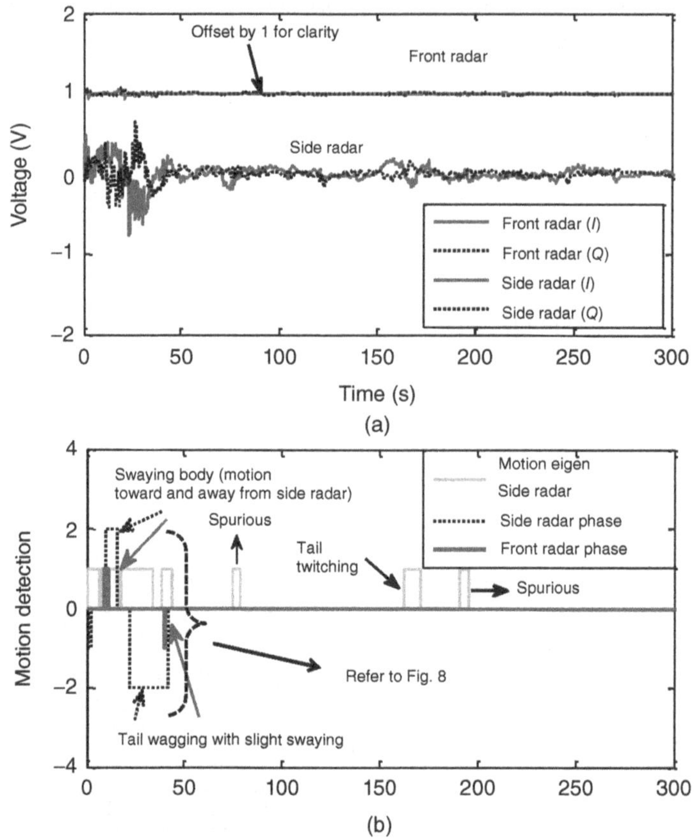

Figure 9.8 (a) Raw data from front and side radar showing changes in amplitude due to motion and (b) result of the detection algorithm for front and side radar. The swaying of the body is detected as locomotion by the side radar and fidgeting by the front radar as expected. A few spurious alerts were generated by the eigen vector algorithm but were revealed as no motion by phase analysis and video reference. © 2012 IEEE. Reprinted, with permission, from Singh et al. [2012].

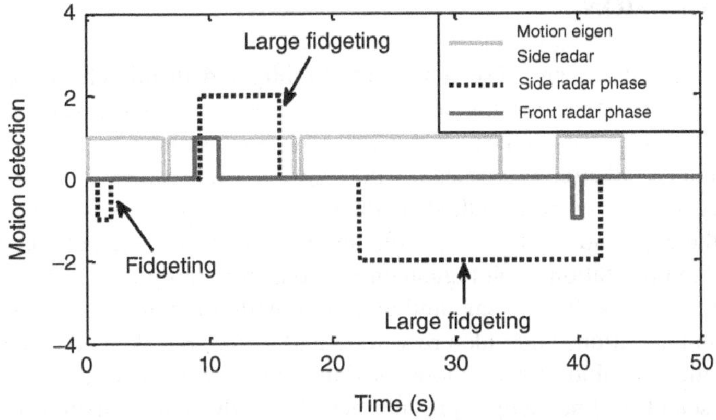

Figure 9.9 A plot of 0–50 s taken from Fig. 9.8 showing a new class of activity (large fidgeting motion) that cannot be considered as locomotion. But the radar has the capability to differentiate between small fidgeting and large fidgeting. © 2012 IEEE. Reprinted, with permission, from Singh et al. [2012].

also be classified based on movement relative to the radar that can be deduced by observing a given amount of radar data samples and calculating the phase angles of the arc transcribed in the *I–Q* plane. If the phase angles are rotating in the clockwise direction, the motion would be classified as moving away from the radar, whereas phase angles in the counterclockwise (CCW) direction would be classified as moving toward the radar. The motion in front of the radar would result in the formation of an arc in the *I–Q* plane. The length of the arc is directly proportional to the amount of the motion (motion component orthogonal to the plane of radar antenna). For a transmitted signal of 24 GHz ($\lambda = 1.25$ cm), a movement of approximately 0.6125 cm results in a complete circle. By counting the number of circles or closed loops in the *I–Q* plane, it is possible to quantify motion as fidgeting or locomotion. For our analysis with different radar modules, different threshold values were used. The phase angles were calculated using MATLAB©; however, to obtain the correct phase values, the *IQ* plots were conditioned to be centered at the origin. After the phase angles were calculated, the algorithm determines the numbers of rotation by counting how many times it passed the initial value of each circular pattern. Two-dimensional movement was calculated by comparing the data obtained from the two sensors (Fig. 9.6).

For characterization of motion of *Chamaeleo jacksonii*, the radar setup is shown in Fig. 9.7. A 10.525 GHz radar module was used, and a stand was used to hold a small branch on which the chameleon was let loose. The experiments were performed under natural light in the morning in a closed room. The sampling rate for radar data acquisition was set to 100 Hz. Measurements were made for 5 min. A standard digital camera was used to record video (640 × 480) as reference. From the inspection of the video, a table was created indicating the type of motion with time. These reference values were then compared with radar data analysis that has been presented in Figs 9.8 and 9.9.

9.3 CONCLUSION

While the applications and research efforts highlighted in this chapter commonly depend on physiological motion sensing, each presents or addresses unique challenges. General advances in the cost, performance, and availability of radio and computational integrated circuits are steadily making it possible for engineers in academia and industry to focus on new methods to distinguish motion of interest from interference, and the exploration of new specific applications where the automated remote sensing and interpretation of biological phenomena provides practical benefits. Modern society is already discovering, and grappling with, the introduction of pervasive mobile communications and video recording and streaming. We can expect that the next generation of ubiquitous remote sensing devices will open doors to new categories of scientific knowledge, provide new biomedical information tools to the general public, and inspire thought and reflection on what society can and should do with this capability and information.

REFERENCES

Baboli M, Singh A, Hafner N, Lubecke V. Parametric study of antennas for long range Doppler radar heart rate detection. Paper presented at the Engineering in Medicine and Biology Society (EMBC), 2012 Annual International Conference of the IEEE; 2012 August 28–2012 Sept 1; 2012.

Boric-Lubecke O, Lubecke VM, Host-Madsen A, Samardzija D, Cheung K. Doppler radar sensing of multiple subjects in single and multiple antenna systems. Paper presented at 7th International Conference on Telecommunications in Modern Satellite, Cable and Broadcasting Services, 2005; 2005 Sept 28–30; 2005.

Haykin S. Adaptive Filter Theory. 2nd ed. Englewood Cliffs, NJ: Prentice Hall; 1991.

Hofsoy, DA, Clauss JF, Wolf B. Monitoring and therapy of sleep-related breathing disorders. Paper presented at the 2009 6th International Workshop on Wearable Micro and Nano Technologies for Personalized Health (pHealth); 2009 June 24–26; 2009.

Lubecke VM, Boric-Lubecke O. Wireless technologies in sleep monitoring. Paper presented at the IEEE Radio and Wireless Symposium, 2009. RWS'09; 2009 Jan 18–22; 2009.

Mack DC, Patrie JT, Felder RA, Suratt PM, Alwan M. Sleep assessment using a passive ballistocardiography-based system: preliminary validation. Paper presented at the Annual International Conference of the IEEE Engineering in Medicine and Biology Society, 2009. EMBC 2009; 2009 Sept 3–6; 2009.

Nobuyuki, A, Yasuhiro N, Taiki T, Miyae Y, Kiyoko M, Terumasa H. Trial of measurement of sleep apnea syndrome with sound monitoring and Spo2 at home. Paper presented at the 11th International Conference on e-Health Networking, Applications and Services, 2009. Healthcom 2009; 2009 Dec 16–18; 2009.

Singh A, Lee SSK, Butler M, Lubecke V. Activity monitoring and motion classification of the lizard *Chamaeleo jacksonii* using multiple Doppler radars. Paper presented at the 2012 Annual International Conference of the IEEE Engineering in Medicine and Biology Society (EMBC); 2012 Aug 28–2012 Sept 1; 2012.

Singh A, Lubecke V. Adaptive noise cancellation for two frequency radar using frequency doubling passive RF tags. IEEE Trans Microwave Theory Tech 2013;61(8):2975–2981.

Wake Up America. A National Sleep Alert. Report of the National Commission on Sleep Disorders Research. Washington (DC): Health and Human Services; 1993.

Widrow B, Glover JR Jr, McCool JM, Kaunitz J, Williams CS, Hearn RH, Zeidler JR, Dong E Jr, Goodlin RC. Adaptive noise cancelling: principles and applications. Proc IEEE 1975;63(12):1692–1716.

Zaffaroni A, de Chazal P, Heneghan C, Boyle P, Mppm PR, McNicholas WT. 2009. Available at: http://ieeexplore.ieee.org/xpl/articleDetails.jsp?arnumber=5332909&newsearch=true&queryText=sleepminder. doi: 10.1109/IEMBS.2009.5332909.

INDEX

Doppler Radar Physiological Sensing, First Edition.
Edited by Olga Boric-Lubecke, Victor M. Lubecke, Amy D. Droitcour, Byung-Kwon Park, and Aditya Singh.